Lecture Notes in Computer Science 9460

Commenced Publication in 1973
Founding and Former Series Editors:
Gerhard Goos, Juris Hartmanis, and Jan van Leeuwen

More information about this series at http://www.springer.com/series/7409

Guido Zuccon · Shlomo Geva
Hideo Joho · Falk Scholer
Aixin Sun · Peng Zhang (Eds.)

Information Retrieval Technology

11th Asia Information Retrieval Societies Conference, AIRS 2015
Brisbane, QLD, Australia, December 2–4, 2015
Proceedings

 Springer

Editors

Guido Zuccon
Queensland University of Technology
Brisbane
Australia

Shlomo Geva
Queensland University of Technology
Brisbane
Australia

Hideo Joho
University of Tsukuba
Ibaraki
Japan

Falk Scholer
RMIT University
Melbourne
Australia

Aixin Sun
Nanyang Technological University
Singapore
Singapore

Peng Zhang
Tianjin University
Tianjin
China

ISSN 0302-9743   ISSN 1611-3349   (electronic)
Lecture Notes in Computer Science
ISBN 978-3-319-28939-7   ISBN 978-3-319-28940-3   (eBook)
DOI 10.1007/978-3-319-28940-3

Library of Congress Control Number: 2015959933

LNCS Sublibrary: SL3 – Information Systems and Applications, incl. Internet/Web, and HCI

Printed on acid-free paper

This Springer imprint is published by SpringerNature
The registered company is Springer International Publishing AG Switzerland

Preface

The 2015 Asian Information Retrieval Societies Conference (AIRS 2015), was the 11th instalment of the conference series, initiated from the Information Retrieval with Asian Languages (IRAL) workshop series back in 1996 in Korea. The conference was held during December 2–4, 2015, at the Queensland University of Technology (QUT), Brisbane, Australia.

The annual AIRS conference is the main information retrieval forum for the Asia-Pacific region and aims to bring together academic and industry researchers, along with developers, interested in sharing new ideas and the latest achievements in the broad area of information retrieval. AIRS 2015 enjoyed contributions spanning the theory and application of information retrieval, both in text and multimedia.

This year we received 92 submissions form all over the world. Submissions were peer reviewed in a double-blind process by at least three international experts, with at least one senior meta-reviewer. The final program of AIRS 2015 featured 29 full papers (32 %) divided in 10 tracks: "Efficiency," "Graphs, Knowledge Bases and Taxonomies," "Recommendation," "Twitter and Social Media," "Web Search," "Text Processing, Understanding, and Categorization," "Topics and Models," "Clustering," "Evaluation," and "Social Media and Recommendation." The program also featured eight short papers and three demonstrations (12 %).

AIRS 2015 featured two keynotes: the first by Peter Bruza (Queensland University of Technology), titled "Quantum Haystacks Revisited," and the second by Peter Bailey (Microsoft Research), titled "The Great Search Bake Off."

The conference and program chairs of AIRS 2015 extend our sincere gratitude to all authors and contributors to this year's conference. We are also grateful to the Program Committee for the great reviewing effort that guaranteed AIRS 2015 could feature a quality program of original and innovative research in information retrieval. Special thanks go to our sponsors for their generosity: the Australian E-Health Research Centre (CSIRO), the Electrical Engineering and Computer Science School and the Information Systems School of QUT, and Springer. We also thank the Special Interest Group in Information Retrieval (SIGIR) for supporting AIRS by granting it in-cooperation status and sponsoring the student travel grant that contributed to pay the travel costs for 10 students presenting their research work at AIRS 2015.

December 2015

Guido Zuccon
Shlomo Geva
Hideo Joho
Falk Scholer
Aixin Sun
Peng Zhang

Organization

General Chairs

Guido Zuccon	Queensland University of Technology, Australia
Shlomo Geva	Queensland University of Technology, Australia

Program Chairs (Full papers)

Hideo Joho	University of Tsukuba, Japan
Falk Scholer	RMIT, Australia

Program Chairs (Short Papers and Demonstrations)

Aixin Sun	Nanyang Technological University, Singapore
Peng Zhang	Tianjin University, China

Student Travel Support Chair

Teerapong Leelanupab	King Mongkut's Institute of Technology Ladkrabang, Thailand
Guido Zuccon	Queensland University of Technology, Australia

Best Paper Awards Chair

Peter Bruza	Queensland University of Technology, Australia

Website, Publicity, and Social Media Chair

Bevan Koopman	Australian E-Health Research Centre, CSIRO, Australia

Program Committee

Akiko Aizawa	National Institute of Informatics, Japan
James Allan	University of Massachusetts Amherst, USA
Omar Alonso	Microsoft, USA
Giambattista Amati	Fondazione Ugo Bordoni, Italy
Timothy Baldwin	The University of Melbourne, Australia
Bodo Billerbeck	Microsoft Research, Cambridge, UK
Martin Braschler	Zurich University of Applied Sciences, Switzerland
Ben Carterette	University of Delaware, USA

Marc Cartright   Google Inc., USA
Yi Chang   Yahoo Labs, USA
Hsin-Hsi Chen   National Taiwan University, Taiwan
Charles Clarke   University of Waterloo, Canada
Fabio Crestani   University of Lugano, Switzerland
Bruce Croft   University of Massachusetts Amherst, USA
Ronan Cummins   National University of Ireland, Ireland
Thomas Demeester   Ghent University, Belgium
Emanuele Di Buccio   University of Padua, Italy
Giorgio Maria Di Nunzio   University of Padua, Italy
Carsten Eickhoff   ETH Zurich, Switzerland
Yi Fang   Santa Clara University, USA
Nicola Ferro   University of Padua, Italy
Ingo Frommholz   University of Bedfordshire, UK
Atsushi Fujii   Tokyo Institute of Technology, Japan
Sumio Fujita   Yahoo!JAPAN Research, Japan
Bin Gao   Microsoft Research Asia, China
Shlomo Geva   Queensland University of Technology, Australia
Cathal Gurrin   Dublin City University, Ireland
Claudia Hauff   Delft University of Technology, The Netherlands
Jiyin He   CWI, The Netherlands
Ben He   University of Chinese Academy of Sciences, China
Liangjie Hong   Yahoo! Labs, USA
Xuanjing Huang   Fudan University, China
Jimmy Huang   York University, UK
Adam Jatowt   Kyoto University, Japan
Shen Jialie   Singapore Management University, Singapore
Jing Jiang   Singapore Management University, Singapore
Hideo Joho   University of Tsukuba, Japan
Gareth Jones   Dublin City University, Ireland
Jaap Kamps   University of Amsterdam, The Netherlands
Min-Yen Kan   National University of Singapore, Singapore
Noriko Kando   National Institute of Informatics, Japan
Makoto P. Kato   Kyoto University, Japan
Diane Kelly   University of North Carolina, USA
Marijn Koolen   University of Amsterdam, The Netherlands
Bevan Koopman   CSIRO, Australia
Udo Kruschwitz   University of Essex, UK
Oren Kurland   Technion, Israel
Wai Lam   The Chinese University of Hong Kong, Hong Kong,
                                SAR China
Yanyan Lan   Chinese Academy of Sciences, China
Birger Larsen   Aalborg University Copenhagen, Denmark
Teerapong Leelanupab   King Mongkut's Institute of Technology Ladkrabang,
                                Thailand
Liz Liddy   Syracuse University, USA

Nut Limsopatham	University of Cambridge, UK
Christina Lioma	University of Copenhagen, Denmark
Yiqun Liu	Tsinghua University, China
Qiaoling Liu	Emory University, USA
Robert Luk	The Hong Kong Polytechnic University, Hong Kong, SAR China
Mihai Lupu	Vienna University of Technology, Austria
Craig Macdonald	University of Glasgow, UK
Massimo Melucci	University of Padua, Italy
Mandar Mitra	Indian Statistical Institute, India
Stefano Mizzaro	University of Udine, Italy
Alistair Moffat	University of Melbourne, Australia
Josiane Mothe	Université de Toulouse, France
Harumi Murakami	Osaka City University, Japan
Douglas W. Oard	University of Maryland, USA
Hiroaki Ohshima	Kyoto University, Japan
Joao Palotti	Vienna University of Technology, Austria
Gabriella Pasi	Università degli Studi di Milano Bicocca, Italy
Benjamin Piwowarski	CNRS/University Pierre et Marie Curie, France
Martin Potthast	Bauhaus University Weimar, Germany
Yongli Ren	RMIT, Australia
Tetsuya Sakai	Waseda University, Japan
Rodrygo Luis Teodoro Santos	Universidade Federal de Minas Gerais, Brazil
Shin'Ichi Satoh	National Institute of Informatics, Japan
Falk Scholer	RMIT University, Australia
Fabrizio Silvestri	Yahoo Labs, Spain
Laurianne Sitbon	Queensland University of Technology, Australia
Kazunari Sugiyama	National University of Singapore, Singapore
Aixin Sun	Nanyang Technological University, Singapore
Anastasios Tombros	Queen Mary University of London, UK
Andrew Trotman	University of Otago, New Zealand
David Vallet	Universidad Autonoma de Madrid, Spain
Bin Wang	Chinese Academy of Sciences, China
Quan Wang	Chinese Academy of Sciences, China
Jun Wang	University College London, UK
Meng Wang	Hefei University of Technology, China
William Webber	University of Melbourne, Australia
Ingmar Weber	Qatar Computing Research Institute, Qatar
Ji-Rong Wen	Renmin University of China, China
Stewart Whiting	University of Glasgow, UK
Alan Woodley	Queensland University of Technology, Australia
Wei Wu	Microsoft Research Asia, China
Jun Xu	Chinese Academy of Sciences, China
Tan Xu	AT&T Research
Takehiro Yamamoto	Kyoto University, Japan

Zheng Ye	York University, UK
Minoru Yoshida	University of Tokushima, Japan
Masaharu Yoshioka	Hokkaido Univeristy, Japan
Haitao Yu	University of Tsukuba, Japan
Xiaojun Yuan	University at Albany, State University of New York, USA
Zheng-Jun Zha	Hefei Institute of Intelligent Machines, China
Peng Zhang	Tianjin University, China
Min Zhang	Tsinghua University, China
Lanbo Zhang	University of California, Santa Cruz, USA
Le Zhao	Carnegie Mellon University, USA
Mianwei Zhou	University of Illinois at Urbana-Champaign, USA
Justin Zobel	University of Melbourne, Australia
Guido Zuccon	Queensland University of Technology, Australia

Steering Committee

Hsin-Hsi Chen	National Taiwan University, Taiwan
Youngjoong Ko	Dong-A University, Korea
Wai Lam	The Chinese University of Hong Kong, Hong Kong, SAR China
Alistair Moffat	University of Melbourne, Australia
Hwee Tou Ng	National University of Singapore, Singapore
Zhicheng Dou	Renmin University of China, China
Dawei Song	Tianjin University, China
Masaharu Yoshioka	Hokkaido Univeristy, Japan

Keynotes

Quantum Haystacks Revisited

Peter Bruza

Queensland University of Technology
Brisbane, Australia
p.bruza@qut.edu.au

Almost ten years ago, Keith van Rijsbergen delivered a Salton Award keynote with the intriguing title "Quantum Haystacks". In this talk he put forward the idea that quantum theory could inspire a "design language for Information Retrieval". As a consequence, this design language relies on quantum logic and quantum probabilities, rather than the classical probabilities that underpin much of Information Retrieval theory. The aim of this talk is to revisit quantum haystacks. It will put forward the idea that quantum theory can inspire a "design language for the Information Retrieval user". By drawing on developments in the emerging field of quantum cognition, theory and empirical evidence will be provided that quantum probabilities more naturally express human conceptual processing and decision making. As both of these pertain to the information seeking user, implications for the development of formal Information Retrieval user models will be presented.

The Great Search Bake Off

Peter Bailey

Microsoft
Canberra, Australia
pbailey@microsoft.com

A major access divide has opened up between the academic and commercial Web search communities over the last 10–15 years due to the availability and importance of large scale search log user data in understanding certain types of search behavior. At the same time, the success and simplicity of Cranfield-inspired Information Retrieval test collection experimentation methods has encouraged our research publication community to chase marginal gains of the latest effectiveness metric. However, experimentation based on test collections remains fundamentally essential in carrying out cost-effective development of search system algorithms. The real challenge and opportunity for all IR researchers, be they academic or industrial, is to improve the external validity of our experimental outcomes. The users of search systems, and their unfailing and surprising variability, have too often been overlooked when developing test collections. Based on a range of research that sits at the intersection of academic and industrial investigations, I hope to persuade you of the centrality of users in search evaluation, and the opportunities to integrate them more within our familiar experimental frameworks.

Contents

Web Search

Text Processing, Understanding and Categorization

Topics and Models

Clustering

Evaluation

Social Media and Recommendation

Short Papers

Demonstrations

Efficiency

On Structures of Inverted Index for Query Processing Efficiency

Xingshen Song[1(✉)], Xueping Zhang[3], Yuexiang Yang[1],
Jicheng Quan[2], and Kun Jiang[1]

[1] College of Computer, National University of Defense Technology,
Changsha, China
{songxingshen,yyx,jiangkun}@nudt.edu.cn
[2] Aviation University of Air Force, Changsha, China
jicheng_quan@126.com
[3] Information Center, PLA University of Science and Technology,
Nanjing, China
university1128@sina.cn

Abstract. Inverted index has been widely adopted by modern search engines to effectively manage billions of documents and respond to users' queries. Recently, many auxiliary index variants are brought up to enhance the engine's compression ratio or query processing efficiency. The most successful auxiliary index structures are Block-Max Index and Dual-Sorted Index, both used for quickening the query processing. More precisely, Block-Max Index is designed for efficient top-k query processing while Dual-Sorted Index introduces pattern matching to solve complex query. There is little work thoroughly analyses and compares the performance of the two auxiliary structures. In this paper, an in-depth study on Block-Max Index and Dual-Sorted Index is presented, with a survey on related top-k query processing strategies. Finally, experimental results on TREC GOV2 dataset with detailed analysis show that Dual-Sorted Index achieves the best query processing performance at the price of huge space occupation, moreover, it sheds light upon the prospect of combining compact data structures with inverted index.

Keywords: Performance evaluation · Inverted index · Block-Max index · Dual-Sorted index · Query processing

1 Introduction

Inverted index is adopted as the core component to fast respond to enormous queries. Given a collection of D documents, an inverted index can be seen as a big table mapping each unique *term* to a *posting list* which contains all the document identifiers (called *docid*) and the number of occurrences in the document (called the *frequency*), and possibly other information like the positions of each occurrence within the documents. We postulate that postings have docids and frequencies, but do not consider other data such as positions or contexts, thus the postings fit in the format (d_i, f_i). The set of terms is called *lexicon*, which is relatively small compared to postings. Its advantages are clear: (i) it removes the redundancy in the documents where the same

© Springer International Publishing Switzerland 2015
G. Zuccon et al. (Eds.): AIRS 2015, LNCS 9460, pp. 3–14, 2015.
DOI: 10.1007/978-3-319-28940-3_1

term occurs more than once, stores only one mapping from term to document instead; (ii) its primitives are composed of docids, document frequencies and positions, these elements can be stored separately or combined arbitrarily; (iii) the posting lists of query terms can be processed in parallel to accelerate the procedure, also, various list intersection and skipping algorithms have been implemented to reach early termination [1–3].

On the other hand, the inverted index may consist of many millions of postings and it can be hardly fit into the main memory, thus compression is needed to save space and reduce the number of disk access. Different ordering schemes and traversal strategies have a large impact on the performance of query response. Also, the inverted index is inferior in searching for substrings and in languages which the terms are not as discrete in English. All these issues have forced variant implementations of inverted index. Different orderings in the lists of documents associated with a term, and different auxiliary information, fit widely different IR tasks. Index designers have to choose the right order for one such task, rendering the index difficult to use for others.

Among various schemes of inverted index, Block-Max Index [4–6] and Dual-Sorted Index [7, 8] are two successful works that have been continuously studied by researchers. Block-Max Index first partitions the sequence of each term into blocks of fixed size(say, 64 or 128 elements), and compresses each block independently with faster list-oriented encoders like simple-X or PFD [9]; then stores the maximum impact value and the head docid for each block in uncompressed form, enabling to skip large parts of the lists. It is simple with little space occupation, but leads to considerable performance gains in conjunctive query and DAAT style pruning approaches. Dual-Sorted Index is a variant of inverted index using wavelet tree, a balanced binary tree-like compact data structure. The wavelet tree can store a sequence (e.g., the posting list) from a symbol universe (e.g., the docid) within asymptotically the same space required by a plain representation of the sequence [10–12]. Dual-Sorted Index allows combining an ordering by decreasing term frequency with an ordering by increasing docid, more importantly, it supports not only typical query scheme, but also sophisticated operations in pattern matching. While researchers keep improving both techniques continuously, missing from the literature is a study that thoroughly measures properties and performances of these two indexes. In this paper, we provide a comprehensive comparison and analysis of the space occupation and response efficiency for different query schemes of these two indexes, using an open source search engine platform-Terrier [13].

This paper is structured as follows: Section 2 provides a background on query processing strategies and wavelet trees; Sect. 3 summarizes both Block-Max Index and Dual-Sorted Index; Sect. 4 describes our experimental setup and comparison results; Conclusions and future work follow in Sect. 5.

2   Background

2.1   Query Processing Strategies

Given a query, the most basic processing form is called *Boolean query processing*, which intersects or merges posting lists of query terms according to their logical relation (AND or OR) [14, 15]. Search engines usually use *ranked query processing*, where a ranking function is used to compute a score for each document passing a simple Boolean filter, and then the k top-scoring documents are returned. To traverse the index structure, there are two basic techniques, DAAT and TAAT. The former searches the posting list of each query term one after another, keeping a temporary accumulator to build the result set; while the latter traverses in an interleaved fashion, keeping aligned by docid. To facilitate different strategies, posting lists are always reordered, in TAAT, lists are sorted by term impact in non-increasing order. Meanwhile, DAAT uses docid-sorted lists instead. DAAT performs well on AND query and supports dynamic pruning algorithms like WAND [16] and Maxscore [17], hence is widely adopted in current search engines.

In WAND, an ingenious pointer movement strategy based on *pivoting* is used, which allows it to skip many documents that would be evaluated by an exhaustive algorithm. It stores for each posting list the highest impact score of any posting in the list, called *maxscore*. There are three steps when the algorithm processes the lists: pivot selection, alignment check and evaluation. Left hand side of Fig. 1 gives a glance into the procedure of WAND, terms are sorted in increasing order by their current docids, maxscores are precalculated and the current minimum score of top-k results is known and used as a threshold, maxscores are accumulated from top to bottom, docid whose sum excesses threshold is chosen as pivot. Thus, we can align the lists above to the position which is no less than pivot docid, if the pivot docid appears in these lists then we evaluate this document, otherwise we sort the lists according to the current docids and pivot again.

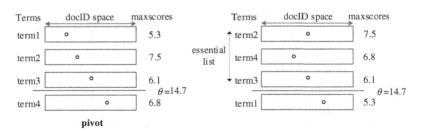

Fig. 1. On the left is the pivot selection phase in WAND. Posting lists are sorted by their current docid and term3 is selected as pivot term. On the right is the selection of essential lists in Maxscore. Lists are sorted by their maximum score and essential lists are those whose sum overpasses current threshold.

Maxscore also precalculates the maximum score of each list, before query processing lists are sorted by their maxscores in decreasing order, after scoring some documents, we again have a threshold that a document must meet to be ranked into current top-k results. According to the threshold lists are divided as essential and non-essential lists. Note that no document can make it into the top-k results just using postings in the non-essential lists, and at least one of the essential terms has to occur in any top-k documents. Each time we pick out the least docid from essential lists, and pointers of other lists are aligned to it, scoring is executed incrementally from top to bottom, once we find current document fail to pass the threshold, scoring is aborted and another docid is picked until essential lists reach their ends.

2.2   Wavelet Tree

Recent years, several researchers have been making efforts to bring the compact data structure to bear on the problems in IR, in particular ranked document retrieval. Brisaboa et al. [18] present an encoding scheme called Directly Addressable Codes (DACs), which enables direct access to any element of the encoded sequences without the need of sampling method, but the symbols must be stored in complete form, making it hard to compress; Culpepper et al. [19, 20] adopt the HSV data structure in combination with a wavelet tree-base representation to retrieve top-k results, but it is not safe; Petri et al. [21] describe a hybrid index consisting of a pruned suffix of document-level posting lists to efficiently handle large intervals, and fast sequential exhaustive processing of smaller sections to create document-level lists on-the-fly. Although quite different in their details, the common vision of these work is to use breakthroughs in compressed pattern matching as an efficient algorithmic base on which the more sophisticated operations required by IR systems can be built.

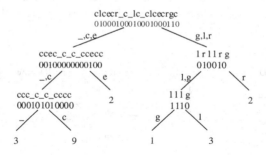

Fig. 2. A wavelet tree over a sequence "clcecr c lc clcecrgc". In each node, the top row shows subsequence $S_v[1, n_v]$ and the second row shows bitmap B_v. Spaces in the sequence is represented by underscores. Since leaf nodes store only one symbol, a sum is saved instead. Here is an example of letters, however, wavelet tree can extend to docid or term frequency sequences easily.

Wavelet tree is a versatile data structure which stores a sequence $S[1, n]$ over an alphabet $\Sigma[0 \cdots \sigma]$, within a space requirement of $n log \sigma (1 + o(1))$ bits. Figure 2 gives an example of the structure of wavelet tree. The tree is a complete balanced binary tree, where each node handles a range of symbols. The root handles $[0 \cdots \sigma)$, its children nodes bisect the range recursively until reach the leaves which only hold a single symbol. The detail procedure is as follows: each node v in the tree handling the range $[a_v., .\omega_v)$ represents the subsequence $S_v[1, n_v]$ of S formed by the symbols in $[a_v., .\omega_v)$, instead of storing S_v, a bitmap B_v is stored, so that $B_v[i] = 0$ if $S_v[i] < a_v + 2^{\lfloor log(\omega_v - \alpha_v) \rfloor - 1}$ and $B_v[i] = 1$ otherwise. Then the alphabet interval breaks into two roughly equal parts:$[\alpha_v, \alpha_v + 2^{[log(\omega_v - \alpha_v)]-1})$ and $[\alpha_v + 2^{[log(\omega_v - \alpha_v)]-1}, \omega_v)$, ditto for S_v and B_v, generating their left and right children nodes. It is easy to figure out the tree has a height of $log\sigma$, and it has exactly σ leaves and σ - 1 internal nodes. The B_v in each level exactly occupies n bits, for a total of at most $n[log\sigma]$. Storing the tree pointers, and the pointers to the bitmaps, requires $O(\sigma logn)$ further bits, if we use the minimum $logn$ bits for the pointers.

Within that space, the wavelet tree is able to return any sequence element $S[i]$, and also to answer another two queries that are fundamental in succinct data structure for text retrieval. Note that searching in bitmap B_v to obtain $B_v[i]$ can be solved in constant time, which enables the following three operations return in $O(log\sigma)$ time.

$access(S, i)$: returns the symbol at the position i in the sequence S.

$rank_c(S, i)$: returns the number of times symbol c appears in the prefix $S[1, i]$.

$select_c(S, i)$: returns the position of the i-th occurrence of symbol c in sequence S.

3 Auxiliary Index Structures

3.1 Block-Max Index

A Block-Max Index (BMI), in a nutshell, augments the commonly used inverted index structure, where for each distinct term t we store a sorted list of the docids of those documents where t occurs, with upper-bound values for blocks of these docids. That is, for every say 64 docids of documents containing a term t, we store the maximum term-wise score of any of these documents with respect to t. This then allows algorithms to quickly skip over blocks of documents whose scores are too low to make it into the top results, as shown in Fig. 3.

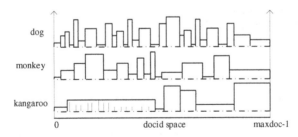

Fig. 3. Three inverted lists are piecewise upper-bounded by the maximum scores in each block. Inside each block are various values in compressed form as shown in the bottom list, one block needs to be decompressed when its maximum score has the potential to be ranked in top-k results, or the whole block will be skipped.

BMI is designed particularly to speed up early termination algorithms like Maxscore and WAND. We refer to the approach in [4] as Block-Max WAND (BMW), and the approach in [22] as Block-Max Maxscore (BMM). BMW defines two functions used in list traversal: *deep pointer movement* refers to receiving a docid from the current list that is equal to or greater than a given one and it usually involves a block decompression; *shallow pointer movement* only moves the current pointer to the corresponding block without decompression. BMW picks a candidate pivot using the list maxscores as in WAND, but then it uses shallow movement to check if it is necessary to decompress the block and evaluate the pivot based on the maxscores of the block, which helps filtering out most of the candidates and achieving much faster query response. Also, if a pivot d fails to make it into the top results, instead of picking the docid next to it, the pointer is moved to the header of the next block, since d is ruled out based on the maxscore of current block. Different from BMW, BMM uses a preprocessing step rather than an online one to detect and align block boundaries. The blocks are repartitioned into intervals with interval boundary, then BMM runs Maxscore within each interval, selecting non-essential lists and doing partial scoring using block maxscores instead of list maxscores.

Shan et al. [5] propose an optimization on BMI which combines the static score such as PageRank and the IR score into one score to give a correct and better estimation of document's upper bound score, they also recalibrate candidate documents using local BMW and BMM to omit invalid scoring. Dimopoulos et al. [6] compare performances of WAND- and Maxscore-based algorithms with and without BMI, then build on their observations by designing and implementing new techniques for exploiting BMI, in particular docid-oriented block selection schemes, on-the-fly generation of BMI, and a new recursive query processing algorithm that uses a hierarchical partitioning of inverted lists into blocks. The core idea of their improvement is to decouple the choice of blocks for storing block maxscores from the choice of blocks for inverted index compression, making the BMI as a structure separate from the inverted lists. In some cases, much smaller blocks are chosen to get better pruning power, however, this also results in a much larger space occupation. To solve the problem, they have defined block boundaries based not on the number of postings in a block, but based on docid space and carefully tuned the block size to achieve good space-time tradeoffs, moreover, on-the-fly Block-Max generation is proposed to further shrink the size of BMI with little time overhead.

3.2   Dual-Sorted Index

The main data structure used in Dual-Sorted Index (DSI) is wavelet tree. To make it suitable for document retrieval, symbols of letters are substituted by docids in the range $[1, D]$, where D denotes the total number of documents in the collection. Let $L_t[1, df_t]$ be the list of docids in which term t appears, in decreasing tf order. Let $N = \sum_t df_t$ be the total number of occurrences of distinct terms in the documents. All the posting lists L_t are concatenated into a unique list $L[1, N]$, and the starting position s_t of list L_t within L is also stored. The sequence L of docids is then represented with a wavelet tree. Note that the structure is a complete balanced binary tree with D leaves. The leaves are labeled left-to-right with the symbols $[1, D]$ in increasing order. For any internal node v of the wavelet tree, let L_v be a subsequence of L containing only the docids on the leaves in the subtree with root v. At each node v of depth ℓ, the docids are split by their most significant bit (*msb*) in the position of $logD - \ell$, for each docid, if $msb_{logD-\ell} = 0$ it will appear below the left child of v, or the right child otherwise. With this property, the docid can be restored by recording its path (e.g. consider alphabet 0, 1, 2, 3, 4 = 000, 001... 100; after 3 times turning left, we descend down to the leaf 0).

Note that L_t of each term is sorted in decreasing tf order, symbols in wavelet tree are sorted in increasing order inherently, thus making the index dual-sorted for different type of retrieval. The tf values are sorted in differential and run-length compressed form in a separate sequence. Finally, the s_t sequence is represented using a bitmap $S[1, N]$, preprocessed for *rank* and *select* queries. Thus $s_t = select_1(S, t)$, and also $Rank_1(S, i)$ tells where $L[i]$ belongs to. Analysis of wavelet tree shows that space occupied by L is $NH_0(L) + o(NlogD)$ bits, here $NH_0(L) = \sum_d dt_d log \frac{N}{dt_d} \leq NlogD$ and dt_d denotes the number of distinct terms in document d. The tf values are stored in a sequence $W[1, N]$ aligned to L, which is also asymptotically similar to the space needed by L. The s_t values are represented to support constant time *rank* and *select* queries, requiring $Vlog\frac{N}{V} + O(V) + o(N)$ bits, which is less than the usual pointers from the vocabulary to the list of each term and here V denotes the number of distinct terms in collection. Roberto improves DSI's time efficiency using a fast implementation from [23] that uses 37.5 % extra space on top of bitmap since L is not expected to be compressible, also s_t is replaced by V pointers from term t to the starting positions of the list L_t in L.

The three fundamental queries are detailed as follows. For $access(L, i)$, a look-up is started from the root node v, if $B_v[i] = 0$ we descend to the left child and i is updated using $i = rank_0(B_v, i)$, or right child with $i = rank_1(B_v, i)$ otherwise. This process is continued recursively until a leaf is reached and $L[i]$ is the concatenated path in binary. For $rank_c(L,i)$, bits in c tell the path to descend, the only thing needs to do is updating i to be the number of times of the current bit in depth ℓ appears up to position i in the node until a leaf is reached. For $select_c(L,i)$, the look-up is processed upwards to the root, as the path is already clear, each level ℓ_i. i is updated by $i = select_0(B_\ell, i)$ if $c[\ell] = 0$, or $i = select_1(B_\ell, i)$ if $c[\ell] = 1$. More complex operations can be implemented by combining the above three queries like retrieving all the values in a range $L[i, j]$ and retrieving the k-th value in a range $L[i,j]$. These algorithms can be easily adopted in Boolean conjunctive and disjunctive queries, when we find the $|q|$ intervals $[s_t, e_t]$ of the query words using pointers from the vocabulary to the inverted lists L_t, we can track all the $|q|$ ranges simultaneously and recursively merge or intersect them to obtain a result. Note that the docids in each list are sorted by term frequencies, so we can immediately compute the documents score and retain the k highest scoring documents. See more details of the algorithms in [7, 8].

4 Experiments

4.1 Experimental Setup

We implement both BMI and DSI for comparison, as an external implementation to compare we choose Terrier as a baseline, which is a highly flexible and effective open source search engine. Terrier supports both disjunctive and conjunctive queries in

DAAT and TAAT traversals with different weighting models, we will show the result that although inferior to the two state-of-the-art indexes, it does gain performance achievement after carefully tuned.

In our experiments, we use the TREC GOV2 collection, which consists of 25.2 million web pages and about 32.8 million terms in the vocabulary crawled from the gov Internet domain. The uncompressed size of these web pages is 426 GB. The collection is indexed using Terrier IR platform, all terms have the Porter stemmer applied, and stopwords have been removed. The docids are assigned by the sequence of their occurrence.

For retrieval, a total of 1000 queries are selected from the TREC2005 Efficiency track queries and classified into different categories according to the number of distinct terms in the query. Unless stated otherwise, the default number of documents retrieved for each query equals to 20 that a result page will contain and BM25 is used as ranking function in order to keep consistent with work in [4, 6].

All the implementations are carried out on an Intel(r) Xeon(r) E5620 processor running at 2.40 GHz with 128 GB of RAM and 12,288 KB of cache. The default physical block size is 16 KB, unless stated otherwise algorithms are implemented using C ++ and compiled with GCC 4.8.1 with –O3 optimizations. In all our runs, the whole inverted index is completely loaded into main memory, in order to warm up the execution environment, each query set is run 4 times for each experiment, and the response times only measured for the last run. Our implementations are available at https://github.com/Sparklexs/Dualsorted-master.

4.2   Index Size Comparison

First we compare the index size of each structure. The baseline is compressed using both Gamma and OptPFD. As depicted in [4], BMI compresses docid-gaps and frequencies using OptPFD, compared with baseline OptPFD, BMI barely expands its size, as it only augments the index with local maxscores and pointers to the header of blocks. For DSI, its docids and frequencies are stored separately, the frequency lists are compressed using Gamma codec, however, the docid lists are hard to compress since they are stored in wavelet trees in primitive form. Moreover, in order to achieve direct access to any symbol in the tree, some additional structures are also stored, all these result in the size of DSI grows nearly 4 times larger than the rest indexes. Table 1 gives the detail of each index. We also show the results after docid reassignment for the collection. The idea of docid reassignment is to reorder the documents in the collection so that similar documents are clustered together, thus shrinking the gap between docids and the index size, it also improves the speed of dynamic pruning for that both potential and invalid documents are batched up to reach an early termination, here we choose to reorder the documents based on an alphabetic sorting of their URLs. As the right hand side of Table 1 shows, baseline and BMI reduce nearly 20 % size after docid reassignment, however, DSI seems insensitive to reassignment as it hardly changes size in both situations, this is mainly due to the fact that DSI separately stores docids in their primitive form and frequencies in decreasing order without relying on docid order. Also note that BMI cuts out the same size as baseline OptPFD, which can be explained by the fact that they adopt the same structure except BMI stores its additional information in its raw format.

Table 1. Size in GB of different inverted index for GOV2

	before reordering	after reordering	Δ
Baseline Gamma	9.86	7.11	2.75
Baseline OptPFD	9.45	7.33	2.12
BMI OptPFD	10.86	8.75	2.12
DSI Gamma	39.2	39.2	0

4.3   Query Processing Efficiency

Next, we compare the timing results of each index for different length of queries. First we use the index without reordering and the processing strategy chosen for baseline index is DAAT, to distinguish different compressions we name the result of baseline Gamma as GDAAT and ODAAT for baseline OptPFD, BMM is short for BMI using Maxscore and BMW for BMI using WAND. For now there isn't any dynamic pruning techniques used in DSI in the literature, as arbitrary symbols in the index are directly accessible, so we simply use DAAT-like strategy for DSI (which is quite different from ordinary DAAT since the documents are not fetched in ascending order). The results are shown in Fig. 4, we omit the result of a single term query, since all the processing strategies degenerate into DAAT with only one posting list and performances of these methods become the same. It is worth mentioning that our implementations achieve a close performance to what has been reported in [21].

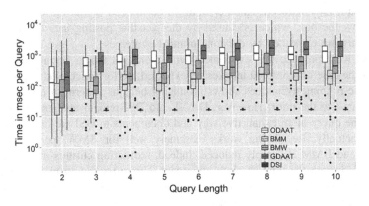

Fig. 4. Efficiency for queries of different length using different indexes. The interquartile range is shown in the box part and solid line in the box represents the median, points represent the outliers. One thing to take notice of is that the box part of DSI shrinks into a solid line.

It is clear that the same structure using different compressions gains performance gap between ODAAT and GDAAT, which can be explained by the fact that shift and concatenation operations are inevitable to decode an integer using bit-oriented codecs, however, list-oriented codecs can decode a batch of integers with a single operation. BMI with dynamic pruning methods further improves processing efficiency, nonetheless, BMM achieves better performance than BMW, which is different from the result in [21], after an inspection into the procedure of both methods, we conclude that the list sorting consumes much time in pivot selection phase of BMW, also BMM selects candidate documents in the essential lists which are likely to belong to more important terms, while BMW selects candidate documents based on their docids without considering the terms' importance, with more query terms, it causes more mis-scoring candidates that slows down the performance of BMW. Here we only implement the basic BMI, some other methods like on-the-fly BMI generation and hierarchical-layered blocks adopted in [6] remain future investigation. When it comes to DSI, we get an interesting result that its time consume is rarely low compared with other strategies, which are an order of magnitude higher than DSI; another surprising phenomenon is that it stays considerably stable when query length grows while other strategies raise their time consume to different extent, the box part is contracted into a solid line which can be hardly noticed. It can be concluded without doubt that DSI outperforms all the others. However, this performance is achieved at the cost that the huge size structure of DSI is fully loaded into the main memory and occupies nearly 50 GB space, while others keep low memory occupation within 6 GB. This result is also consistent with the inference before, we postulate that scoring a document and maintaining a priority queue for the result cost constant time, then the only factor that influences time efficiency is the size of the collection, as the three basic operations can be implemented in $O(logD)$ time, so is the intersection and merge of the posting lists. Also the documents in DSI are sorted by term frequency in decreasing order, the potential candidates to be ranked in top-k lie in the front of each posting list. With this in mind, we can just evaluate few documents rather than the whole list and the query time is significantly reduced. Another thing to be noticed is that the outliers always appear below the box, which is caused by the fact that posting lists for uncommon terms are missing or pruned by dynamic pruning strategies.

Second we reproduce this experiment with reordered index. Figure 5 shows the results for the case of reordered indexes, as expected, performance of structures which use dynamic pruning techniques shows promising benefits in term of query response time compared to the prior. However, DAAT and DSI remain unaffected, as DAAT processes the query in an exhaustive way and documents in DSI are ordered by term frequency actually. Reordering works extremely well for BMW, performance gap between it and BMM is sharply reduced. Indeed, reordering clusters the related documents and shorten the time of pivot alignment. For query length no longer than 5, BMM and BMW are about the same efficiency, their average response times fall below 10^2 ms, as query length grows, the performance gap rises again.

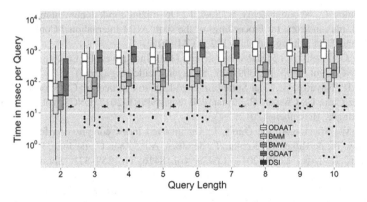

Fig. 5. Efficiency for queries of different length using reordered indexes.

5 Conclusions

In this paper we have presented an explicit study of two state-of-the-art inverted index structures used in safe retrieval in disjunctive queries. Our main contribution is an experimental comparison of these two indexes combined with different compression techniques and traversal strategies, we also discuss the effect on query processing efficiency brought by docid reassignment. As shown in the result, DSI reveals its superiority over other techniques in query response, however, a serious disadvantage is obvious that it occupies too much storage and memory space. Overall, DSI is a promising structure taking its feature of bridging the gap between IR problems and pattern matching data structures into account.

There are still many open problems and opportunities for future research, including mitigating the space issue using on-the-fly index generation and extending DSI with dynamic pruning techniques and list-oriented compression codecs.

References

1. Zobel, J., Moffat, A.: Inverted files for text search engines. ACM Comput. Surv. (CSUR) **38** (2), 6 (2006)
2. Lemire, D., Boytsov, L., Kurz, N.: SIMD Compression and the Intersection of Sorted Integers (2014). arXiv:1401.6399
3. Barbay, J., López-Ortiz, A., Lu, T., Salinger, A.: An experimental investigation of set intersection algorithms for text searching. J. Exp. Algorithmics (JEA) **14**, 7 (2009)
4. Ding, S., Suel, T.: Faster top-k document retrieval using block-max indexes. In: Proceedings of the 34th international ACM SIGIR Conference on Research and Development in Information Retrieval, pp. 993–1002. ACM (2011)
5. Shan, D., Ding, S., He, J., Yan, H., Li, X.: Optimized top-k processing with global page scores on block-max indexes. In: Proceedings of the fifth ACM International Conference on Web Search and Data Mining, pp. 423–432. ACM (2012)

6. Dimopoulos, C., Nepomnyachiy, S., Suel, T.: Optimizing top-k document retrieval strategies for block-max indexes. In: Proceedings of the sixth ACM International Conference on Web Search and Data Mining, pp. 113–122. ACM (2013)

7. Navarro, G., Puglisi, S.J.: Dual-sorted inverted lists. In: Chavez, E., Lonardi, S. (eds.) SPIRE 2010. LNCS, vol. 6393, pp. 309–321. Springer, Heidelberg (2010)

8. Konow, R., Navarro, G.: Dual-Sorted Inverted Lists in Practice. In: Calderón-Benavides, L., González-Caro, C., Chávez, E., Ziviani, N. (eds.) SPIRE 2012. LNCS, vol. 7608, pp. 295–306. Springer, Heidelberg (2012)

9. Catena, M., Macdonald, C., Ounis, I.: On inverted index compression for search engine efficiency. In: de Rijke, M., Kenter, T., de Vries, A.P., Zhai, C., de Jong, F., Radinsky, K., Hofmann, K. (eds.) ECIR 2014. LNCS, vol. 8416, pp. 359–371. Springer, Heidelberg (2014)

10. Navarro, G.: Wavelet trees for all. J. Discrete Algorithms **25**, 2–20 (2014)

11. Gagie, T., Navarro, G., Puglisi, S.J.: New algorithms on wavelet trees and applications to information retrieval. Theoret. Comput. Sci. **426**, 25–41 (2012)

12. Claude, F., Navarro, G.: The wavelet matrix. In: Calderón-Benavides, L., González-Caro, C., Chávez, E., Ziviani, N. (eds.) SPIRE 2012. LNCS, vol. 7608, pp. 167–179. Springer, Heidelberg (2012)

13. Ounis, I., Amati, G., Plachouras, V., He, B., Macdonald, C., Lioma, C.: Terrier: A high performance and scalable information retrieval platform. In: Proceedings of the OSIR Workshop, pp. 18–25 (2006)

14. Li, X., Wang, Y., Li, X., et al.: Parallelizing skyline queries over uncertain data streams with sliding window partitioning and grid index. Knowl. Inf. Syst. **41**(2), 277–309 (2014)

15. Ma, D., Rao, L., Wang, T.: An empirical study of SLDA for information retrieval. In: Salem, M.V.M., Shaalan, K., Oroumchian, F., Shakery, A., Khelalfa, H. (eds.) AIRS 2011. LNCS, vol. 7097, pp. 84–92. Springer, Heidelberg (2011)

16. Petri, M., Culpepper, J.S., Moffat, A.: Exploring the magic of WAND. In: Proceedings of the 18th Australasian Document Computing Symposium, pp. 58–65. ACM (2013)

17. Turtle, H., Flood, J.: Query evaluation: strategies and optimizations. Inf. Process. Manage. **31**(6), 831–850 (1995)

18. Brisaboa, N.R., Ladra, S., Navarro, G.: DACs: bringing direct access to variable-length codes. Inf. Process. Manage. **49**(1), 392–404 (2013)

19. Culpepper, J., Navarro, G., Puglisi, S.J., Turpin, A.: Top-*k* Ranked document search in general text databases. In: Berg, M., Meyer, U. (eds.) ESA 2010, Part II. LNCS, vol. 6347, pp. 194–205. Springer, Heidelberg (2010)

20. Culpepper, J.S., Petri, M., Scholer, F.: Efficient in-memory top-k document retrieval. In: Proceedings of the 35th International ACM SIGIR Conference on Research and Development in Information Retrieval, pp. 225–234. ACM (2012)

21. Petri, M., Moffat, A., Culpepper, J.S.: Score-safe term-dependency processing with hybrid indexes. In: Proceedings of the 37th International ACM SIGIR Conference on Research & Development in Information Retrieval, pp. 899–902. ACM (2014)

22. Chakrabarti, K., Chaudhuri, S., Ganti, V.: Interval-based pruning for top-k processing over compressed lists. In: IEEE 27th International Conference on Data Engineering (ICDE), 2011, pp. 709–720. IEEE (2011)

23. González, R., Grabowski, S., Mäkinen, V., Navarro, G.: Practical implementation of rank and select queries. In: Proceeding of WEA, pp. 27–38 (2005)

Access Time Tradeoffs in Archive Compression

Matthias Petri, Alistair Moffat[(✉)], P.C. Nagesh, and Anthony Wirth

Department of Computing and Information Systems,
The University of Melbourne, Victoria 3010, Australia
ammoffat@unimelb.edu.au

Abstract. Web archives, query and proxy logs, and so on, can all be
very large and highly repetitive; and are accessed only sporadically and
partially, rather than continually and holistically. This type of data is
ideal for compression-based archiving, provided that random-access to
small fragments of the original data can be achieved without needing to
decompress everything. The recent RLZ (relative Lempel Ziv) compression approach uses a semi-static model extracted from the text to be compressed, together with a greedy factorization of the whole text encoded
using static integer codes. Here we demonstrate more precisely than
before the scenarios in which RLZ excels. We contrast RLZ with alternatives based on block-based adaptive methods, including approaches that
"prime" the encoding for each block, and measure a range of implementation options using both hard-disk (HDD) and solid-state disk (SSD)
drives. For HDD, the dominant factor affecting access speed is the compression rate achieved, even when this involves larger dictionaries and
larger blocks. When the data is on SSD the same effects are present, but
not as markedly, and more complex trade-offs apply.

1 Introduction

Large data archives are often retained for long periods. Examples include web
crawls; site edit histories for resources such as the Wikipedia; query, proxy, and
click logs; and many other forms of meta-data associated with the way we store
and access information. Such archives are rarely decoded in full, and even partial-
access operations may be infrequent. Moreover, the data might be highly repetitive, with occasional very long repeated strings, and repeated strings that are
widely separated. There is thus considerable interest in specialized compression
techniques that provide a high level of space saving for such data, plus the ability
to support random access to small fragments of it.

The Relative Lempel-Ziv (RLZ) compression approach is designed for
archives like these [5]. It involves a plain-text *dictionary* extracted from the collection of documents via fixed-interval sampling across their concatenation. The
documents are then factored against the dictionary using the standard Lempel-
Ziv greedy parsing approach, and factor descriptions consisting of copy offsets
and copy lengths are represented with static integer codes. Because the dictionary and encodings are both static, decoding is possible from any point in the
encoded stream, provided only that a corresponding code-aligned byte or bit

© Springer International Publishing Switzerland 2015
G. Zuccon et al. (Eds.): AIRS 2015, LNCS 9460, pp. 15–28, 2015.
DOI: 10.1007/978-3-319-28940-3_2

address is given for the document that is required. Moreover, decoding is fast – during decoding operations the dictionary is stored in memory uncompressed, allowing rapid access to factors that can then be copied directly to the output stream as required. More details of the RLZ approach are given in Sect. 2.

While the approach provided by RLZ is indeed a good solution to the question of archive compression, other methods based on *adaptive* compression mechanisms are available. For example, standard tools like GZip and xz can be applied on a per-block basis. The block size then becomes an important parameter that trades compression effectiveness against access speed. The larger the block size, the better the compression rate, but the longer it takes for a fragment of text to be reconstructed, since decompression must start at the beginning of a block.

Our purpose in this paper is to provide detailed evidence of RLZ's capability in archive compression. Our analysis includes the effects of the storage device chosen, and both hard-disk drives (HDD) and solid-state disk (SSD) storage are employed. We analyze the factors that determine the time required to access a fragment of text from an arbitrary location in a large corpus, and show how different compression techniques can be evaluated. The approaches explored include making use of a facility provided by the standard ZLIB library in which a "priming" text enhances compression effectiveness during the start-up phase of GZip's Lempel-Ziv implementation. The various options are compared on the 426 GiB GOV2 crawl of the .gov domain, which contains a broad mix of HTML, PDF, and other document formats.

Based on those experiments, we conclude that for HDD the dominant factor affecting access speed for random decoding is compression effectiveness, with block size a secondary factor; whereas for SSD decompression speed is also a factor. Our results confirm and extend those of Hoobin et al. [5], providing additional insights into the behavior of this important archiving technique. Our new implementation of RLZ will be made available on completion of the project, so that other compression approaches can also be incorporated as they are developed.

2 RLZ Compression

We now provide a brief description of the RLZ archive compression mechanism [5].

Forming a Dictionary. The collection of documents to be stored are concatenated to make a single large file; we let C denote that single string, and $|C|$ be its length in bytes. Two parameters are then identified: the *dictionary size*, denoted $|D|$ (with D to be used for the dictionary); and the sample size s, chosen to be a factor of $|D|$. The dictionary is formed by taking $|D|/s$ samples, each s bytes long, from C, extracting them at regular $|C|/(|D|/s)$-byte intervals. For example, if $|C| = 64\,\text{GiB}$ and $s = 1\,\text{kiB}$, then a dictionary of $|D| = 64\,\text{MiB}$ would be formed by concatenating a total of 65,536 samples, extracted every 1,048,576 bytes of C. Figure 1 shows the process of extracting regular samples from C to form the dictionary D, regardless of the underlying document boundaries.

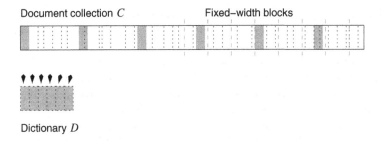

Fig. 1. Constructing the RLZ dictionary D by selecting regular samples from the document collection C. Document boundaries in C are shown by dotted lines; block boundaries (over part of the collection) by dashed lines.

Factoring the Collection. Once D has been formed, C is broken into a sequence of *blocks*, and each block independently *factored* against D, using a left-to-right greedy approach. The blocks might be variable-length and formed by considering individual documents in the collection; might be variable-length and formed by taking groups of documents to reach some minimum size; or might be fixed-length and formed by taking some exact number of bytes. In our implementation we adopt the latter approach, meaning that access to any byte range or to any particular document requires that the corresponding block or blocks be identified and retrieved.

To generate the factorization for each of the blocks, D is indexed via a suffix array or similar structure, so that for an arbitrary string S, the set of longest-matching prefixes of S that appear in D can be identified. Starting at the beginning of each block, factors relative to D are identified and represented by a pair of integer values: the *length* of the factor, and its *offset* in D. If the next character in the block does not appear in D, a *literal* is generated – a factor length of zero, and then an ASCII character code rather than a dictionary offset. There are a range of ways in which the presentation of literals can be optimized, including the application of a minimum match length, or separating them into a distinct third stream. These alternatives are explored in Sect. 4; Hoobin et al. [5] assume that literals are sufficiently rare that intermingling them in the stream of offsets will not adversely affect compression effectiveness. Except when specifically described otherwise, references to factor offsets below include any literals that may have been required. The last factor in each block is truncated so that it finishes at the block boundary. The compressed equivalent of each fixed-length block is then the fundamental access unit for decoding, with higher-level operations such as document retrieval and byte-range retrieval implemented on top of the block access routines.

Compression Rate. The total cost of storing C is the cost of storing D, plus the cost of storing all of the $\langle \textit{offset}, \textit{length} \rangle$ pairs. The dictionary can be stored using any desired compression mechanism, and is fully decoded into memory prior to subsequent access operations. Even stored uncompressed, it is typically a small fraction of the original collection. Continuing the previous example, $|D|/|C| = 0.1\,\%$, and a compressed representation of D should occupy well under $0.03\,\%$ of $|C|$.

The majority of the space required is in the $\langle \textit{offset}, \textit{length} \rangle$ pairs. As already noted, they are separated into two streams on a per-block basis, with each stream coded using a static method such as 32-bit or minimal-width binary integers, or the variable-width byte-oriented vbyte approach [9]. The two coded streams are then typically padded to a byte or word boundary, concatenated to make a single unit, and a small prelude added that includes a count of the number of factors contained. Continuing with the same example, suppose that C is partitioned uncompressed into blocks of 16 kiB; that the average factor length is 20 bytes; that each $\textit{offset}$ is coded in $\log_2 |D| = 26$ bits; and that almost all factor lengths are coded in one byte each (vbyte codes for factor lengths of up to 127). Then each factor requires 34 bits, and the offsets and lengths for a block are stored in around 3.4 kiB, a compression rate of approximately $3.4/16 \approx 22\,\%$. Previous experimental results with RLZ suggest that all these various estimates are reasonable [5], and they are further confirmed in the experiments described in Sect. 4.

Random Access Decoding. To provide random-access decoding, index pointers to each block in the compressed integer stream are maintained in an auxiliary structure. The block size determines the number of index points and hence the size of the index, which is important because the index must also be retained in memory during access operations. In the same example, with blocks of 16 kiB, a set of 4,194,304 indexing pointers into the compressed stream is required, with each pointer 34 bits long to address a compressed file of approximately 16 GiB. That is, in the example an index to allow random access to blocks consumes 17 MiB, a further overhead.

To decode a fragment of C specified by an uncompressed byte range (for example, if one document is required, and a mapping from document identifiers to byte addresses is available) standard mod/div arithmetic is performed to determine the ordinal numbers of the block or blocks that are required. The block index (required to be memory resident) is then used to determine the address of the bundle of de-interleaved $\langle \textit{offset}, \textit{length} \rangle$ pairs for that block, and a file operation undertaken to fetch the relevant data from secondary storage. The dictionary D (also memory resident) is then used, with $D[\textit{offset}]$ to $D[\textit{offset} + \textit{length} - 1]$ copied to a decode buffer for each $\langle \textit{offset}, \textit{length} \rangle$ factor extracted from the compressed blocks. The required range of bytes from within the block can then be written to the output stream once the block decode buffer is filled. That is, after a compressed block has been fetched into main memory, reconstructing a fragment of C consists of decoding two sequences of integers using static integer codes, and then copying strings. Both operations are fast. Further blocks are fetched and decoded if required, until the byte range specified in the query has been delivered.

Ferrada et al. [2] have also considered random access in RLZ mechanisms.

Memory Footprint. Compression effectiveness is in part determined by the amount of space used for the dictionary, as another dimension of effectiveness-efficiency trade-off. For example, if the memory required ($64\,\mathrm{MiB} + 17\,\mathrm{MiB}$ in the example scenario) must be reduced for some reason, either the block size can be increased, potentially affecting access speed; or the dictionary size decreased, potentially affecting compression rate. If the block size is increased to $64\,\mathrm{kiB}$, the index reduces to $4.3\,\mathrm{MiB}$. The drawback, of course, is that four times as much data must be transferred into main memory to fulfill a request, and more of it is likely to be required to be decoded as well, unless internal structure is added within each block. As is demonstrated in the experiments below, transfer and decoding times are usually small, and block sizes in the tens of kilobyte range are acceptable. The uncompressed dictionary D is then the dominant memory requirement during random-access decoding. To mitigate this cost, methods have been developed for pruning the dictionary to remove unused or under-used strings [7].

Access Time. In a memory-to-memory context, string-copy decoders similar to RLZ generate text at around $250\,\mathrm{MiB}$–$300\,\mathrm{MiB}$ per second.[1] A compressed block derived from $64\,\mathrm{kiB}$ of C can thus be decoded in around $0.25\,\mathrm{ms}$. But that can only happen once it has been fetched from secondary memory. Table 1 provides indicative performance figures for mechanical (HDD) and solid-state (SSD) secondary memory devices. In a mechanical disk, there is a non-trivial startup time for each data transfer, involving (with high probability) a seek operation to move the read head, followed by a delay resulting from rotational latency. Solid-state disks achieve higher data transfer rates, and commence the data transfer relatively quickly after the request is received.

Table 1. Performance of different storage media. Extracted from product specifications of current devices: Seagate ST3000DM001 (HDD), Intel SSD 750 Series (SSD).

Medium	Random read latency	Sequential transfer rate
Hard disk (HDD)	$8.5\,\mathrm{ms}$	150 MiB/s
Solid-state disk (SSD)	$0.12\,\mathrm{ms}$	1000 MiB/s

If compressed blocks are stored on HDD, the seek-plus-latency cost of approximately $8.5\,\mathrm{ms}$ dominates the cost of transferring the data (around $0.15\,\mathrm{ms}$ for the compressed equivalent of a block of, say, $64\,\mathrm{kiB}$ of C), and the cost of decoding that block once it is in memory (around $0.25\,\mathrm{ms}$). Based on this arithmetic, and assuming that each query consists of accessing a 16 kiB segment of C, a throughput of around 110 random-access queries per second should be possible. Of that time, decoding activity occupies less than 3 %. On the other hand, if the whole collection is decoded sequentially (meaning that seek and latency times are

---

[1] https://github.com/Cyan4973/lz4, accessed 27 July 2015.

amortized to zero), and if compression effectiveness of 30 % or better is achieved (meaning that decoding cost completely subsumes transfer cost) then data can be handed to another process at the measured peak output rate. Continuing the same example, a rate of 300 MiB decoded per second correspond to up to 5,000 64 kiB-blocks, or 20,000 16 kiB-blocks.

If SSD is used, the situation for random access changes markedly. Now the transfer initialization time is around 0.1 ms, meaning that something like 2,900 64 kiB blocks per second can be fetched and decoded, with the decoding taking around 60 % of the total time. Sequential access continues to be dominated by decoding cost, and remains capped at around 20,000 16 kiB-blocks per second. All of these estimated access time and throughput rates are validated empirically in Sect. 4.

3   Block-Based Adaptive Alternatives

We now consider additional options for archive compression.

Standard Compression Libraries. Standard compression tools such as GZip, BZip2, and xz, are *adaptive*, in that they use dynamic models and codes, so as to be versatile across file types. For example, the well-known GZip compressor adopts the same Lempel-Ziv factorization approach as RLZ, starting each compression run with an empty dictionary, and then adding each parsed factor's text for possible use in subsequent factorizations. If GZip is applied independently to blocks, its "always-start-from-zero" approach puts it at a disadvantage compared to RLZ, because the global RLZ dictionary allows identification of long factors right from the beginning of every block.

On the other hand, adaptive compression techniques build models that are focused on exactly the content being compressed, and hence have an ability to be locally sensitive in a way that RLZ does not. Adaptive methods are also able to exploit encodings for factor offsets and lengths that are adaptive rather than static, further enhancing their ability to provide locally sensitive compression. That is, while RLZ's use of a global dictionary and static encodings for factor offsets and lengths gives it an advantage on very short blocks, localized adaptive methods may obtain better compression as the block size is increased. Part of our purpose in this investigation is to explore the options provided by these alternatives.

Block Size. A second area for exploration is the effect of block size. The connection between block size and the size of the block index was discussed above. In the case of RLZ, because it typically uses static integer codes, increasing block size has no effect on compression effectiveness. But if large blocks are passed to an adaptive compression utility, average compression effectiveness is likely to improve, because the start up cost of the model is amortized over a longer section of text. This then raises an interesting trade-off – at what block size does an adaptive dictionary provide better compression than a static RLZ-style dictionary of some given size.

For random-access operations using mechanical disk, the added decoding cost due a large block size may not matter. Even with a block size of 512 kiB, decoding of half a block, to reach a given byte address within it, takes around 0.8 ms; transfer of a full block takes approximately 1.1 ms, assuming a 25 % compression rate; and the seek-plus-latency time of around 8.5 ms is unchanged. That is, it should be possible to extract fragments from a block representing 512 kiB of text in around 11 ms, or at an estimated rate of approximately 90 queries per second.

Batch-Mode Operation. If queries are batched and processed "elevator" style, higher query throughput rates can be achieved, because average disk-seek times are likely to be smaller when the access requests are sorted. For example, if 110 random-access queries per second can be supported without batching, and if batches of sufficient size can be accumulated so that the average seek-plus-latency time drops from 8.5 ms to say 4.5 ms then the same hardware configuration should support approximately 200 queries per second. The drawback is that on average the queries will have much greater latencies before being processed – perhaps measured in tens or hundreds of seconds, rather than tens of milliseconds. In applications that fetch small fragments of a large archive, this mode of operation may still be acceptable.

4 Experiments

A New Implementation. To allow precise characterization of the performance of RLZ compression, we have created a new implementation based on fixed-length data blocks, each compressed independently, with a block index maintained in memory so that random-access queries can be supported. The system is written using ≈4000 lines of C++11 code with the help of the sdsl library [4]. We use gcc 4.9.2 running on Ubuntu 15.04 in our experiments, with all optimizations enabled.

We have explored five variants, including three RLZ versions:

- RLZ-UV, using unsigned 32-bit integers for factor offsets, and vbyte for factor lengths, as described by Hoobin et al. [5];
- RLZ-PV, using packed $\log_2 |D|$-bit integers for factor offsets, and vbyte for factor lengths; and
- RLZ-ZZ, using ZLIB (the basis of the standard GZip compression utility) version 1.2.8 (http://zlib.net) to represent each of the streams of 32-bit factor offsets and the stream of 32-bit factor lengths, on a block-by-block basis.

Each of these three methods makes use of a sampled dictionary. We also applied each of ZLIB and LZ4 (https://github.com/Cyan4973/lz4) to independent blocks, without use of a dictionary, following preliminary experimentation that included BZip2 and xz. The latter two were slower, and gave less interesting trade-offs between access speed and compression effectiveness. Finally, as a sixth system and a further baseline, we measured the performance of a COPY mechanism that does no compression at all.

Datasets. Our experiments focus on the GOV2 collection, a crawl of the .gov domain undertaken in early 2004, with documents stored in as-crawled order. This collection contains around 25 million documents as a mixture of PDF, HTML, text, and other formats, averaging 18 kiB each, and totaling 426 GiB.[2] We use both the full collection and a 64 GiB prefix of it.

Query Streams. We explore three modes of retrieval: FULL, in which the archive is decoded sequentially; RANDOM, in which a set of 10,000 random unaligned locations is accessed and a 16 kiB fragment retrieved from each; and BATCH, in which those same 10,000 locations are accessed, but with the queries sorted by address. The "Sequential" mode explored by Hoobin et al. [5] most closely matches our FULL mode, in that they measured retrieval of 100,000 consecutive GOV2 documents. Similarly, their "Query Log" mode corresponds broadly to our RANDOM mode, but with 100,000 document requests in the query stream, and hence more possibility of caching affecting throughput.

Hoobin et al. [5] also make use of a second URL-sorted GOV2 collection. They obtain notably different query throughput results for the two orderings, particularly with regard to decoding speed, differences that we were unable to reproduce with our implementation. An examination of their code suggests that the differences arise from a mode in their software that because of compiler optimization inadvertently results in no decoded output being generated. As a result, we believe that the "Sequential" retrieval speeds shown in their Table 5 (including decoding rates as high as 80,000 documents per second) should be discounted; and (for other reasons) possibly some of their other speed results too.[3] That is, our work here can be seen in part as representing re-measurement of the techniques Hoobin et al. [5] describe.

Dictionary Size and Formation. The effectiveness of the RLZ mechanism is heavily affected by the dictionary size. In their GOV2 experiments Hoobin et al. [5] work with dictionary sizes between 0.5 GiB and 2 GiB. Here we focus on smaller dictionaries, and explore the range from 16 MiB to 256 MiB for the 64 GiB test file, and the range 64 MiB to 1024 MiB for the full GOV2 collection. As described in Sect. 2, we followed the "standard" approach of selecting fixed-interval samples from the collection, presuming it to have been concatenated into a single large file. Other dictionary construction methodologies have been shown to result in small compression effectiveness gains [7]; we also explored a range of other heuristics, but found the simple interval-based sampling approach to be relatively robust. We used samples of length $s = 1024$ throughout, matching (when $|D| = 1$ GiB) some of the experiments carried out by Hoobin et al. [5]. We tested block sizes of 16 kiB, 64 kiB, and 256 kiB. All compression rates include the cost of storing the dictionary, compressed as a character stream using ZLIB, and the cost of the index table for block access, also stored using ZLIB.

---

[2] http://ir.dcs.gla.ac.uk/test_collections/gov2-summary.htm, 27 July 2015.

[3] Our concerns in this regard have been communicated to the authors of [5].

Hardware Configuration. All experiments were run on a server equipped with two Intel Xeon E5640 CPUs running at 2.67 GHz using 144 MiB RAM, a Western Digital 5 TiB (WD50EFRX-68MYMN1) HDD and a 500 GiB Samsung 850 EVO SSD. Before each experiment the operating system caches were cleared to minimize caching effects using `echo 3 > /proc/sys/vm/drop_caches`. We also took care with file placement on the HDD, noting the effect that fragmentation and track assignment can have on disk-based experimentation [8]. In some cases this meant deleting and re-copying indexes, so as to ensure that measurements were made in a fair and consistent manner. The SSD did not suffer from this variability.

High-Level View. Figure 2 presents an overview of the six methods, measured using the 64 GiB file, and shows the gross relative performance across the three querying modes and two hardware configurations. Each pane plots the relationship between compression rate, as a percentage of the original file size, on the

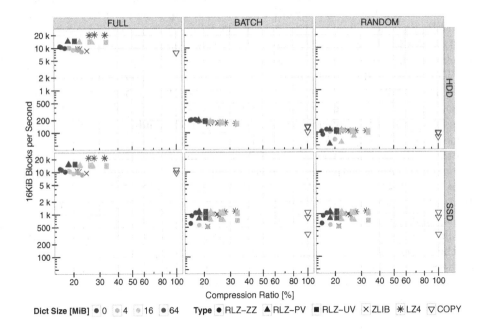

Fig. 2. Query processing rates measured as 16 kiB units retrieved per second, for three different processing modes, two types of secondary storage, block sizes of 16 kiB, 64 kiB, and 256 kiB (not individually identified in the plots), and a 64 GiB prefix of GOV2. In the FULL mode, throughput rates are for aligned 16 kiB units; for the BATCH and RANDOM modes, for unaligned 16 kiB units. The COPY, LZ4, and ZLIB methods do not use a dictionary, and are shown as 0 MiB. In general, larger block sizes lead to better compression effectiveness; together with faster access in the case of FULL operation, and slower access in the case of BATCH and RANDOM operation. (Color figure online)

horizontal axis; and access speed, measured by the number of 16 kiB blocks accessed per second. Each pane contains 36 plotted points: three RLZ variants, each with three different dictionary sizes and three different block sizes (27 data points); plus two blocked adaptive methods using the same three different block sizes (6 data points); plus the COPY method using the three block sizes. Each color corresponds to a dictionary size, and each point shape corresponds to a method. Within each method, the larger the dictionary size and/or the larger the block size, the better the compression. But increased block sizes also correspond to slower decoding. All six panes show the absolute advantage of using virtually any compression method, with the COPY approach the slowest in several cases, and never the fastest. Data compression often pays for itself. Note also that for each method, dictionary, and block size combination the compression rate is the same across all six panes.

The two left panes confirm that sequential decoding is very fast, with the LZ4, RLZ-UV and RLZ-PV approaches having a moderate speed advantage over the other mechanisms, but with all of the compressed approaches delivering 10,000+ documents (each a 16 kiB unit in these experiments) per second, or 160 MiB+/second. There is little measurable difference in performance between HDD and SSD. Unsurprisingly, the larger the dictionary and/or the larger the block size, the better the compression.

The BATCH and RANDOM modes are much slower. In the two middle panes, depicting BATCH access, there is a clear trend on the HDD for better compression to correspond to higher query throughput, with query rates of between 100 documents (unaligned 16 kiB units in this querying mode) and 200 documents per second, and relatively little differentiation between the compression techniques. On the SSD, much faster rates of 800–2,000 documents per second result, with throughput more sensitive to the choice of compression technique. Finally, the right two panes show the further slowdown arising from RANDOM access. On the HDD, query rates are around 100 documents/second; and on the SSD querying throughput is the same as for BATCH retrieval.

The SSD RANDOM and BATCH querying rates are around half those predicted by the model described in Sect. 2. Measurement of the operating characteristics of the SSD used in the experiments indicate that its mean latency is higher than is shown in Table 1, approximately 0.25 ms per access, explaining the difference between predicted and measured querying rates.

Detailed View – Random Access. Figure 3 shows a focused view corresponding to the two right-hand panes in Fig. 2, measured using the full 426 GiB GOV2 collection, and with the COPY method omitted. It considers only the RANDOM queries, using correspondingly larger dictionaries of 64 MiB, 256 MiB, and 1 GiB, and unchanged block sizes of 16 kiB, 64 kiB, and 256 kiB. At the increased scale of these graphs, it is possible to identify a Pareto frontier for each different dictionary size, and quantify the tension between compression and throughput that is controlled by block size.

For random access, the raw speed of LZ4 is less of an advantage, and it is part of the trade-off frontier only when no dictionary can be used, and when

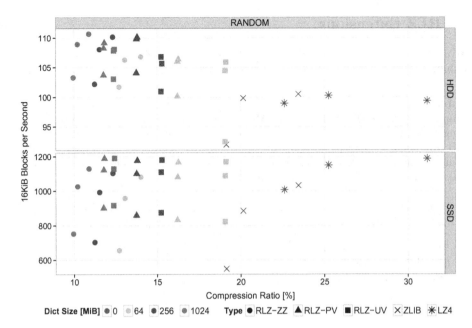

Fig. 3. Query processing rates for the RANDOM processing mode, measured as unaligned 16 kiB units retrieved per second, for two types of secondary storage, block sizes of 16 kiB, 64 kiB, and 256 kiB (not individually identified in the plots), and the full GOV2 collection. Note that the upper and lower panes have different vertical scales. (Color figure online)

the fast data rates of SSD are available. If dictionary space is not a restriction, then the RLZ-ZZ methods dominate absolutely for HDD retrieval, and for much of the frontier with SSD retrieval. The remaining part of the SSD frontier is pinned on the RLZ-PV method, highlighting that unaligned bit-wise integers can be processed just as efficiently as can the aligned 32-bit integers preferred by Hoobin et al. [5], and give better compression.

Comparing our results with those of Hoobin et al. [5], we have measured very similar throughput rates for RANDOM queries, and by adding blocking to the RLZ-ZZ approach, have slightly improved its compression effectiveness. That small gain, and the reduction in transfer and decoding time that accompanies it, gives the RLZ-ZZ approaches the upper hand, and dictionaries as small as 256 MiB are sufficient to attain high RANDOM query throughput even compared to RLZ-PV, and also compact storage. On SSD, the situation is similar, but if query throughput is the primary goal, the RLZ-PV represent the best combination of attributes.

5   RLZ Extensions

We briefly describe two different ways in which RLZ compression can be enhanced.

Table 2. Use of ZLIB priming with the 64 GiB prefix of GOV2. In the ZLIB′ method, a uniform sampled dictionary of 256 MiB is employed. In the RLZ-ZZ′ method, the same 256 MiB dictionary is used, plus two fixed pre-computed integer sequences of 64 kiB containing factor lengths and factor offsets respectively. The two values for each combination are the compression rate, as a percentage of the original collection, and the measured RANDOM-mode throughput, in documents per second using SSD.

Block size	ZLIB		ZLIB′		RLZ-ZZ		RLZ-ZZ′	
	Comp.	Thrpt.	Comp.	Thrpt.	Comp.	Thrpt.	Comp.	Thrpt.
16 kiB	24.83 %	990	22.64 %	955	17.56 %	1043	17.37 %	946
64 kiB	22.29 %	840	21.53 %	825	16.56 %	905	16.47 %	866
256 kiB	21.53 %	513	21.33 %	508	16.26 %	599	16.21 %	581

Priming in RLZ-ZZ. The ZLIB compression library offers the ability to "prime" the compression process, by providing data that is considered to precede the sequence that is to be compressed, thereby providing a model to initialize the dictionary. In the same way that RLZ employs a dictionary, so too can a ZLIB′ approach, in which a uniform sampled dictionary is created, and then each block of data is ZLIB-compressed using priming text drawn from the dictionary in the vicinity of the block being compressed. A similar approach has been demonstrated to be effective when compressing Yahoo email archives [1]. A primed variant of RLZ-ZZ can also be constructed, using pre-computed sequences of factor offsets and factor lengths. Table 2 shows that when the block size is small, priming achieves a worthwhile benefit, but that the gain for larger block sizes is smaller. Priming causes a small decrease in query throughput rates.

Three Streams. Using a full factor – requiring 30+ bits – to represent a literal is expensive, and it is not actually necessary for literals to be mingled with the stream of dictionary offsets. If a third stream is added, containing only the sequence of literals, it can be compressed separately. Once a separate stream is allowed, it also makes sense to force any short factors in to it too – if the next match in the dictionary is of length less than some value *min_literal*, then the entire factor is coded as literals. Similar optimizations are used in many Lempel-Ziv implementations; see, for example, Fiala and Greene [3]. The third stream can be coded using any of the mechanisms already discussed, or any other coding method [6]; here we use of ZLIB for all three.

Table 3 provides a detailed comparison between RLZ-ZZ and RLZ-ZZZ. The gain in compression is larger with a small dictionary than with a large dictionary, since the bigger the dictionary, the less likely it is that short factors will get

Table 3. Use of a three-way split of streams, using $min\_literal = 4$, a 64 GiB prefix of GOV2, and three different dictionary sizes. Values reported are compression rates, as a percentage of the original collection. The final column shows the measured RANDOM-mode throughput, as unaligned 16 kiB accesses per second using SSD secondary storage, for the RLZ-ZZZ method with a dictionary of 256 MiB, and can be compared with the values in Table 2.

Block size	RLZ-ZZ			RLZ-ZZZ			
	16 MiB	64 MiB	256 MiB	16 MiB	64 MiB	256 MiB	Thrpt.
16 kiB	22.89 %	20.03 %	17.56 %	22.42 %	19.80 %	17.47 %	1029
64 kiB	21.58 %	18.89 %	16.57 %	20.99 %	18.54 %	16.39 %	896
256 kiB	21.18 %	18.54 %	16.27 %	20.57 %	18.17 %	16.06 %	591

generated. That is, the use of three streams can be viewed as being a way of making slightly better use of a small dictionary. Decoding speed is only marginally affected.

6   Summary and Conclusion

We have extended the experimentation of Hoobin et al. [5] to SSD memory, and undertaken a systematic study of blocking effects and access time trade offs in archive compression. The RLZ-ZZ static-dictionary method provides an outstanding balance between random access query throughput and compression effectiveness, for both HDD devices and SSD devices. We have also measured the effect of two simple techniques that provide small additional compression gains, without any great loss of throughput.

Acknowledgments. This work was supported under the Australian Research Council's *Discovery Projects* scheme (project DP140103256).We have had access to the code of Hoobin et al. while working on this project, and we thank them for making it available.

References

1. Bergman, A., Zohar, E.: Compressing Yahoo mail. In: Proceedings of the DCC, pp. 223–232 (2015)
2. Ferrada, H., Gagie, T., Gog, S., Puglisi, S.J.: Relative Lempel-Ziv with constant-time random access. In: Moura, E., Crochemore, M. (eds.) SPIRE 2014. LNCS, vol. 8799, pp. 13–17. Springer, Heidelberg (2014)
3. Fiala, E.R., Greene, D.H.: Data compression with finite windows. Commun. ACM **32**(4), 490–505 (1989)
4. Gog, S., Beller, T., Moffat, A., Petri, M.: From theory to practice: plug and play with succinct data structures. In: Gudmundsson, J., Katajainen, J. (eds.) sea 2014. LNCS, vol. 8504, pp. 326–337. Springer, Heidelberg (2014)

5. Hoobin, C., Puglisi, S.J., Zobel, J.: Relative Lempel-Ziv factorization for efficient storage and retrieval of web collections. PVLDB **5**(3), 265–273 (2011)
6. Moffat, A., Turpin, A.: Compression and Coding Algorithms. Kluwer, Boston (2002)
7. Tong, J., Wirth, A., Zobel, J.: Principled dictionary pruning for low-memory corpus compression. In: Proceedings of the SIGIR, pp. 283–292 (2014)
8. Webber, W., Moffat, A.: In search of reliable retrieval experiments. In: Proceedings of the 10th Australasian Document Computing Symposium, pp. 26–33 (2005)
9. Williams, H.E., Zobel, J.: Compressing integers for fast file access. Comput. J. **42**(3), 193–201 (1999)

Large Scale Sentiment Analysis with Locality Sensitive BitHash

Wenhao Zhang$^{(\boxtimes)}$, Jianqiu Ji, Jun Zhu, Hua Xu, and Bo Zhang

State Key Lab of Intelligent Technology and Systems,
Tsinghua National TNLIST Lab, Department of Computer Science and Technology,
Tsinghua University, Beijing 100084, China
{zhangwenhao10,jijq10}@mails.tsinghua.edu.cn,
{dcszj,xuhua,dcszb}@mail.tsinghua.edu.cn

Abstract. As social media data rapidly grows, sentiment analysis plays an increasingly more important role in classifying users' opinions, attitudes and feelings expressed in text. However, most studies have been focused on the effectiveness of sentiment analysis, while ignoring the storage efficiency when processing large-scale high-dimensional text data. In this paper, we incorporate the machine learning based sentiment analysis with our proposed Locality Sensitive One-Bit Min-Hash (BitHash) method. BitHash compresses each data sample into a compact binary hash code while preserving the pairwise similarity of the original data. The binary code can be used as a compressed and informative representation in replacement of the original data for subsequent processing, for example, it can be naturally integrated with a classifier like SVM. By using the compact hash code, the storage space is significantly reduced. Experiment on the popular open benchmark dataset shows that, as the hash code length increases, the classification accuracy of our proposed method could approach the state-of-the-art method, while our method only requires a significantly smaller storage space.

Keywords: Sentiment analysis · Locality Sensitive Hashing · Large scale

1 Introduction

With the rapidly growing of new media data such as Twitter, Facebook, Weibo, more and more people express their opinions or attitudes towards a topic on the Internet. The large-scale text data was quite useful for a commercial company to identify the users' attitudes with their products and services, and develop a better marketing strategy and product [5]. Besides, the administrations could find quick insights about public opinions for an event, and even have more insightful conclusions for psychology and sociology. As the volume of text data becomes larger and larger, large-scale sentiment analysis will play an important role in the field of natural language processing.

G. Zuccon et al. (Eds.): AIRS 2015, LNCS 9460, pp. 29–40, 2015.
DOI: 10.1007/978-3-319-28940-3_3

Sentiment classification (i.e., whether the sentiment orientation of the text is positive/negative) is one of the most important tasks in sentiment analysis. Many machine learning approaches have been applied to this task, where the representation of a document plays a key role in classification. Bag of words, N-grams are simple and effective ways to build language models. However, these representations need large amounts of memory for large-scale classification. As the scale of text data on the Internet becomes larger, there is an emerging need to scale up sentiment analysis methods. In the filed of sentiment analysis, researchers are usually aiming at improving the classification accuracy [7,13], but there is little literature about reducing the storage for large-scale corpus. In this paper, we design a method which incorporates sentiment analysis with our proposed Locality Sensitive One-Bit Min-Hash (BitHash) method. BitHash compresses each data sample into a compact binary hash code while preserving the pairwise similarity of the original data. The binary code can be used as a compressed and informative representation in replacement of the original data for subsequent processing. By using the compact hash code, the storage space can be significantly reduced.

As we know, finding the nearest neighbor and measuring the distances between the instances are fundamental steps for machine learning methods. For large scale dataset, comparing the query with each sample in the dataset is infeasible, because the linear complexity is not scalable in practical setting [11]. The applications such as for natural language processing and computer vision will also suffer from curse of dimensionality, because the words and visual descriptors might have millions of dimensions. So there exist many works about approximate nearest neighbors methods which can significantly reduce the complexity of the exact nearest neighbors. Min-hash is one of such methods, which is simple and has been largely used in the search engine and clustering tasks. Inspired by [4], which demonstrates a hashing learning method. It proves that B-Bit Min-Hash method's estimators could be naturally integrated with learning algorithms such as SVM.

In our paper, we propose a new approach BitHash, which can compress each data sample into a compact binary hash code while preserving the pairwise similarity of the original data. We rigorously analyze the variance of BitHash, showing that as pairwise Jaccard similarity increases, the variance ratio of BitHash over the original min-hash decreases. BitHash could easily be integrated with linear learning machine like SVM. It could significantly reduce feature dimensions as a new representation method, and help reduce the storage for large-scale sentiment analysis substantially.

We have three key contributions in this paper. Firstly, we are the first to combine Locality Sensitive Hashing technique to scale up sentiment analysis; Secondly, we propose One Bit Min-Hash (BitHash) method, which provides an unbiased estimate of pairwise Jaccard similarity, and the estimator is a linear function of Hamming distance. Finally, we apply BitHash into sentiment analysis, which can significantly reduce the feature dimensions as a more compressed and informative representation method, and help reduce the storage for large-scale text substantially.

In Sect. 2, we introduce some related work on sentiment analysis methods and the fundamental Locality Sensitive Hashing (LSH) technique Min-Hash method. After brief reviewing Min-Hash, we introduce BitHash in Sect. 3, and rigorously analyze its variance. We conduct Jaccard similarity estimation experiment in Sect. 3.3, to verify the variance analysis in Sect. 4. In Sect. 5, we describe how to Integrate Locality Sensitive BitHash representation with SVM to deal with the large-scale text sentiment classification problem. In Sect. 6, we show our experimental results on IMDB movie reviews dataset. Finally, Sect. 7 concludes this paper.

2   Related Work

2.1   Sentiment Analysis

Much research about sentiment analysis has been done in the past years, and [5,9] provide a comprehensive overview about sentiment analysis methods, and we know that machine learning methods are widely used and achieve state-of-the-art experiment results.

Notations. Sentiment classification could be viewed as a machine learning problem. More formally, the problem can be formulated as follows, given a dataset of pairs $\{x(i), y(i)\}_{(i=1,...,N)}$ where $x(i)$ is the i-th document in N training samples, and $y(i) \in \{+1, -1\}$ is the sentiment orientation label of $x(i)$. We train two models: $p^+(x|y = +1)$ for $\{x(i)$ subject to $y(i) = +1\}$ and $p^-(x|y = -1)$ for $\{x(i)$ subject to $y(i) = -1\}$. For an input x at test time, the sentiment orientation could be classified by computing the ratio $r = p^+(x|y = +1)/p^-(x|y = -1) \times p(y = +1)/p(y = -1)$. If $r < 1$, then x is assigned to negative class, otherwise to the positive class.

Representation and Classification Methods. Ngram is a simple but effective language model, and easy to be combined with classifiers. [9] claimed that SVM method with Unigrams could achieve the best result for sentiment analysis, when comparing with Naive Bayes and other Ngram features. [7] compared different Ngrams patterns with SVM, which shows Unigrams + Bigrams + Trigrams achieves the best performance when comparing with less Ngram combinations. However, the Ngram representation suffers from the curse of dimensionality, which is sparse and needs large memory space. Naive Bayes Support Vector Machine (NB-SVM) [12] applied NB log-count ratio as feature with LibLinear SVM classifier. It yields strong baseline result and the output even can beat many intricate approaches.

Recurrent neural network [8] provides a better word vector representation which has better performance than Ngram. However, For the individual model, NB-SVM with Trigram outperforms other representation and classifier [7], while the paper ensembles different model together and achieves state-of-the-art performance. However, those approaches focus on the effectiveness of sentiment analysis methods, and the curse of the dimensionality has not been solved when meeting with large-scale text datasets.

For the dimensionality reduction, we think of an idea which incorporating hash technique into sentiment analysis feature representation. [4] shows that hashing technique could be used in machine learning area theoretically. In this paper, we exploit One-Bit Min-Hash (BitHash) to generate binary hash code, and apply it to sentiment analysis. By using the BitHash codes representation, the storage space for strong hash codes is largely reduced, and our method provides an unbiased estimate of pairwise Jaccard similarity, and the estimator is a linear function of Hamming distance, which is very simple.

2.2   Min-Hash Review

Before introducing BitHash method, we take a brief review of Min-Hash, which is a building block of our proposed BitHash.

Fig. 1. The general framework of sentiment classification with BitHash technique

Min-hash [1] is a popular hashing method for Jaccard similarity, which is the most widely-used pairwise similarity between two sets A and B, defined as

$$J(A, B) = \frac{|A \bigcap B|}{|A \bigcup B|}. \tag{1}$$

For data of bag-of-words representation, e.g. texts can be represented as sets of words or Ngrams, denote the dictionary (the whole set of elements) by W of which the cardinality is $|W| = d$. We assign each word with a unique ID from the non-negative integer set $I = \{0, 1, 2, \ldots, d-1\}$, and thus any set of words S is represented by a subset of non-negative integers in I.

Min-Hash outputs a hash value by first generating a random permutation $\pi : I \rightarrow I$, and then taking the smallest permuted ID, i.e. $min(\pi(S)) := min_{i \in S} \pi(i)$. It is proven [1] that for any two non-empty sets A and B,

$$Pr[min(\pi(A)) = min(\pi(B))] = \frac{|A \bigcap B|}{|A \bigcup B|} = J(A, B). \tag{2}$$

Define random variable

$$X_\pi = \begin{cases} 1, & min(\pi(A)) = min(\pi(B)) \\ 0, & otherwise \end{cases}.$$

Then
$$\mathbb{E}[X_\pi] = Pr[X_\pi = 1] = J(A, B). \tag{3}$$

Generate K random permutations $\pi_1, \pi_2, \ldots, \pi_K$ independently, and define correspondingly $X_{\pi_1}, X_{\pi_2}, \ldots, X_{\pi_K}$. The estimator of Min-Hash is

$$X = \frac{1}{K} \sum_{i=1}^{K} X_{\pi_i} = \frac{1}{K} \sum_{i=1}^{K} 1_{min(\pi_i(A))=min(\pi_i(B))}, \tag{4}$$

which is an unbiased estimator of $J(A, B)$, and its variance is

$$Var[X] = \frac{1}{K} J(A, B)(1 - J(A, B)). \tag{5}$$

However, the feature representation by using Min-Hash is consisting with hash integers, which couldn't be adapted directly with linear learning method such as SVM. So we propose to use a new approach called BitHash to deal with this problem.

3   BitHash

3.1   Sentiment Classification Framework with BitHash

The general framework of the sentiment classification with BitHash is shown in Fig. 1. Documents are first represented as sets of N-grams, then each N-gram is replaced by its index (an nonnegative integer) in the dictionary, which is built based on a training set of documents. Now each document is represented by a set of nonnegative integers. To get a compact and similarity-preserving representation of the set, we transform it into a binary string using BitHash. Finally, we feed the (Extended) BitHash code into a classifier which predicts sentiment orientations.

3.2   BitHash

BitHash is short for One-Bit Min-Hash, which is based on Min-Hash, while producing more compact hash code. One shortage of Min-Hash is that each of its hash value is an integer represented by multiple bits (e.g. 32 bits or 64 bits in modern architecture). It would be desirable to find a family of hash functions for Jaccard similarity that produce 1 bit per hash value. Fortunately, theoretical result shows that such a family of binary hash functions exists. Charikar [2] proves that any Locality-Sensitive Hash (LSH) family $\mathcal{H}$ that has the following property

$$Pr_{h \in \mathcal{H}}[h(x) = h(y)] = sim(x, y) \tag{6}$$

induces a binary LSH family $\bar{\mathcal{H}}$ s.t.

$$Pr_{\bar{h} \in \bar{\mathcal{H}}}[\bar{h}(x) = \bar{h}(y)] = \frac{1 + sim(x, y)}{2}. \tag{7}$$

And

$$\bar{h}(x) = b(h(x)) \tag{8}$$

$h \in \mathcal{H}$ and $b \in \mathcal{B}$, where $\mathcal{B}$ is a pairwise independent family of hash functions that map the elements in the integer set I to $\{0, 1\}$. Formally,

$$Pr_{b \in \mathcal{B}}[b(u) = b(v)] = \begin{cases} \frac{1}{2}, & u \neq v \\ 1, & u = v \end{cases}.$$

This construction of $\bar{\mathcal{H}}$ is given in the proof of the Lemma 2 in [2]. Since Min-Hash satisfies (2), and thus it satisfies the property (6). Therefore the theory guarantees that it is possible to construct a binary hash function family for Jaccard similarity.

Without loss of generality, we assume that the cardinality of W is even. In practice, we may construct a hash function $\bar{h} \in \bar{\mathcal{H}}$ using two random permutations π and ϕ in the following way:

- Apply π to data and generate a Min-Hash value y, which is an integer;
- Use ϕ to map the integer y to another integer z;
- if z is even, output 1, otherwise output 0.

Formally, the construction of a BitHash function $\bar{h}$ is

$$\bar{h}(S) = Mod(\phi(min(\pi(S))), 2), \tag{9}$$

where

$$Mod(x, 2) = \begin{cases} 1, & x \ is \ odd \\ 0, & x \ is \ even \end{cases}.$$

Generate independently $\pi_1, \pi_2, \ldots, \pi_K$ and $\phi_1, \phi_2, \ldots, \phi_K$, then we may construct K independent BitHash functions $\bar{h}_1, \bar{h}_2, \ldots, \bar{h}_K$. Denote $\hat{h} = (\bar{h}_1, \bar{h}_2, \ldots, \bar{h}_K)$, and thus $\hat{h}$ outputs K-bits. For two data A and B, we have that

$$\mathbb{E}[d_{Hamming}(\hat{h}(A), \hat{h}(B))] = \frac{K(1 - J(A, B))}{2}, \tag{10}$$

where $d_{Hamming}(\cdot, \cdot)$ measures the Hamming distance between the two input binary strings. Therefore the estimator with respect to K functions given by BitHash is

$$E_{1-bit, K} = 1 - \frac{2 \times d_{Hamming}(\hat{h}(A), \hat{h}(B))}{K}. \tag{11}$$

3.3   Variance of Bit-Hash

In this subsection we analyze the variance of the estimator $E_{1-bit, K}$ given by BitHash. The variance of $d_{Hamming}(\hat{h}(A), \hat{h}(B))$ is

$$\begin{aligned} Var[d_{Hamming}(\hat{h}(A), \hat{h}(B))] &= K\frac{1 - J(A, B)}{2}(1 - \frac{1 - J(A, B)}{2}) \\ &= K\frac{1 - J(A, B)}{2}(\frac{1 + J(A, B)}{2}). \end{aligned} \tag{12}$$

Table 1. Jaccard similarity estimation results when $p = 0.95$

	Min-Hash	BitHash
K	MSE	MSE
100	8.615e-4	1.829e-3
200	4.362e-4	9.364e-4
300	2.834e-4	5.932e-4
400	2.169e-4	4.492e-4
500	1.695e-4	3.627e-4

Table 2. Jaccard similarity estimation results when $p = 0.8$

	Min-Hash	BitHash
K	MSE	MSE
100	2.207e-3	5.397e-3
200	1.115e-3	2.759e-3
300	7.400e-4	1.958e-3
400	5.572e-4	1.399e-3
500	4.551e-4	1.156e-3

Table 3. Jaccard similarity estimation results when $p = 0.65$

	Min-Hash	BitHash
K	MSE	MSE
100	2.527e-3	7.612e-3
200	1.237e-3	3.863e-3
300	8.321e-4	2.522e-3
400	6.154e-4	1.867e-3
500	4.947e-4	1.506e-3

Table 4. Jaccard similarity estimation results when $p = 0.5$

	Min-Hash	BitHash
K	MSE	MSE
100	2.203e-3	8.916e-3
200	1.120e-3	4.466e-3
300	7.442e-4	3.003e-3
400	5.493e-4	2.203e-3
500	4.421e-4	1.753e-3

Therefore the variance of the estimator $E_{1-bit,K}$ given by BitHash is

$$Var[E_{1-bit,K}] = \frac{1}{K}(1 - J(A,B))(1 + J(A,B)). \qquad (13)$$

Thus the variance ratio of BitHash over Min-Hash is

$$r = \frac{\frac{1}{K}(1 - J(A,B))(1 + J(A,B))}{\frac{1}{K}J(A,B)(1 - J(A,B))}$$
$$= \frac{1 + J(A,B)}{J(A,B)}. \qquad (14)$$

It is easy to see that $r \geq 2$. When $J(A,B)$ is very close to 1, the variance of BitHash is around 2 times that of the original Min-Hash. The fact that the variance of BitHash is larger than that of Min-Hash is not surprising since each hash value produced by BitHash only contains a single bit, while each Min-Hash value usually has 32 or 64 bits. In fact, the variance gap can be compensated by using $2K$ BitHash functions, when K hash functions are required by Min-Hash. More concretely, the variance of BitHash with $2K$ hash functions is about the same as that of Min-Hash with K hash functions, when the pairwise Jaccard similarity $J(A,B)$ to estimate is very close to 1, while the space for storing the hash values is reduced by a factor of 16 or 32 by using BitHash. Another advantage of BitHash is that its estimator is very simple, of which the main part involves computing a Hamming distance which can be computed very fast. This is because the computation of Hamming distance can be accomplished with a bitwise exclusive-or (XOR) operation followed by a non-zero bit counting operation, both of which can be executed very efficiently by modern CPUs. The framework of generating BitHash codes from text is shown in Fig. 1.

4   Jaccard Similarity Estimation

In this section we conduct experiments to show the accuracy of BitHash and Min-Hash in estimating pairwise Jaccard similarity. We use synthetic datasets in this experiment. The data are generated as binary strings of fixed length $d = 1000$. Each bit of the binary string has a constant probability p to be 1 or otherwise 0. As p approaches 1, more and more pairs of data will highly overlap and have high Jaccard similarity. For each p, we generate a set of 100 data with fixed length d, and compute the true pairwise Jaccard similarity as the ground truth. Then we apply both methods, namely, BitHash and Min-Hash, with fixed number K of hash values, to approximate the similarity using their estimators respectively. We test $p = 0.95, 0.8, 0.65, 0.5$ respectively, and vary K from 100 to 500. For each set of p and K, we repeat the test 100 times. We measure the estimation error by mean-squared error (MSE).

Table 1 shows that when the pairwise Jaccard similarity is very high, the estimation MSE of BitHash is about 2 times that of Min-Hash, which verifies the variance analysis in Sect. 3.3. Tables 1, 2, 3 and 4 show that as p decreases, the MSE ratio of BitHash over Min-Hash increases, because the pairwise Jaccard similarity to estimate decreases. With p fixed, as the number of hash values K gets larger, the MSEs of both methods get smaller proportionally.

5   Integrating BitHash with Machine Learning Algorithms

Machine Learning algorithms like SVM and logistic regression are extremely popular. There are a lot of open source softwares which provides the tools for us, such as LibLinear [3], Pegasos [10], etc.

The L_2-regularized SVM solves the following optimization problem:

$$\min_{\mathbf{w}} \frac{1}{2}\mathbf{w}^T\mathbf{w} + C\sum_{i=1}^{n} \log(1 + e^{-y_i\mathbf{w}^T\mathbf{x_i}}) \tag{15}$$

Given a dataset $\{(x_i, y_i)\}_{i=1}^{n}, x_i \in \mathbb{R}^D, y_i \in \{-1, +1\}$, where D is the dimension of the dataset. $C > 0$ is a penalty parameter to avoid overfitting.

Since the computation of inner product in SVM is different from Hamming distance, BitHash code cannot be directly fed to SVM as input. To incorporate the BitHash code with SVM, after representing each data sample with BitHash code, we extend it into another binary string to feed into SVM, as inspired by [4]: if a BitHash value is 0, we extend it to 01, otherwise we extend it to 10. For example, if a BitHash code is 011001, then the extended binary code to feed into SVM is 011010010110. Table 5 shows some examples illustrating the process of transforming the BitHash codes to extended codes. There is a simple relation between the inner product of two extended binary codes and the Hamming distance of two corresponding BitHash codes: their sum equals K, the BitHash code length.

Table 5. BitHash code to extended code: an example

BitHash code	Extended code
1011000	10011010010101
0010110	01011001101001
1101101	10100110100110

6 Experiments

In this section, the effectiveness of our proposed framework, linear machine learning method with BitHash, is demonstrated. The experiments are conducted on a well-known IMDB movie reviews dataset [6].

6.1 Experiment Setting

Data Set and Experiment Setup. The Stanford IMDB movie reviews dataset [6] has been widely used in sentiment analysis. The dataset consists 50,000 full-length movie reviews. And their sentiment orientation have already been labelled as either positive or negative. For the supervised learning methods, the dataset needs to be splitted into training and test sets. In order to remove the uncertainty of the data split, we randomly select 30,000 reviews out of the datasets, and a five-fold validation method is applied in our experiments. Four folds are used for training and the rest one fold is for testing.

The framework of our experiments is shown in Fig. 1, and each review is represented by a set of its Unigram, Bigrams and Trigrams tokens. The average unique N-grams (Unigram, Bigrams and Trigrams) dictionary length (feature dimensions) is about 4.3 million. For the traditional N-gram representations, each N-gram is a float or integer number, will take 4 bytes (32 bit) space. So the average storage cost for each review is about 140 million bits. In our experiments, we test the results when using BitHash representations of various lengths K (number of bits), ranging from $K = 100$ to $K = 20,000$. The L_2-regularized SVM in LibLinear [3] is used as the linear learning algorithm in our experiments.

We use state-of-the-art sentiment classification method Naive Bayes Support Vector Machine (NB-SVM) as our baseline. NB-SVM applies NB log-count ratio for each N-gram as feature with SVM, and the output even can beat other intricate approaches [7,12].

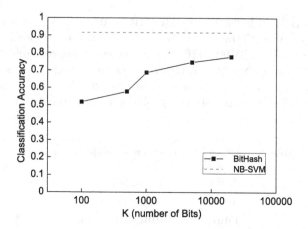

Fig. 2. Sentiment analysis accuracy achieved by using BitHash representations of various length K (number of bits) on IMDB movie reviews dataset, comparing with state-of-the-art baseline NB-SVM classification result

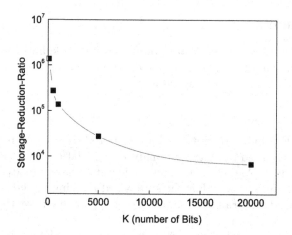

Fig. 3. Storage-Reduction-Ratio achieved by using BitHash representations of various length K (number of bits) on IMDB movie reviews dataset

6.2 Experiment Results

The accuracy of sentiment classification and the Storage-Reduction-Ratio (SRR) are two key metrics in our experiments. We use them to make a comparison from the algorithm effectiveness and storage consumption reduction respectively. SRR is the ratio between the storage requirement of original Ngrams representation and BitHash codes.

The experiments by using linear SVM with BitHash have been conducted. The accuracy and SRR of sentiment analysis using various length of BitHash representation are shown in Figs. 2, 3 and Table 6. Under different length of

Table 6. Storage Reduction Ratio (SRR) Result by using BitHash representations of various length K (number of bits)

K	100	500	1000	5000	20000
SRR	1,371,916	274,383	137,192	27,438	6,860

BitHash codes, as the length K increases, the accuracy of sentiment classification gradually grows and approaches state-of-the-art, NB-SVM, while the Storage-Reduction-Ratio decreases (shown in Table 6).

Particularly, when $K = 20,000$, the accuracy is still satisfiable, while the Storage-Reduction-Ratio is more than 6,000, or in other words, the storage requirement of BitHash is at most $\frac{1}{6000}$ that of Ngrams-like representation which NB-SVM is based on. If Ngram token is stored with 32 bits float or integer, then the storage for a review needs $D \times 32$ bits, where D is the size of dictionary. In our experiment, $D = 4,287,237$, so the storage requirement for a review is $13,719,584$ bits. Therefore, by using BitHash with $K = 20,000$, the storage requirement is reduced by a factor of $6,860$. Table 6 shows the Storage-Reduction-Ratio results for various BitHash code length K.

7 Conclusion and Future Work

In this paper, we scale up sentiment analysis with our proposed Locality Sensitive BitHash method. BitHash could compress each data sample into a compact binary hash code while preserving the pairwise similarity of the original data. The binary code can be used as a compressed and informative representation in replace of the original data, and it can be easily integrated with linear learning classifier. Our experiments results show that BitHash method reduces the storage requirement by a factor of more than $6,000$, while achieving satisfiable accuracy result. We believe our framework may provide an inspiring insights for large-scale sentiment analysis researches.

In the future, we would like to consider more text features and rich side-information to our framework, as well as make a distributed version, which is easy to be used as a fundamental work for large-scale sentiment analysis researchers.

Acknowledgments. The work was supported by the National Basic Research Program (973 Program) of China (No. 2013CB329403), National Natural Science Foundation of China (Nos. 61322308, 61332007), and the Tsinghua National Laboratory for Information Science and Technology Big Data Initiative.

References

1. Broder, A.: On the resemblance and containment of documents. In: Proceedings of the Compression and Complexity of Sequences 1997, p. 21 (1997)

2. Charikar, M.: Similarity estimation techniques from rounding algorithms. In: ACM Symposium on Theory of Computing, pp. 380–388 (2002)
3. Hsu, C.W., Chang, C.C., Lin, C.J., et al.: A practical guide to support vector classification (2003)
4. Li, P., Shrivastava, A., Moore, J.L., König, A.C.: Hashing algorithms for large-scale learning. In: Advances in Neural Information Processing Systems, pp. 2672–2680 (2011)
5. Liu, B.: Sentiment analysis and subjectivity. Handb. Nat. Lang. Process. **2**, 627–666 (2010)
6. Maas, A.L., Daly, R.E., Pham, P.T., Huang, D., Ng, A.Y., Potts, C.: Learning word vectors for sentiment analysis. In: Proceedings of the 49th Annual Meeting of the Association for Computational Linguistics: Human Language Technologies, pp. 142–150. Association for Computational Linguistics, Portland (2011). http://www.aclweb.org/anthology/P11-1015
7. Mesnil, G., Ranzato, M., Mikolov, T., Bengio, Y.: Ensemble of generative and discriminative techniques for sentiment analysis of movie reviews (2014). arXiv preprint arXiv:1412.5335
8. Mikolov, T., Karafiát, M., Burget, L., Cernocký, J., Khudanpur, S.: Recurrent neural network based language model. In: INTERSPEECH, pp. 1045–1048 (2010)
9. Pang, B., Lee, L.: Opinion mining and sentiment analysis. Found. Trends Inf. Retrieval **2**(1–2), 1–135 (2008)
10. Shalev-Shwartz, S., Singer, Y., Srebro, N., Cotter, A.: Pegasos: primal estimated sub-gradient solver for SVM. Math. Program. **127**(1), 3–30 (2011)
11. Wang, J., Kumar, S., Chang, S.F.: Semi-supervised hashing for large-scale search. IEEE Trans. Pattern Anal. Mach. Intell. **34**(12), 2393–2406 (2012)
12. Wang, S., Manning, C.D.: Baselines and bigrams: simple, good sentiment and topic classification. In: Proceedings of the 50th Annual Meeting of the Association for Computational Linguistics: Short Papers, vol. 2, pp. 90–94. Association for Computational Linguistics (2012)
13. Zhu, J., Xing, E.P.: Conditional topic random fields. In: Proceedings of the 27th International Conference on Machine Learning (2010)

Graphs, Knowledge Bases and Taxonomies

Knowledge-Based Query Expansion in Real-Time Microblog Search

Chao Lv, Runwei Qiang, Feifan Fan, and Jianwu Yang$^{(\boxtimes)}$

Institute of Computer Science and Technology,
Peking University, Beijing 100871, China
{lvchao,qiangrw,fanff,yangjw}@pku.edu.cn

Abstract. Since the length of microblog texts, such as tweets, is strictly limited to 140 characters, traditional Information Retrieval techniques usually suffer severely from the vocabulary mismatch problem such that they cannot yield good performance in the context of microblogosphere. To address this critical challenge, in this paper, we propose a new language modeling approach for microblog retrieval by inferring various types of context information. In particular, we expand the query using knowledge terms derived from Freebase so that the expanded one can better reflect the information need. Besides, in order to further answer users' real-time information need, we incorporate temporal evidences into the expansion methods so that the proposed approach can boost recent tweets in the retrieval results with respect to a given topic. Experimental results on two official TREC Twitter corpora demonstrate the significant superiority of our approach over baseline methods.

Keywords: Query expansion · Microblog search · Freebase

1 Introduction

Information Retrieval (IR) in the microblogosphere such as Twitter has attracted increasing research attention along with the fast development of social media. To explore the information seeking behavior in microblogoshpere, TREC first introduced a Real-Time Search Task (RTST) in 2011 [15], which can be summarized as "At time T, give me the most relevant tweets about topic X".

However, it is inherently quite challenging to develop an effective real-time IR platform in the context of microblogosphere. First, in contrast to traditional web search techniques, real-time search task usually has to face to the problem of severe vocabulary mismatch. Since the tweets are very short, there is a large risk that query terms fail to match any word observed in relevant tweets. This problem is extremely severe especially when people search certain entities, which usually have several alternative aliases. Besides, real-time search usually indicates the information need concerning something that is happening right now. Thus, it is very crucial for the IR approach to boost the recent tweets relevant to the given topic. This real-time information need requires search engines to trade off between the recency and relevance score computed between the query and tweet.

© Springer International Publishing Switzerland 2015
G. Zuccon et al. (Eds.): AIRS 2015, LNCS 9460, pp. 43–55, 2015.
DOI: 10.1007/978-3-319-28940-3_4

Query Expansion (QE) methods based on pseudo-relevance feedback (PRF) [11,13,18] are widely used in microblog search to mitigate the problems mentioned above. However, these methods rely much on the assumption that the top ranked documents in the initial search are relevant and include good words for query expansion. Nevertheless, in real world, this assumption does not always hold in microblogosphere [4,14], considering the example that the query contains proper nouns difficult to understand. What's more, even if the top ranked documents are highly relevant to the topic, it is still very likely that they contain numerous topic-unrelated words due to the informality of the tweet content [14].

To overcome the limitations of existing methods, we utilize Freebase[1] as the knowledge source to infer more topic-related context information for each query. Freebase is a practical, scalable tuple database used to organize general human knowledge [3], covering a large amount of knowledge in different aspects (domains), into a hierarchical structure. In this paper, we propose a knowledge query generation method, in which we first match related concepts in Freebase with respect to the query, and then extract useful terms from different properties of the concepts to generate the knowledge query. By interpolating the original query with the knowledge query, we can better reflect the users' information need.

To further incorporate the temporal evidence in microblogosphere, we follow the work of Li and Croft [9] and incorporate a prior distribution regarding to the recency of documents into the language modeling frameworks. More specifically, while selecting top knowledge terms from Freebase using an association based method, we assign each top ranked pseudo-relevance document with a time prior so that the words appearing in more recent documents are associated with higher probability.

The main contributions of this paper include: (1) we propose a novel approach to generate knowledge terms from Freebase to expand the original query, which can result in better understanding of information need; (2) the temporal evidence is incorporated in our QE method to trade off between relevance and recency; (3) we perform a set of experiments on two official twitter test collections published by TREC, to compare our proposed method with the state-of-the-art baseline methods. And, the experimental results demonstrate that our proposed approach can give rise to significant better retrieval performance.

2   Related Work

QE methods based on PRF assume that most frequent terms in the pseudo-relevance documents are useful, which may not hold in practice. Cao et al. [4] then integrated a term classification process to predict the effectiveness of expansion terms. Miyanishi et al. [14] proposed a manual tweet selection feedback to improve the retrieval performance. However, this method sometimes fails due to the content redundancy of tweets, which contain meaningless words that may degrade search results. Several approaches have been proposed to use the external resource such as Wikipedia, WordNet and ConceptNet to improve

---

[1] http://www.freebase.com

query expansion [7,17]. Medelyan *et al.* [10] explored the possibilities of using Wikipedia's articles as an external corpus to expand ad-hoc queries and demonstrated Wikipedia especially useful to improve weak queries which PRF is unable to improve. Pan *et al.* [16] proposed using Dempster-Shafer's Evidence Theory to measure the certainty of expansion terms from the Freebase structure. To the best of our knowledge, query expansion based on Freebase knowledge in microblog search is novel and effective.

Previous works showed that temporal evidence can be incorporated into IR [5,6]. Li and Croft [9] adopted a prior distribution regarding to the recency of documents into language modeling frameworks for retrieval. Liang *et al.* [11] proposed a temporal re-ranking component to evaluate the temporal aspects of documents. Miyanishi *et al.* [14] assumed that similar temporal models share similar temporal property and proposed a query-document dependent temporal relevance model. Albakour *et al.* [1] introduced a decay factor to balance the short-term and long-term interests for a given topic.

3   Proposed Methods

Given the RTST, we assume that a query Q is obtained as a sample from a generative model $\hat{\theta}_Q$, while the document D is generated by model $\hat{\theta}_D$. If $\hat{\theta}_Q$ and $\hat{\theta}_D$ are the estimated query and document language model respectively, according to [8], the relevance score of D with respect to Q can be computed by the following negative KL-divergence function:

$$S(Q, D) = -D(\hat{\theta}_Q || \hat{\theta}_D) \propto \sum_{w \in V} P(w|\hat{\theta}_Q) \cdot \log P(w|\hat{\theta}_D) \tag{1}$$

Within this ranking formula, the retrieval problem is essentially equivalent to the problem of estimating $\hat{\theta}_Q$ and $\hat{\theta}_D$. In principle, we can use any language model for the query and document, which is very flexible.

The start point of our study is to infer more topic-related context for the query with the help of Freebase. In this section, we first elaborate on why we choose Freebase as our knowledge base. Based on the characteristics of Freebase, we describe our proposed method of generating knowledge query in detail. Further improvements can be obtained by combining the knowledge-based query expansion with model-based pseudo-relevance feedback method.

3.1   Why We Choose Freebase

Freebase is a large collaborative knowledge base consisted of data harvested from sources such as the Semantic Web and Wikipedia, as well as individually contributed data from community members. In Freebase, human knowledge is described by structured categories, which are also known as *types* and each type has a number of defined *properties*. In this way, Freebase merges the scalability of structured databases with the diversity of collaborative wikis into a structured general human knowledge [3]. This structured knowledge shows two superiorities compared with the semi-structured or plain contents: (1) when searching in the Freebase (with API), different types can be integrated for a more accurate concept, and some types such as name and alias are more important; (2) when generating knowledge terms, we can treat different types and the corresponding properties as different evidence sources.

3.2   Generation of Knowledge Query

We generate the knowledge query based on types, aiming extracting terms from different properties for a given query. The basic procedures of our proposed method include:

- **Concept Match.** We select the topic-related concepts with the help of Free-base API. Taking the query "Mila Kunis in Oz Movie" as an example, we match two concepts "Mila Kunis" and "The Wizard of Oz" in Freebase.
- **Term Selection.** Freebase describes the human knowledge of a given concept using types and properties. For some important meta types such as alias, name and notable_for, we directly add terms from these type properties to the knowledge query. For other types (i.e. description and domain-specific types), we adopt an association based term selection method to extract the topic-related top K terms.
- **Domain Filtering.** The concept in Freebase may be involved in several domains, such as TV, book and celebrities. When searching microblog about an entity (or concept) during a specific time period, users usually focus on one specific domain of that entity. Thus, aside from the common domain, we only use Freebase knowledge from one specific domain.

Then, we view the top ranked knowledge terms from selected properties equally to form a new knowledge query Q_{fb}. After that, the knowledge query model $\hat{\theta}_{Q_{fb}}$ is interpolated with the original query model $\hat{\theta}_Q$:

$$P(w|\hat{\theta}_{Q_1}) = (1 - \alpha) \cdot P(w|\hat{\theta}_Q) + \alpha \cdot P(w|\hat{\theta}_{Q_{fb}}) \qquad (2)$$

where $\alpha \in [0, 1]$ is the weighting parameter to control the influence of the knowledge query. Both $\hat{\theta}_Q$ and $\hat{\theta}_{Q_{fb}}$ are estimated according to the maximum likelihood estimator.

Concept Match. We then describe our concept match algorithm in detail, which can be concluded as two steps:

1. **Noun Phrase Detection.** For a given query Q, we first split Q by space and receive a sequence of words $q_1, q_2, \cdots q_n$. Part-of-speech Tagging is then performed on each word, and all the noun phrases are extracted with rule-based method [2] from the original query.

2. **Maximum Match.** For each noun phrase, we regard it as a new query and get the related concepts as described in **Algorithm 1**. The FreebaseSearch function searches the given query in the Freebase and returns the top ranked concept if found. The match process ends if a related concept is found or none of the separate words can find a match.

Algorithm 1. GetConcept(NQ)

Require:
  Noun Phrase Query $NQ = q_1q_2 \cdots q_n$.
Ensure:
  Candidate Concept Set $CSet$.
1: $CSet = \text{FreebaseSearch}(NQ)$
2: **if** $CSet$ is empty **then**
3:    **if** n $==$ 1 **then**
4:       **return** $\emptyset$
5:    **end if**
6:    $NQ_1 \leftarrow q_1q_2...q_{n-1}$
7:    $NQ_2 \leftarrow q_2q_3...q_n$
8:    $CSet_1 \leftarrow \text{GetConcept}(NQ_1)$
9:    $CSet_2 \leftarrow \text{GetConcept}(NQ_2)$
10:   **return** $CSet_1 \cup CSet_2$
11: **else**
12:   **return** $CSet$
13: **end if**

Term Selection. For each returned concept from Freebase API, different types and corresponding properties which reflect the different aspects of the concept are provided by the search result. Some types (i.e. meta types) are very general and precise, such as alias, name and notable_for in the common domain, we directly add the property terms to the knowledge query for these types. When it comes to other types such as description and domain specific ones whose properties contain a lengthy text, an association based term selection method is utilized to extract the topic-related knowledge terms. Effective term selection is an important issue for an automatic query expansion technique. In microblog retrieval, a good expansion term should satisfy the following criteria:

1. The term should be semantically associated with the concept from the original query;
2. The term extracted from Freebase should be suitable for the local corpus and it should provide useful information for the query as well;
3. As the user's intent may change and events related to the given topic will develop over the time, the ranking function should favor the short-term words that are mostly used in recent tweets.

The candidate terms extracted from Freebase have met the first criterion to some extent. In order to satisfy the second criterion, we score the candidate terms with an association based method on the basis of the top ranked N pseudo-relevance documents (PRD):

$$Score(w) = \sum_{D \in PRD} P(D) \cdot P(w|D) \cdot \prod_{i=1}^{n} P(q_i|D) \qquad (3)$$

where $P(D)$ is the document prior which is usually assumed to be uniform, and $\prod_{i=1}^{n} P(q_i|D)$ is the query likelihood given the document model, which is traditionally computed using Dirichlet smoothing. To further meet the third criterion, we follow the work of [9] and incorporate the temporal evidence into the document prior in Eq. 3 by using an exponential distribution:

$$P(D|T_D) = r \cdot e^{-r(T_Q - T_D)} \qquad (4)$$

where r is the exponential parameter that controls the temporal influence, T_Q is the query issue time and T_D is the tweet post time. Both T_Q and T_D are measured in fractions of days. Note that T_D is constantly less than T_Q as we cannot use the future evidence. Finally, we select the top scored K words from the common description and domain specific properties, to form the knowledge query Q_{fb} along with the terms extracted from meta properties.

Domain Filtering. One advantage of Freebase is that it provides the domain information of a given concept. A domain is a collection of types which share a namespace (e.g. TV, book and celebrities). As mentioned above, only properties from common domain and top ranked domain are used for term selection. Hence, a domain ranking function should be defined to rank all the domains of a given concept. We first define the domain language model $\hat{\theta}_{dm}$ as follows:

$$P(w|\hat{\theta}_{dm}) = \frac{c(w, dm)}{\sum_{w' \in dm} c(w', dm)} \qquad (5)$$

where $c(w, dm)$ is the count of word w occurred in the domain related properties. Then, we compute the KL-divergence score of the domain language model with the document language model estimated on the top ranked N PRD using empirical word distribution.

$$Score(dm) = -D(\hat{\theta}_{dm} || \hat{\theta}_{PRD}) \propto \sum_{w \in dm} P(w|\hat{\theta}_{dm}) \cdot \log P(w|\hat{\theta}_{PRD}) \qquad (6)$$

Note that we use Dirichlet smoothing method while estimating $\hat{\theta}_{PRD}$. With this formula, we can rank all the domains and choose terms from the top ranked domain for term selection.

3.3   Mixture Feedback Model

With the knowledge query environment, we believe the information need is more understandable, which could lead to a high precision in the top retrieved documents. Based on this hypothesis, we further utilize a model-based feedback to update the query representation. More specifically, we update the $\hat{\theta}_{Q_1}$ with the simple mixture model $\hat{\theta}_F$ which is widely used in microblog retrieval [11,18].

$$P(w|\hat{\theta}_{Q_2}) = (1 - \beta) \cdot P(w|\hat{\theta}_{Q_1}) + \beta \cdot P(w|\hat{\theta}_{Q_F}) \tag{7}$$

where $\beta \in [0, 1]$ is a weighting parameter to control the amount of model-based feedback.

The model-based feedback model generates a feedback document by mixing the query topic model $\hat{\theta}_F$ with the collection language model $\hat{\theta}_C$. Under this simple mixture model, the log-likelihood of feedback documents F is:

$$\log P(F|\hat{\theta}_F) = \sum_w c(w, F) \cdot \log((1 - \lambda) \cdot P(w|\hat{\theta}_F) + \lambda \cdot P(w|\hat{\theta}_C)) \tag{8}$$

where $c(w, F)$ is the count of word w occurred in the set of feedback documents F. Then we follow the work of [18] and implement the EM algorithm with the fixed smoothing parameter $\lambda = 0.5$. No matter whether or not the query finds its knowledge terms in Freebase, the query environment will be updated by the model-based feedback.

4   Evaluation

4.1   Experimental Setup

Data Set. Two corpora (i.e. Tweets11 and Tweets13 collection) are used in our experiments, which are crawled via the official API released by TREC organizers [12]. Tweets11 collection has a sample of about 16 million tweets, ranging from January 24, 2011 to February 8, 2011 while Tweets13 collection contains about 259 million tweets, ranging from February 1, 2013 to March 31, 2013. Tweets11 is used for evaluating the effectiveness of the proposed real-time Twitter search systems over 50 official topics in the TREC'11 Microblog track as well as 60 official topics in the TREC'12 Microblog track, respectively. And, Tweets13 is used in evaluating the proposed real-time Twitter search systems over 60 official topics in the TREC'13 Microblog track. In our experiments, topics for Tweets13 are used for tuning the parameters and then we use the best parameter settings to evaluate our methods with topics for Tweets11.

The tweets and their corresponding topic information were preprocessed in several ways. We first discarded the non-English tweets using a language detector, named *ldig*[2]. Second, in conformance with the track's guidelines, all simple retweets were removed by deleting documents beginning with the string 'RT'. Moreover, we crawled all the shortened URLs and add the web title to the origin tweet. In the end, each tweet was stemmed using the Porter algorithm and stopwords were removed using the InQuery stopwords list.

---

[2] http://github.com/shuyo/ldig

Evaluation Metric. In TREC Microblog Track, tweets were judged on the basis of the defined information using a three-point scale [15]: irrelevant (labeled as 0), minimally relevant (labeled as 1), and highly relevant (labeled as 2). The main evaluation metric is Mean Average Precision (MAP) for top 1000 documents and Precision at N (P@N), which are widely used in IR. MAP and P@30 with respect to *allrel* (i.e. tweet set judged as highly or minimally relevant) are used in this paper.

Baselines. To demonstrate the performance of our proposed method, we compare our knowledge-based query expansion methods with several baseline methods. The simple KL-divergence retrieval model (denoted as **SimpleKL**) [19] is used as our first baseline. That is, we estimate $\hat{\theta}_Q$ and $\hat{\theta}_D$ with empirical word distribution, and we choose Dirichlet smoothing method for document model estimation. Throughout this paper, we set the Dirichlet smoothing parameter $\mu = 100$. We use the Simple Mixture Model [18] (denoted as **QESMM**) as our second baseline, and optimize the number of feedback documents to 5 and the number of terms in the feedback model to 100. The smoothing parameter β is set as 0.9. We also compare our method with the state-of-the-art real-time ranking model (denoted as **RTRM**) under language modeling framework, proposed by [11]. **RTRM** approach utilized a two-stage pseudo-relevance feedback query expansion to estimate the query language model and expand documents with shortened URLs in microblog. We tune all the parameters of these models with TREC'13 topics on Tweets13 corpus.

4.2 Experimental Results

We conduct several experiments to measure the effects of our query expansion methods. For our knowledge-based query expansion method, we label the method with query model $\hat{\theta}_{Q_1}$ as **QEFB**, and the one with $\hat{\theta}_{Q_2}$ as **QEFB+SMM**. When selecting knowledge terms from Freebase description and domain specific property, we set the top ranked PRD number N to 100 and the expanded term number K to 10. Following Li and Croft [9], unless otherwise specified, we set the parameter r as 0.01. α in Eq. 2 is set as 0.5, which means we regard the original query and the knowledge query equally important. The query expansion parameters in the mixture feedback model are set like **QESMM**. All the parameters are tuned with TREC'13 topics. Then we test the optimized models on Tweets11 corpus with TREC'11 and TREC'12 topics.

Table 1 shows the performance comparison of different query expansion methods. To test for statistical significance, we used a paired t-test. † and ‡ indicate that the corresponding improvements over **SimpleKL** and **QESMM** are statistically significant ($p < 0.05$), respectively. Note that all the methods listed in the table estimate the document model as **SimpleKL**. As we can see, all of the query expansion methods have significant MAP and P@30 improvements compared with the **SimpleKL** method, which indicates the importance of query expansion in microblog retrieval. When the query is expanded with the Freebase knowledge query, our approach can retrieve more relevant documents in

Table 1. The performance comparison of different query expansion methods. The best performances are marked in bold.

Topics	TREC'13		TREC'12		TREC'11	
Method	MAP	P@30	MAP	P@30	MAP	P@30
SimpleKL	0.2926	0.4939	0.2681	0.3872	0.3572	0.3773
QESMM	0.3260	0.5117	0.2955	0.3989	0.3878	0.4133
RTRM	0.3486	0.5567	0.3282	0.4439	0.3975	0.4300
QEFB	0.3149 †	0.5117	0.3144†	0.4262	0.4125†	0.4207
QEFB+SMM	0.3652 †‡	0.5750 †‡	0.3320 †‡	0.4450 †‡	0.4241 †	0.4300 †

the top results. Thus, we can further improve the retrieval performance by combining the knowledge-based expansion method with mixture feedback model. Our knowledge-based query expansion method **QEFB+SMM** achieves the best retrieval performance in the three topic sets. More specifically, for TREC'12 topics, our method **QEFB+SMM** improves the MAP over **SimpleKL** and **QESMM** by 23.83 % and 12.35 %, respectively; while the corresponding increments in terms of P@30 are 14.93 % and 11.56 %, respectively. For TREC'11 topics, the **QEFB+SMM** raises the MAP over **SimpleKL** and **QESMM** by 18.73 % and 9.36 %,respectively; while the corresponding P@30 improvements are 13.97 % and 4.04 %, respectively. Our method also beats the state-of-the-art baseline **RTRM**, which has a good performance in Microblog Track.

4.3   Discussion

Many parameters in our proposed approach can affect the system performance. In this section, we analyze the robustness of the parameter settings in knowledge-based query expansion method. All these experiments in this section are run on TREC'13 topics, which are used for parameter selection.

Effects of Knowledge Query. For the query modeling, we propose using knowledge query to make the information need more comprehensible. Many factors affect the quality of the knowledge terms: (1) whether the maximum match algorithm can get topic-related concept from the Freebase; (2) whether the Freebase structured information (i.e. type information and domain information used in term selection and domain filtering) is helpful; (3) how the number of knowledge terms K affects the retrieval performance and (4) how the number of pseudo-relevance documents N used for term selection and domain selection affects the retrieval performance.

To answer the first question, we create the run **QEManualFB**, which means we manually select the concept from Freebase for each query. For the second question, we dismiss the structured information of Freebase and select terms from all the properties and regard all the properties equally. We label this run as **QEWiki**.

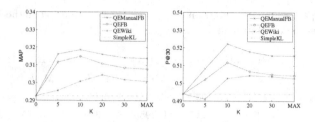

Fig. 1. Sensitivity to the selected knowledge term number K.

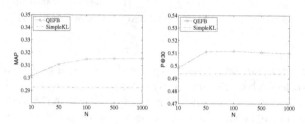

Fig. 2. Sensitivity to the PRD number N for knowledge term selection.

Figure 1 shows the MAP and P@30 scores of all the models for different K and fixed $N = 100$. MAX means all the candidate terms that satisfy $Score(w) > 0$ in Eq. 3 are selected. We can see that though **QEFB** is not better than **QEManualFB**, the performance gap between them is not large, which verifies the effectiveness of our concept match algorithm. **QEFB** is consistently better than **QEWiki**, which proves the importance of the structured information in Freebase. Moreover, when K is set around 10, **QEFB** can get its optimal retrieval performance and is significantly better than that of **SimpleKL**, which indicates the effectiveness of the association based term selection method.

We then fix the term number K to 10 and change the PRD number N for term selection and domain filtering. Figure 2 shows the MAP and P@30 scores of our **QEFB** model aganist different values of N. We can observe that when N is larger than 100, the performance changes slightly. It means that top 100 pseudo-relevance documents can provide enough information for selecting good knowledge terms from description property.

Effects of the Interpolation Coefficients. Recall that we first expand the query with knowledge query, and further expand the updated query $\hat{\theta}_{Q_1}$ with model-based feedback. The first-stage query expansion is controlled by a coefficient α, while the second-stage expansion is controlled by β. Figure 3 (left) shows the performance changing of **QEFB** ($N = 100, K = 10$) against different values of α. When $\alpha = 0$, the **QEFB** reduces to the baseline method **SimpleKL**. When $\alpha = 1$, we completely ignore the original query and only use the knowledge query. We can observe that the performance of **QEFB** is better than **SimpleKL** when α is no greater than 0.7. The optimal performance can be obtained when α is set

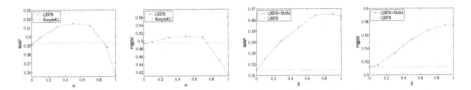

Fig. 3. Sensitivity to the first-stage knowledge query expansion coefficient α (left) and second-stage mixture feedback interpolation coefficient β (right).

around 0.5. Figure 3 (right) shows the performance changing of **QEFB+SMM** against different values of β. When $\beta = 0$, the **QEFB+SMM** reduces to **QEFB**. We can see the second-stage expansion seems to be more robust and constantly better than **QEFB**. After knowledge-based query expansion, the query can be more comprehensible and get more top related tweets, which lead to a further improvement with traditional model-based feedback.

5   Conclusion

In this study, we proposed using knowledge-based query expansion to solve the problems in microblog search. With the knowledge terms derived from the Freebase, the queries in microblogosphere can be more comprehensible and thus more relevant documents can be retrieved. The knowledge terms from Freebase should also co-occur with query terms in PRD, which has the potential to alleviate topic drift often induced by knowledge-based QE. Freebase's structured information is also well utilized in knowledge query generation procedure. Moreover, we incorporated the temporal evidence into query representation. Thus the proposed method favors recent tweets which satisfy the real-time information need in microblog retrieval. Our thorough evaluation, using two standard TREC collections, demonstrates the effectiveness of the proposed method.

Acknowledgments. The work reported in this paper was supported by the National Natural Science Foundation of China Grant 61370116.

References

1. Albakour, M., Macdonald, C., Ounis, I., et al.: On sparsity and drift for effective real-time filtering in microblogs. In: Proceedings of the 22nd ACM International Conference on Information & Knowledge Management, pp. 419–428. ACM (2013)
2. Bird, S., Klein, E., Loper, E.: Natural Language Processing with Python. O'Reilly Media, Inc., Newton (2009)

3. Bollacker, K., Evans, C., Paritosh, P., Sturge, T., Taylor, J.: Freebase: a collaboratively created graph database for structuring human knowledge. In: Proceedings of the 2008 ACM SIGMOD International Conference on Management of Data, pp. 1247–1250. ACM (2008)

4. Cao, G., Nie, J.Y., Gao, J., Robertson, S.: Selecting good expansion terms for pseudo-relevance feedback. In: Proceedings of the 31st Annual International ACM SIGIR Conference on Research and Development in Information Retrieval, pp. 243–250. ACM (2008)

5. Dakka, W., Gravano, L., Ipeirotis, P.G.: Answering general time-sensitive queries. IEEE Transactions on Knowledge and Data Engineering $24(2)$, 220–235 (2012)

6. Dong, A., Zhang, R., Kolari, P., Bai, J., Diaz, F., Chang, Y., Zheng, Z., Zha, H.: Time is of the essence: improving recency ranking using twitter data. In: Proceedings of the 19th International Conference on World Wide Web, pp. 331–340. ACM (2010)

7. Kotov, A., Zhai, C.: Tapping into knowledge base for concept feedback: leveraging conceptnet to improve search results for difficult queries. In: Proceedings of the fifth ACM International Conference on Web search and Data Mining, pp. 403–412. ACM (2012)

8. Lafferty, J., Zhai, C.: Document language models, query models, and risk minimization for information retrieval. In: Proceedings of the 24th Annual International ACM SIGIR Conference on Research and Development in Information Retrieval, pp. 111–119. ACM (2001)

9. Li, X., Croft, W.B.: Time-based language models. In: Proceedings of the Twelfth International Conference on Information and Knowledge Management, pp. 469–475. ACM (2003)

10. Li, Y., Luk, W.P.R., Ho, K.S.E., Chung, F.L.K.: Improving weak ad-hoc queries using wikipedia asexternal corpus. In: Proceedings of the 30th Annual International ACM SIGIR Conference on Research and Development in Information Retrieval, pp. 797–798. ACM (2007)

11. Liang, F., Qiang, R., Yang, J.: Exploiting real-time information retrieval in the microblogosphere. In: Proceedings of the 12th ACM/IEEE-CS Joint Conference on Digital Libraries, pp. 267–276. ACM (2012)

12. Lin, J., Efron, M.: Overview of the TREC-2013 microblog track. In: Proceedings of TREC. vol. 2013 (2013)

13. Lv, Y., Zhai, C.: A comparative study of methods for estimating query language models with pseudo feedback. In: Proceedings of the 18th ACM Conference on Information and Knowledge Management, pp. 1895–1898. ACM (2009)

14. Miyanishi, T., Seki, K., Uehara, K.: Improving pseudo-relevance feedback via tweet selection. In: Proceedings of the 22nd ACM International Conference on Information and Knowledge Management, pp. 439–448. ACM (2013)

15. Ounis, I., Macdonald, C., Lin, J., Soboroff, I.: Overview of TREC-2011 microblog track. In: Proceeddings of the 20th Text REtrieval Conference (TREC 2011) (2011)

16. Pan, D., Zhang, P., Li, J., Song, D., Wen, J.-R., Hou, Y., Hu, B., Jia, Y., De Roeck, A.: Using Dempster-Shafer's evidence theory for query expansion based on freebase knowledge. In: Banchs, R.E., Silvestri, F., Liu, T.-Y., Zhang, M., Gao, S., Lang, J. (eds.) AIRS 2013. LNCS, vol. 8281, pp. 121–132. Springer, Heidelberg (2013)

17. Xu, Y., Jones, G.J., Wang, B.: Query dependent pseudo-relevance feedback based on wikipedia. In: Proceedings of the 32nd International ACM SIGIR Conference on Research and Development in Information Retrieval, pp. 59–66. ACM (2009)
18. Zhai, C., Lafferty, J.: Model-based feedback in the language modeling approach to information retrieval. In: Proceedings of the Tenth International Conference on Information and Knowledge Management, pp. 403–410. ACM (2001)
19. Zhai, C., Lafferty, J.: A study of smoothing methods for language models applied to ad hoc information retrieval. In: Proceedings of the 24th Annual International ACM SIGIR Conference on Research and Development in Information Retrieval, pp. 334–342. ACM (2001)

Enrichment of Academic Search Engine Results Pages by Citation-Based Graphs

Shuhei Shogen, Toshiyuki Shimizu[✉], and Masatoshi Yoshikawa

Graduate School of Informatics, Kyoto University, Kyoto 606-8501, Japan
shogen@db.soc.i.kyoto-u.ac.jp, {tshimizu,yoshikawa}@i.kyoto-u.ac.jp

Abstract. Researchers' readings of academic papers make their research more sophisticated and objective. In this paper, we describe a method of supporting scholarly surveys by incorporating a graph based on citation relationships into the results page of an academic search engine. Conventional academic search engines have a problem in that users have difficulty in determining which academic papers are relevant to their needs because it is hard to understand the relationship between the academic papers that appear in the search results pages. Our method helps users to make judgments about the relevance of papers by clearly visualizing the relationship. It visualizes not only academic papers on the results page but also papers that have a strong citation relationship with them. We carefully considered the method of visualization and implemented a prototype with which we conducted a user study simulating scholarly surveys. We confirmed that our method improved the efficiency of scholarly surveys through the user study.

Keywords: Scholarly survey · Academic search engine · Citation graph

1 Introduction

Researchers read academic papers related to their own research to acquire knowledge and/or cite them when they write their own papers. Doing so makes their research more sophisticated and objective. Here, we call researchers' examinations of papers related to their work, "scholarly surveys." Scholarly surveys are an important means for researchers, especially those who have recently changed their field or who do not have much experience, to get an understanding of their field of study. However, it is very difficult for researchers to read all the papers related to their research because their information processing ability is limited. Therefore, there is a need for automated techniques to make scholarly surveys easier.

Progress in Web technologies has reached a point where we can now download and use many kinds of publications such as books and pictures electronically. The trend in digitization of publications is especially evident in academic journals. Researchers used to conduct scholarly surveys by reading journals directly. However, since the advent of search engines that are customized to search academic literature (called "academic search engines" hereafter), they can retrieve

© Springer International Publishing Switzerland 2015
G. Zuccon et al. (Eds.): AIRS 2015, LNCS 9460, pp. 56–67, 2015.
DOI: 10.1007/978-3-319-28940-3_5

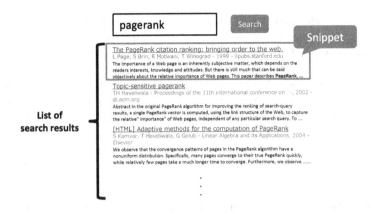

Fig. 1. Conventional interface of an academic search engine.

and read papers on the Web. Popular academic search engines include Google Scholar[1] and Microsoft Academic Search[2]. After a user inputs a query (a sequence of keywords), the academic search engine searches for and displays a list of relevant papers. The search results are ranked according to the relevance to the input query, the number of citations of the paper, etc. Results deemed more relevant have a higher rank and are located at a higher position on the list. In addition, they include bibliographic information such as title, author, published year, and journal as a snippet, along with links to the papers. Figure 1 illustrates the user interface of a search engine results page of a conventional academic search engine.

Inevitably, some papers on the search results page will be judged by the user as relevant, i.e., the targets of the survey, while other papers will be judged irrelevant. The user makes such judgments by checking the snippets of the search results. However, in some cases, the user cannot do so because the search results page has limited space to display snippets. To address this problem, we propose a method that considers the citing and cited relationship between academic papers.

With regard to two papers that are in a citing and cited relationship, if a user surveys one paper, he/she tends to survey the other paper. If we could visualize such a relationship and display it on the search results page, users could easily distinguish the papers relevant to their research from the non-relevant ones, and this would improve the efficiency of conducting scholarly surveys. Conventional search engines produce rankings according to the relevance to the query and display lists of search results; however, they do not explicitly consider the relationship between search-result papers.

In the following sections, we describe a method of incorporating a graph based on citing and cited relationships between papers into the results page of

---

[1] http://scholar.google.com/.

[2] http://academic.research.microsoft.com/.

an academic search engine. Our method incorporates a citation graph, whose nodes represent papers in the search results or the papers that cite or are cited by them and whose edges represent citation relationships between them, and it displays this graph to users in an understandable fashion. Representing the citation relationship as a graph allows users to understand the relationship between papers intuitively. Citation graphs may help newcomers to a research field to find papers useful for their initial scholarly surveys. Also, they may help experts of their fields find recently published papers easily. To the best of our knowledge, no other study has proposed a method of incorporating a graph that aggregates search results into an academic search engine's results pages. The contributions in this paper are twofold:

(1) We describe the idea and merits of incorporating a graph based on citation relationships into the search results page of an academic search engine. We also propose a method of doing so and implemented it.
(2) We describe the results of a user study examining the effectiveness of the proposed method.

The remainder of this paper is organized as follows. Section 2 introduces previous work related to this research. Section 3 describes the features of scholarly surveys using academic search engines and the merits of incorporating citation relationships into the interface of an academic search engine. Section 4 explains the method of incorporating a citation graph into the search results page of an academic search engine and the interface of the academic search engine we implemented. Section 5 describes the content and results of the experiment performed to examine the effectiveness of the proposed method. Section 6 concludes this paper and describes future work.

2 Related Work

An academic search engine is a search engine that has been customized to search academic literature; representatives are Google Scholar [1], Microsoft Academic Search, and CiteSeerX[3]. Beel and Gipp [2,3] reported that popular academic search engines rank their search results based on the relevance between the query and the retrieved paper, number of citations of the paper, and names and reputations of the authors and the journals. Verberne et al. [4] proposed a method of suggesting additional terms to specify the topic when reformulating the input query.

Our research visualizes the relationship between academic papers by using network graphs [5]. Nanba et al. [6] proposed a system that visualizes the statements on why the citing paper cites the cited paper along with the citing and cited relationships between papers and displays them to users. Here, Microsoft Academic Search introduced a similar function called the "Citation Graph." It displays a set of papers that cite the target paper appearing in the search results

---

[3] http://citeseerx.ist.psu.edu/index.

as a graph. He et al. [7] extracted and visualized the time-series process of the evolution of a research field by analyzing the abstracts and citation information of papers of the field. Dunne et al. [8] proposed a system that visualizes a summary of collected papers by using the citations and automatic summarization to help researchers explore papers rapidly. The goal of the above research is to support browsing of papers, whereas our goal is to apply a graphical method to the results pages.

The proposed method improves the usability of an academic search engine by visualizing the search results understandably. One of its visualization methods improves the interface of snippets. "Snippets" in Web search engines are summary sentences displayed on the search results page. Users check each snippet and decide whether they should follow the link to the corresponding Web page or not. Some researchers have attempted to improve the content of summary sentences in snippets [9], while others have surveyed the influence of snippets on user behavior [10]. These researches are relevant to ours in that they discuss the information contained in search engine results pages. Apart from these, other studies have proposed methods of displaying clustered search results to users [11,12]. Scaiella et al. [13] categorized the set of search results obtained from a Web search engine by using the categories of Wikipedia and displayed them with labels (the representative name of each cluster). Mirylenka and Passerini [14] proposed a method of organizing papers in the search results of an academic search engine and displaying them as a graph whose node represents a research category. These researches are relevant to ours in terms of their organizing the search results of an academic search engine; however, they are different in that their goal is to extract research topics by clustering papers in search results, while ours is to support users' judgment of the relevance of search results.

3   Advantages of Incorporating Citation into Interfaces of Academic Search Engines

In this section, we describe the characteristics of scholarly surveys using academic search engines and summarize their search process and problems. Then we explain the advantages of incorporating citing and cited relationships into the results pages.

3.1   Characteristics of Scholarly Surveys Using Academic Search Engines

Current scholarly surveys are usually performed using an academic search engine. Different from general Web searches, we think that scholarly surveys using an academic search engine have the following characteristics:

(1) They need to retrieve various relevant documents.
(2) They can determine the papers that they should survey next with reference to the citing and cited relationship.

With regard to (1), some researchers have claimed that recall is an important factor in a scholarly survey [15]. For example, in a normal Web search, when users want to know tomorrow's weather, they are usually satisfied after checking only a few search results. On the other hand, a scholarly survey needs to search and check a large number of papers ranging from famous papers in the field to the papers relevant to the user's own research because he/she needs to acquire enough knowledge about the field. With regard to (2), we think that two papers having a citing and cited relationship may be similar in content to each other. This allows users of an academic search engine to select the papers that they should survey next from the papers that cite or are cited by the relevant paper. In this way, they can efficiently survey the studies related to their own research.

3.2 Problems in the Process of Scholarly Surveys on Academic Search Engines

The user looks for the information they need by inputting a sequence of words, i.e., a query, and checks the snippets of the corresponding search results from the top of the list to the bottom. When a user judges a paper to be related to his or her own research, they follow the link to it and read it. They repeat the process until they are satisfied with the results of the survey, or else they reformulate the input query because the search results are not relevant to their information needs.

However, search results pages have limited space for displaying snippets, and thus, they may not be able to show enough information for users to be able to make good judgments on the relevance of the papers. Therefore, users may miss the papers that they should survey or conduct unnecessary surveys because of misjudgments they make about the relevance of the search results.

3.3 Incorporating Citation into Search Engine Results Pages

To address the problem mentioned in Sect. 3.2, It is possible to exploit the idea of relevance feedback. For example, we can devise a system that reranks the search results when users notify the search engine that the paper is relevant to their information needs. However, this method sometimes forces users to provide feedback, which may be a burden.

As mentioned in Sect. 3.1, with regard to two papers that have a citing and cited relationship, if one is a target of the users' survey, the other also tends to be a target. We thought the method of aggregating citing and cited relationships whose heads or tails are papers in the search results (called "proximate citing and cited relationships" after this) is effective and considered displaying it to users understandably. This method allows users to grasp the relationships between papers in the search results and distinguish the ones relevant to their needs before they follow the links to them. If we suppose there is a search-result paper Z that is relevant to the users' needs and they can see that Z and X have a citing and cited relationship, they can recognize X as a target and not overlook it. Likewise, if users can see that Z and Y have no citing and cited relationship,

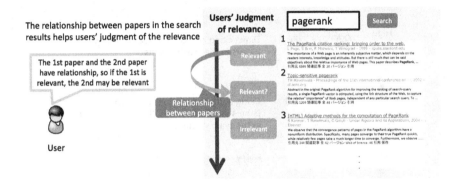

Fig. 2. Example of advantage of displaying the relationship between papers in search results.

they perhaps can exclude Y from the targets of the survey. An example of this process is illustrated in Fig. 2. In this way, displaying proximate citing and cited relationships can help users make relevance judgments and improve the efficiency of scholarly surveys.

4   Proposed Method

In this section, we describe the method of incorporating a graph based on citing and cited relationships into the search results page of an academic search engine. First, we consider what kind of information should be included in the citation graph. Then we explain the features and functions of the graph we have implemented.

4.1   Visualization Using a Citation Graph

One possible method of visualizing proximate citing and cited relationships is to incorporate a list of the citing papers and the cited papers into the snippet of each search result. However, this method makes it difficult for users to understand the citing and cited relationship because it requires them to check the citing papers and the cited papers for each search result. Instead, we propose a more understandable visualization that works by summarizing the citing and cited relationships between these papers as a graph whose nodes represent individual papers and whose edges represent citing and cited relationships.

In composing a citation graph from proximate citing and cited relationships, if we incorporate all the citing and cited relationships into a graph, the amount of displayed information would be excessive. This causes various problems in that it takes up too much space, imposes a heavy cognitive load, etc. Thus, we categorize citing and cited relationships from the viewpoint of whether they support the users' judgments of relevance and consider what type of relationship should be displayed to them. We propose three types of relationship:

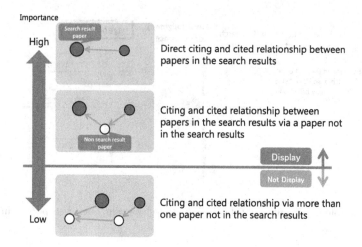

Fig. 3. Categories of citing and cited relationships in the proposed method.

(1) Direct citing and cited relationships between papers in the search results.
(2) Citing and cited relationships between papers in the search results via a paper not in the search results.
(3) Citing and cited relationship except for (1), (2) (via more than one paper not in the search results).

Figure 3 summarizes this categorization. With regard to type (1), two papers in the search results that have a citing and cited relationship are likely to be similar in content. Displaying such relationships is important for supporting users' relevance judgments. Next, type (2) means that papers A or B in the search results cite or are cited by paper C, which is in or not in the search results. We consider that although this type of relationship is less important than type (1), it is still important. In addition, papers in search results tend to be similar in content to the papers that cite or are cited by them. This tendency implies that such papers can be new targets of the users' surveys. For these reasons, it is important for users to display this type of relationship. Finally, type (3) has a problem in that it is much less important than type (2), and too many nodes and edges satisfy this rule.

The above consideration led us to include only types (1) and (2); that is, we excluded type (3).

4.2   Features and Functions of Citation Graph

In this section, we explain our implementation of the interface with citation graph. The interface receives query inputs by users and makes retrievals from the literature databases. It displays the graph based on the criteria in Sect. 4.1, along with a list of papers that are ranked based on their citation count, relevance to the query, and other indicators. Figure 4 shows examples of the citation

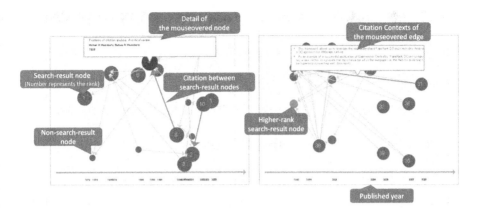

Fig. 4. Citation graph in the proposed interface.

graph obtained by the proposed method. We create the list by using Microsoft Academic Search and display the top ten search results on the first page. In addition, this interface has links to other search results pages at the bottom of the page. For example, if users want to check the page that contains the 11th to the 20th search results, they can click link "2." Each snippet contains the title of the paper, a link to the page including the details of the paper in Microsoft Academic Search, and summary sentences of the paper.

Determination of the Size of a Node According to Its Importance. Changing the size of a node displayed in a graph can highlight the paper corresponding to the node. Hence, it is likely that changing the size according to the importance of the paper will help users to pay more attention to important papers. Moreover, although there are various indicators that represent the importance of papers such as the journals that published the paper, the authors of the paper, etc., one of the most important is the citation count (i.e., the number of citations) [16]. Thus, in our method, if the paper has a large citation count, the size of the corresponding node becomes large, and vice versa. This function allows the interface to express the importance of a single paper on the graph as well as the relationship between papers. For example, it helps users to understand that the relationship between two papers must be important because the individual papers are likely to be important.

In addition, when the citation count of a paper that cites and is cited by a paper in the search results is less than a threshold, we decided not to display the node corresponding to that paper and the surrounding edges. We suppose the threshold is given by users.

Arrangement of Nodes. If the nodes and edges of the citation graph are disordered, users looking at the interface of the graph may experience significant cognitive load. Here, arranging the nodes in the citation graph on the basis of time would be a familiar way to relieve the load because the cited paper always

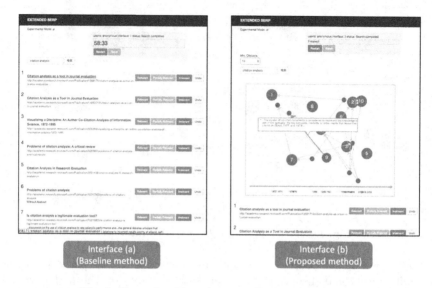

Fig. 5. Example of interfaces used in the experiment.

precedes the citing paper. Hence, we decided to position the nodes according to the published year. For example, the node of a paper published in 1990 is located on the left of another paper published in 2000. This arrangement eases the users' cognitive load by keeping the graph in order. Apart from this, arranging the papers as a time series allows users to understand the developmental history of their research topic. It enables them to improve their understanding of their research field.

5   User Study

Here, we explain the user study examining the effectiveness of the proposed method. We asked actual users of an academic search engine to participate the experiment, and to conduct scholarly surveys using the interface that we implemented on the existing academic search engine (Microsoft Academic Search) and using the same engine but without our interface. Figure 5 shows the interfaces used in the experiment. The two interfaces were as follows:

(a) Academic search engine with no additional information
    This interface was the baseline in this experiment; it was a conventional academic search engine.
(b) Academic search engine with a graph based on citing and cited relationship
    This was the interface based on the method described in Sect. 4.

5.1   Procedure

We asked nine students in our laboratory, eight graduate students and an undergraduate student, to participate in the experiment. They had their own research

Table 1. Number of papers judged to be relevant or partially relevant.

	(a)	(b)
Topic 1	4.83 (2.14)	7.83 (4.17)
Topic 2	5.67 (4.72)	7.00 (4.20)
Topic 3	6.50 (2.17)	5.50 (2.88)
All topics	11.33 (5.66)	13.56 (6.56)

Table 2. Ratio of papers judged to be relevant or partially relevant.

	(a)	(b)
Topic 1	0.42 (0.04)	0.70 (0.08)
Topic 2	0.54 (0.25)	0.90 (0.01)
Topic 3	0.71 (0.08)	0.52 (0.27)
All topics	0.55 (0.18)	0.71 (0.21)

fields related to data mining or text processing and surveyed related papers if needed. The procedure of this experiment is listed below:

(1) We divided the participants into three groups so that each group had three participants. The groups were named "group 1," "group 2" and "group 3."
(2) A participant was selected from each group and asked to prepare a research topic for a survey; there was a total of three research topics. The prepared research topics were named "topic 1," "topic 2", and "topic 3" in correspondence with the groups. The participants were asked to write about the title, abstract (about 500 characters in Japanese) of the research topic, and the situation in which they actually surveyed the research topic. The titles of the topics were "bias analysis of news articles," "causal relationship mining," and "privacy protection of trajectory data."
(3) Before the surveys, all of the participants were asked to use the two interfaces to get accustomed to them.
(4) All the participants in group 1 were asked to complete their surveys according to the description and situation of topic 2 in (2) using interface (a). If they found a paper relevant to the predefined information need, they judged the relevance of the paper after they followed the link to the paper. The relevance of papers judged by the participants had three levels, i.e., "relevant," "partially relevant", and "irrelevant." The time limit of each survey was 60 min.
(5) Using the same procedure as (4), all participants in group 1 were asked to survey topic 3 using interface (b).
(6) After the participants had performed each survey, they were asked to answer a post-survey questionnaire. Also, when they had finished all assigned surveys, they were asked to answer an exit questionnaire.
(7) The participants in groups 2 and 3 were also asked to survey the research topics and answer the questionnaires as well as (4)–(6). We balanced the assignment between the groups so that all the topics were equally surveyed.

5.2 Results

Efficiency of Survey. First, we show the results of the study related to the efficiency of surveys obtained from the log data. Table 1 shows the average number of papers judged to be "relevant" or "partially relevant" by the participants.

Numbers in parentheses represent standard deviations. The results suggest that the participants using interface (b) judged more papers to be relevant compared with those using interface (a), excluding the surveys of topic 3. This indicates that the relationships displayed in the citation graph helped them choose relevant search results. The surveys of topic 3 with interface (b) obtained fewer relevant papers because the scope of the topic was narrow, and the participants could not find relevant papers easily.

Next, we describe the ratio of papers judged to be relevant by the participants to all papers whose links they followed and judged the relevance of during the surveys. This indicator represents how many unnecessary surveys the proposed method could reduce, which is one of the problems of scholarly surveys mentioned in Sect. 3.2. We can also consider it as a precision based on the papers the participants read. The average numbers of all papers that a participant judged the relevance (followed the link) per topic in the survey were about 20. Table 2 shows the average ratio of papers judged to be "relevant" or "partially relevant" to all papers of which he/she judged the relevance. Numbers in parentheses represent standard deviations. We can observe that the participants using interface (b) could find a higher ratio of papers relevant to their needs compared with the participants using interface (a), excluding the surveys of topic 3. The reason is that displaying a graph enabled them to understand the relationship between the papers in the search results easily and thus made it easier to find only the relevant papers from a large number of papers.

Subjective Preferences. Next, we briefly review the participants' subjective preferences from the results of the post-survey and exit questionnaires for each interface. We asked the participants about the usability of the interfaces, the cognitive load when they used the interface, the confidence on the relevance judgments before they read a paper, and so on. The results indicated that interface (b) was better than interface (a) as the confidence in the relevance judgments was higher with reasonable cognitive load.

Also, we got some comments which reveal that the citation graph improved the satisfaction and usability of the interface. On the other hand, some participants pointed out the possibility that the citation graph had a bad influence on the interface when the research topic of the survey was limited to a narrow one or they had little knowledge on the topic.

6 Conclusion and Future Work

We proposed a method of supporting researchers' scholarly surveys using academic search engines. This method extracts citing and cited relationships between papers and incorporates them into the search results pages in an intuitive fashion. We explained the implementation of an interface based on the proposed method and described a user study in which actual users simulated scholarly surveys.

There is still a need to consider how best to deal with citing and cited relationships as there are some patterns on the relationships. We also need to refine the definition of scholarly surveys so that they can express the users' information needs more clearly, conduct another user study that includes more participants and research topics, and analyze the results deeply.

Acknowledgments. We thank Associate Professor Hidetsugu Nanba at Hiroshima City University for his valuable comments on our research.

References

1. Jacsó, P.: Google Scholar: the pros and the cons. Online Inf. Rev. **29**(2), 208–214 (2005)
2. Beel, J., Gipp, B., Wilde, E.: Academic Search Engine Optimization (ASEO): optimizing scholarly literature for Google Scholar and co. J. Sch. Publ. **41**(2), 176–190 (2010)
3. Beel, J., Gipp, B.: Academic search engine spam and Google Scholar's resilience against it. J. Electron. Publ. **13**(3) (2010)
4. Verberne, S., Sappelli, M., Kraaij, W.: Query term suggestion in academic search. In: de Rijke, M., Kenter, T., de Vries, A.P., Zhai, C.X., de Jong, F., Radinsky, K., Hofmann, K. (eds.) ECIR 2014. LNCS, vol. 8416, pp. 560–566. Springer, Heidelberg (2014)
5. Viégas, F.B., Donath, J.: Social network visualization: can we go beyond the graph? In: Workshop on Social Networks, CSCW, vol. 4, pp. 6–10 (2004)
6. Nanba, H., Abekawa, T., Okumura, M., Saito, S.: Bilingual PRESRI - integration of multiple research paper databases. In: Proceedings of RIAO, pp. 195–211 (2004)
7. He, Q., Chen, B., Pei, J., Qiu, B., Mitra, P., Giles, C.L.: Detecting topic evolution in scientific literature: how can citations help? In: Proceedings of CIKM, pp. 957–966 (2009)
8. Dunne, C., Shneiderman, B., Gove, R., Klavans, J., Dorr, B.J.: Rapid understanding of scientific paper collections: integrating statistics, text analytics, and visualization. JASIST **63**(12), 2351–2369 (2012)
9. Varadarajan, R., Hristidis, V.: A system for query-specific document summarization. In: Proceedings of CIKM, pp. 622–631 (2006)
10. Clarke, C.L.A., Agichtein, E., Dumais, S.T., White, R.W.: The influence of caption features on clickthrough patterns in web search. In: Proceedings of SIGIR, pp. 135–142 (2007)
11. Nguyen, T., Zhang, J.: A novel visualization model for web search results. IEEE Trans. Vis. Comput. Graph. **12**(5), 981–988 (2006)
12. Giacomo, E.D., Didimo, W., Grilli, L., Liotta, G.: Graph visualization techniques for web clustering engines. IEEE Trans. Vis. Comput. Graph. **13**(2), 294–304 (2007)
13. Scaiella, U., Ferragina, P., Marino, A., Ciaramita, M.: Topical clustering of search results. In: Proceedings of WSDM, pp. 223–232 (2012)
14. Mirylenka, D., Passerini, A.: Navigating the topical structure of academic search results via the wikipedia category network. In: Proceedings of CIKM, pp. 891–896 (2013)
15. Kim, Y., Seo, J., Croft, W.B.: Automatic boolean query suggestion for professional search. In: Proceedings of SIGIR, pp. 825–834 (2011)
16. Yan, R., Tang, J., Liu, X., Shan, D., Li, X.: Citation count prediction: learning to estimate future citations for literature. In: Proceedings of CIKM, pp. 1247–1252 (2011)

Quality-Aware Review Selection Based on Product Feature Taxonomy

Nan Tian[1]([✉]), Yue Xu[1], Yuefeng Li[1], and Gabriella Pasi[2]

[1] Faculty of Science and Engineering, Queensland University of Technology,
Brisbane, Australia
{n.tian,yue.xu,y2.li}@qut.edu.au
[2] Department of Informatics, Systems and Communication,
University of Milano Bicocca, Milano, Italy
pasi@disco.unimib.it

Abstract. User-generated information such as online reviews has
become increasingly significant for customers in decision making
processes. Meanwhile, as the volume of online reviews proliferates, there
is an insistent demand to help users in tackling the information overload
problem. A considerable amount of research has addressed the problem
of extracting useful information from overwhelming reviews; among the
proposed approaches we remind review summarization and review selection. Particularly, to address the issue of reducing redundant information,
researchers attempt to select a small set of reviews to represent the entire
review corpus by preserving its statistical properties (e.g., opinion distribution). However, a significant drawback of the existing works is that
they only measure the utility of the extracted reviews as a whole without considering the quality of each individual review. As a result, the set
of chosen reviews may consist of low-quality ones even if its statistical
property is close to that of the original review corpus, which is not preferred by the users. In this paper, we propose a review selection method
which takes the reviews' quality into consideration during the selection
process. Specifically, we examine the relationships between product features based upon a domain ontology to capture the review characteristics
based on which to select reviews that have good quality and to preserve
the opinion distribution as well. Our experimental results based on real
world review datasets demonstrate that our proposed approach is feasible
and able to improve the performance of the review selection effectively.

Keywords: Review selection · Review quality · Product feature
taxonomy

1 Introduction

The advent of Web 2.0 has determined the generation of a huge amount of user
generated information such as online reviews; this phenomenon has increased
in recent years, as more and more online users tend to write textual reviews to

© Springer International Publishing Switzerland 2015
G. Zuccon et al. (Eds.): AIRS 2015, LNCS 9460, pp. 68–80, 2015.
DOI: 10.1007/978-3-319-28940-3_6

share their personal experiences and opinions on the purchased products. Online reviews are an invaluable information resource, and they have become increasingly important in decision making processes [2] and promoting business [15]. However, the volume of review data is overwhelming while the quality of reviews varies greatly. Hence, some researchers have made efforts to evaluate the review quality/helpfulness automatically [2,3,6]. In detail, existing approaches examine a number of data features about the reviews such as writing quality and writer's expertise to find the hidden association between the review quality and these defined features. However, such review quality prediction approaches usually require considerable time and resources such as training and review data labelling. Meanwhile, another set of studies aim to extract a subset of useful reviews as a representation of the original review corpus [4,11]. For instance, Lappas et al. [4] proposed to extract a set of reviews that accurately reflect the proportion of opinions of each product feature (called opinion distribution) in a collection of reviews. One significant drawback of this method is that the utility of the selected reviews is only measured as a whole in terms of opinion distribution, while the quality of each individual review has been largely ignored, which could allow some low-quality reviews to be included in the selected review set.

In recent years, ontology learning has attracted significant attention. Researchers have made a lot of efforts to find relationship between different terms or concepts more effectively and accurately. By making use of various techniques such as text mining, we are now able to generate product ontology or taxonomy about product features and relationships between features from data about products or even from user generated information such as tags and review text [1,10]. In this paper, we propose a method, called Quality-Aware Review Selection (QAS), to assess the review quality by using a product feature taxonomy to improve the performance of review selection. More specifically, we propose not only to select a small set of reviews which retain the same opinion distribution (i.e., positive and negative opinion proportion) as in the original review collection, but also we ensure that each of the selected reviews is of good quality by examining the comprehension of the features mentioned in the review.

The remainder of the paper is organized as follows. The related works will be discussed in Sect. 2. Section 3 provides an insight about our proposed quality-aware review selection approach. The evaluation of our approach is reported in Sect. 4. Finally, we conclude the paper in Sect. 5.

2   Related Work

2.1   Review Helpfulness Prediction

A number of studies focused on identifying data features related to review text or review content, on which a review's quality or helpfulness can be learnt [3,6,9]. In detail, these approaches formulated the review helpfulness prediction as a regression or classification problem. For instance, Liu et al. [7] made use of radial basis functions to predict the helpfulness of movie reviews based upon three factors: reviewer expertise, writing style and timeliness. O'Mahony et al. [9]

investigated a number of structural features related to review texts such as the percentage of uppercase characters and the number of sentences of reviews. They made use of a classifier to distinguish helpful and unhelpful reviews based upon identified structural features. In addition, Lu et al. [8] proposed to exploit and utilize contextual information about the writers' identities and social networks to improve review quality estimation.

Meanwhile, some researchers attempted to value the reviews by examining mentioned features and opinions from users. For instance, Liu et al. [6] considered the number of product features in the review one of the most important indicators of the review helpfulness. They employed SVM (Support Vector Machines) as the model of classification to filter out low-quality reviews so that a more accurate opinion summarization (e.g., average feature rating score) can be achieved. Zhang et al. [13] proposed a supervised method to predict the utility of the reviews by taking sentiments into account. Specifically, the subjective words that express users' opinions were considered one factor for learning the utility of the reviews.

2.2 Review Selection

A number of recent works have focused on extracting reviews for the users based upon reviews' utility. The aim is to preserve the integrity and immediacy of actual reviews. For instance, Tsaparas et al. [11] and Xu et al. [12] attempted to generate a subset of reviews which provide comprehensive information of the reviewed product by covering all product features and both positive and negative opinions toward each feature. They formulated it as a maximum coverage problem. However, this formulation fails to capture the proportion of opinions in the review corpus since each opinion polarity on features is treated equally. In order to overcome this shortcoming, Lappas et al. [4] proposed to select a subset of reviews that emulate the opinion distribution in the original review collection as accurately as possible. They formulated this task as a combinatorial optimization problem. In detail, the utility of the generated review set is measured based upon how the proportion of opinions in the selected reviews is close to that in the original review corpus. One major problem of this approach is that the utility of selected reviews is only benchmarked as a whole, while the quality of each individual review has been omitted.

2.3 Taxonomy or Ontology Learning

Meanwhile, product classifications or taxonomies are often available, provided by product manufacture organizations or companies for promotion or marketing purposes. Moreover, ontology learning is a wide studied area. In recent years, some researchers seek to create a hierarchical structure about products or items from user generated content. Djuana et al. [1] proposed both to construct a tag ontology from folksonomy based on WordNet, and to personalize the tag ontology based on user clusters. Furthermore, Tian et al. [10] presented an approach to construct a hierarchical product profile which contains product features and relationships between them. Specifically, the proposed model identifies a group of frequent patterns as the potential features to assist selecting useful association rules for finding the relationships between features.

3 The Proposed Approach

Prior work has basically employed supervised learning approaches such as classification to estimate the quality of the reviews based on statistical features related to review texts, which ignores the usefulness of information buried in reviews. For instance, a review written in a very professional style is still useless to a user if it lacks the discussion of the features concerned by the user. According to Liu et al. [6], a review of good quality should be a rather complete and detailed comment on a product, by presenting several aspects of a product. According to this observation, we attempt to find a way to formulate this characteristic for determining the quality of reviews.

3.1 Product Feature Taxonomy

Reviews vary in terms of coverage and focus by considering different product features. More specifically, some reviews talk about a number of unrelated features, while others may focus on one or several specific features only by analysing them from different angles. We believe that a review's helpfulness or quality can be better predicted if considering how it covers the product features than when analyzing its textual features or the writer's reputation.

In order to measure the aforementioned review characteristics, we first need a structural profile of the product that provides the relationships between its different features. It could be a standard ontology provided by domain experts or an ontology automatically generated from domain data such as reviews by using ontology learning methods. In this work, we make use of the product profile called product feature taxonomy proposed in [10], defined below, for assisting the review analysis. Figure 1 shows part of a feature taxonomy for a product (i.e., digital camera) generated from a collection of reviews. As shown, it is a tree structure describing the relationships between the product features.

Definition 1 (Feature Taxonomy): We formally represent a feature taxonomy as a set of features and their relationships, denoted as $FT = \{F, L\}$, F is a set of

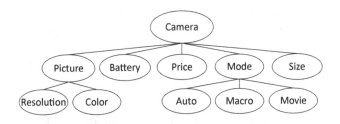

Fig. 1. Product Feature Taxonomy

features where $F = \{f_1, f_2, \ldots, f_n\}$ and L is a set of links. The feature taxonomy has the following constraints and terms that are used in this paper:

(1) The link which connects a pair of features represents the sub-feature relationship. For $f_i, f_j \in F$, if f_j is a direct sub-feature of f_i, then (f_i, f_j) is a link in the taxonomy and $(f_i, f_j) \in L$, which indicates that f_j is more specific than f_i. f_i is called the parent feature of f_j denoted as $P(f_j)$.
(2) Except for the root, each feature has only one parent feature. This means that the taxonomy is structured as a tree.
(3) The root of the taxonomy represents the product itself.

3.2 Review Comprehension

The product feature taxonomy provides a thorough picture of the hierarchical relation between different features of a certain product, which can be used as an indicator of review quality. For instance, if a user describes *vivid color* and *high resolution* of the pictures captured by a camera, the quality of *picture* is discussed in depth in this review. Based on the observation in [6], we believe such review is of good quality. Therefore, we first need to identify all features mentioned in the review. Due to the fact that the same feature may be represented in different ways, we make use of WordNet to find synonyms of a certain feature to address this problem. A feature is considered included in the review if a similar word to the feature name, according to a semantic similarity, is found in the review instead of the exact feature name. In order to measure how comprehensive a review is, we propose to examine the coverage of the features in the review. The coverage of a feature can be determined based on its sub features in the product feature taxonomy. In Definition 2 below, a notion of maximum sub tree is defined. In this paper, we propose to generate all maximum sub trees in a review according to its identified features, then measure the comprehension of the review based on its maximum sub trees.

Definition 2 (Maximum Sub Tree): Let $FT = \{F, L\}$ be a product feature taxonomy, $F_r \subseteq F$ be a set of features identified from review r and $f \in F_r$ be a specified feature. The maximum sub tree rooted at f is defined as $MST_{r,f} = \{SF_f, SL_f\}$ which satisfies the following constraints:

- f is the root of the sub tree, $f \in SF_f$
- $SF_f \subseteq F_r$, $SL_f \subseteq L$
- $\forall g \in SF_f$, there must be a path $< f, f_1, f_2, \ldots, f_n, g >$ between f and g in the feature taxonomy FT, $f_i \in SF_f$, $i = 1, \ldots n$, and $(f, f_1), \ldots, (f_i, f_{i+1}), (f_n, g) \in SL_f$, $j = 1, \ldots, n-1$
- $\forall g \in F_r$ and $g \notin SF_f$, there is no path $< f, f_1, f_2, \ldots, f_n, g >$ between f and g in the feature taxonomy FT and $f_i \in SF_f$, $i = 1, \ldots, n$.

These maximum sub trees are disjoint, i.e., there is no overlap between any two sub trees. The features in one sub tree are considered related in terms of F_r since they are linked in the sub tree, while the features from different sub trees are not considered related since there is no path or link between these features. Let $MST_r = \{MST_{r,f_1}, ..., MST_{r,f_m}\}$ be a set of m maximum sub trees generated for review r, $f_1, ..., f_m \in F_r$ are the root for the sub trees respectively, then $SF_{f_1} \cap SF_{f_2} \cap ... \cap SF_{f_m} = \emptyset$, that is, $\{SF_{f_1}, SF_{f_2}, ..., SF_{f_m}\}$ is a partition to F_r. For example, if a review contains *"picture"*, *"color"*, *"mode"*, *"auto"*, *"macro"*; according to the feature taxonomy shown in Fig. 1, we can generate two maximum sub trees: one consists of *"picture"* and *"color"*, the other consists of *"mode"*, *"auto"* and *"macro"*.

We employ the term comprehension to indicate how comprehensively a review discusses the features of the considered product. We employ the sub-feature relation of the feature taxonomy to measure the degree of comprehension of a review. Specifically, the more numerous are the sub features of a given feature in a review, the more comprehensive with respect to that feature the review is. Therefore, we calculate the ratio between the number of the feature's direct sub features appearing in the review and the total number of direct sub features in the feature taxonomy based upon the generated maximum sub trees. Let $MST_{FT,f} = \{SF_f, SL_f\}$ be a maximum sub tree of which feature f is the root; we can derive all direct sub features of f in the taxonomy: $DSF_f(FT) = \{sf | sf \in F \text{ and } (f, sf) \in L\}$. Hence, the comprehension of f can be derived by the following equation:

$$comp_{r,f} = \begin{cases} \frac{|DSF_f(FT) \cap F_r|}{|DSF_f(FT)|} & |DSF_f(FT)| \neq 0 \\ 1 & otherwise \end{cases} \qquad (1)$$

As shown in Eq. 1, some features in the feature taxonomy may not have any sub feature. In such case, the comprehension degree with resepct to that feature is 1. As to the comprehension of a review r, we calculate the average comprehension value of the features in r as the comprehension value of the review:

$$COMP_r = \frac{\sum_{f \in F_r} comp_{r,f}}{|F_r|} \qquad (2)$$

The comprehension score is used to indicate how comprehensive a review is so that it can be employed to assess the review quality. This measure allows to assess the quality of both reviews that provide in-depth information on various product features and reviews that focus on one single feature. For instance, if a review discusses of a single feature by covering all or most of its sub features, the comprehension value of this review will be still quite good. The higher the comprehension value a review obtains, the better quality this review has.

3.3   Characteristic Review Selection

Lappas et al. [4] believe that selecting a subset of reviews which preserves the opinion distribution statistics in the underlying review corpus brings a great balance between review summarization and review selection. Given a collection of reviews R and a list of features F, a feature-opinion vector $\pi(R)$ is defined as $< f_{1P}, f_{1N}, f_{2P}, f_{2N}, ..., f_{mP}, f_{mN} >$, where $f_{iP} = \frac{\#Positive\_f_i}{|R|}$ and $f_{iN} = \frac{\#Negative\_f_i}{|R|}$ measure the proportion of positive and negative opinion for each product feature in R. $\pi(R)$ can be constructed by using sentiment analysis techniques such as [5]. To find the best set of n reviews SR as the representation of R, the feature-opinion vector in SR, i.e., $\pi(SR)$, should be close to $\pi(R)$ as much as possible. $D(\pi(SR), \pi(R))$ is used to indicate the distance between two feature-opinion vectors. A Greedy algorithm based on opinion distribution distance has been introduced in their work.

The selected reviews as a whole reflect the characteristics of the original review corpus and also avoid redundant information as other prior work does. Inspired by this motivation, we aim to combine our work and this idea together to improve the performance of the review selection task.

3.4   Quality-Aware Review Selection

There are two obvious drawbacks for the above characteristic review selection method. Firstly, the utility of a review is only determined based on if it can minimize the opinion distribution distance between the extracted review set SR and the review corpus R without examining the review's actual quality. Therefore, a low-quality review could be selected if it can make the SR obtain a smaller opinion distribution distance value. Secondly, when multiple reviews make the current SR achieve the same minimum $D(\pi(SR), \pi(R))$, this method fails to distinguish them to determine which review should be the most appropriate one to be chosen. In such case, the review with the top position according to the order of the reviews in the original corpus will be selected. In other words, the same review dataset with different order might lead to a different result.

To overcome the aforementioned shortcomings, we propose to select the reviews based on both their review comprehension score and their opinion distribution. In this work we employ the sentiment analysis algorithm proposed by Lau et al. [5] to identify all feature-sentiment pairs and to determine the orientation (positive or negative) of each sentiment of a feature so that the opinion distribution can be generated. Then, in the ith iteration of the selection process, we aim to select one review to form a set SR of $i$reviews that can make $\pi(SR)$ close to $\pi(R)$. In each iteration, the strategy is to first find a set of candidate reviews, then from the candidate reviews to select the most appropriate review. The number of candidate reviews is dynamically determined based on the opinion distribution distance between all potential review set SR and the original review set R.

In the ith iteration, let SR^{i-1} be the current selected review set, for all remaining reviews in $R \backslash SR^{i-1}$, the minimum opinion distance between all potential review set SR^i and the original review set R is defined as:

$$D_{min}^i = min_{r \in R \backslash SR^{i-1}} D(\pi(SR^{i-1} \cup \{r\}), \pi(R)) \tag{3}$$

According to D_{min}^i, we extract a number of reviews whose derived opinion distribution distance value is close to D_{min}^i as the candidates for further selection. The following equation is defined to measure how close the opinion distribution distance $D(\pi(SR^{i-1} \cup \{r\}), \pi(R))$ is to the minimum distance D_{min}^i:

$$C(r, i) = \frac{D(\pi(SR^{i-1} \cup \{r\}), \pi(R)) - D_{min}^i}{D_{min}^i} \quad (4)$$

$C(r, i)$ is called the closeness of r in the ith iteration. Based on $C(r, i)$, we can generate a set of candidate reviews as defined below in the ith iteration:

$$Candidates^i = \{r | r \in R \backslash SR^{i-1} \text{ and } C(r, i) < \sigma\} \quad (5)$$

where σ is a threshold. In the experiments reported in Sect. 4, $0 <= \sigma <= D_{min}^i *$ 5 %. Among the identified candidate reviews, we calculate their comprehension value based upon the given product feature taxonomy and choose the one which obtains the highest comprehension score as the best review to be added to the selected review set: $r_{selected} = argmax_{r \in Candidates^i}\{COMP_r\}$. Here below the algorithm employed for the reviews selection is sketched.

Algorithm 1. Quality-Aware Review Selection Algorithm

Input: $R, FT = \{F, L\}$, and n.
Output: $SR \subseteq R, |SR| = n$
1: $SR = \emptyset$
2: **for** $i = 1...n$ **do**
3:    $D_{min}^i = min_{r \in R \backslash SR^{i-1}} D(\pi(SR^{i-1} \cup \{r\}), \pi(R))$
4:    $Candidates^i = \{r | r \in R \backslash SR^{i-1} \text{ and } C(r, i) < \sigma\}$
5:    $r_{selected} = argmax_{r \in Candidates^i}\{COMP_r\}$
6:    $SR = SR \cup \{r_{selected}\}$
7: **end for**
8: **return** SR

---

4 Experiment and Evaluation

In this section we present the experiment that we performed to evaluate the effectiveness of our proposed approach. Our experiment is carried out using real data collected from one of the most popular e-commerce websites: www.amazon.com. On Amazon, users are able to vote each review as helpful or unhelpful from their perspective, and the form *"m of n people found the following review helpful"* indicates the helpfulness voting. Therefore, we use the ratio between the number of positive votes and the total number of votes as the gold standard of the review quality, which indicates if the review is preferred by users. For instance, the helpfulness score of a review is *0.8*, if *8* out of *10* online users rate this review as helpful. We use six datasets of reviews for testing (5 cameras and 1 laptop). Each dataset consists of all online users' reviews for a certain product. In addition, there are more than 350 reviews in each dataset averagely.

Table 1. Average helpfulness comparison

Product	Model	n = 10	n = 20	n = 30
CAM1	Baseline	0.54	0.56	0.60
	QAS	0.64	0.65	0.64
CAM2	Baseline	0.42	0.49	0.53
	QAS	0.62	0.57	0.64
CAM3	Baseline	0.47	0.52	0.49
	QAS	0.62	0.59	0.55
CAM4	Baseline	0.50	0.56	0.59
	QAS	0.55	0.66	0.69
CAM5	Baseline	0.55	0.66	0.60
	QAS	0.68	0.67	0.72
PC	Baseline	0.52	0.53	0.55
	QAS	0.57	0.64	0.62

Table 2. Opinion distribution distance comparison

Product	Model	n = 10	n = 20	n = 30
CAM1	Baseline	0.263	0.169	0.113
	QAS	0.310	0.161	0.125
CAM2	Baseline	0.283	0.187	0.120
	QAS	0.335	0.198	0.129
CAM3	Baseline	0.309	0.184	0.129
	QAS	0.332	0.166	0.130
CAM4	Baseline	0.219	0.140	0.089
	QAS	0.227	0.141	0.091
CAM5	Baseline	0.365	0.207	0.136
	QAS	0.346	0.197	0.138
PC	Baseline	0.216	0.128	0.082
	QAS	0.220	0.113	0.079

We use the review selection method proposed in [4] as the baseline for comparison. In detail, their proposed Greedy algorithm has a satisfactory performance in terms of running time and capturing the opinion distribution of the original review corpus. In order to provide a comprehensive and objective evaluation, we benchmark the effectiveness of our proposed method from three aspects: quality of the selected reviews, the opinion distribution distance between the selected reviews and the review corpus, and product rating results derived from the selected reviews. For each review dataset, both selection models are employed to generate the result set of n reviews ($n = 10, 20, 30$) for 10 runs. Particularly, in each run, we shuffle the review dataset to change the order of the reviews. All experimental results are derived from the average of all runs.

First of all, we compare the average helpfulness score of n selected reviews in the generated review set to check if our method is able to find reviews of better quality. More specifically, the ratio between helpfulness votes and total votes provided by the review is used as the helpfulness score to indicate the review quality. The average helpfulness score of all reviews in generated review set from both models is calculated. The higher the average helpfulness score obtained, the more the reviews of better quality have been selected. The experimental results are reported in Table 1 below.

Table 1 illustrates the average helpfulness scores of 10, 20, and 30 reviews selected by the baseline and by our proposed method, respectively. From the results, we can see that the review set produced by our method always obtains better helpfulness score than the baseline does. It is because that in each iteration of the review selection process in which a new review is to be added into the existing review set, our method always chooses the review of best quality according to the comprehension value from a number of reviews instead of the one which achieves the minimum opinion distribution distance. As a result, this

Table 3. Standard deviation comparison

Product	Model	n = 10	n = 20	n = 30
CAM1	Baseline	0.026	0.013	0.020
	QAS	0.007	0.004	0.002
CAM2	Baseline	0.026	0.013	0.009
	QAS	1.11E-16	1.11E-16	0.0
CAM3	Baseline	0.028	0.017	0.011
	QAS	1.11E-16	1.11E-16	1.11E-16
CAM4	Baseline	0.055	0.023	0.015
	QAS	0.033	0.016	0.011
CAM5	Baseline	0.059	0.029	0.020
	QAS	0.021	0.010	0.007
PC	Baseline	0.101	0.051	0.030
	QAS	0.095	0.046	0.021

Table 4. Average product rating comparison

	Golden Standard	QAS	Baseline
CAM1	9.6	9.6	9.4
CAM2	9.0	9.0	9.5
CAM3	8.0	7.7	8.7
CAM4	8.4	8.7	9.2
CAM5	8.0	8.3	8.5
PC	8.2	8.4	8.6

proves that making use of structural relationships between features to assess review quality is effective.

In addition, we are also concerned if the review set produced by our method can still preserve the characteristics of the original review corpus with respect to the opinion distribution. Thus, we calculate the opinion distribution distance between the selected reviews and the review corpus for both the proposed model and the baseline model. The experimental results are shown in Table 2.

Our method aims to find a combination that optimizes both review quality and opinion distribution. Based upon the comparison provided by Table 2, we can see that our method selects reviews of much better quality without sacrificing the characteristic of opinion distribution in the original review collection, which is a satisfactory result. Thus, it is evident that our method could achieve a great balance between considering individual review quality and optimizing the utility of selected reviews as a whole.

As discussed in the previous section, one significant drawback of the baseline model is that it cannot obtain a consistent performance if the order of the reviews in the review dataset changes. Thus, we also undertake an experiment to test if our method has a stable performance on review selection. In detail, we calculate the standard deviation of the average helpfulness scores of selected reviews in all runs. The derived results are provided in Table 3 below.

From Table 3 we can see that the standard deviation values derived by our method are much lower than those of the baseline. This indicates that the review set generated by the baseline model is quite different in different runs with different orders. This is because the baseline selection method is heavily related to the review order in the dataset. In contrast, relatively low standard deviation scores of our method show that the selected reviews in each run tend to be quite similar, which indicates that our method is much less related to the review order in the dataset.

To further demonstrate the impact of our approach, we aim to justify the performance with respect to product rating. We assume that a better selected review set should provide a more accurate product rating, which is in accordance with the objective evaluation of the products. Therefore, we seek to use the product rating score provided by authors of the reviews as the measurement in this regard. For the ground truth data, we look for the objective evaluation reference from external authoritative sources. A number of websites provide a professional "expert review" for the products we test, which gives a score in the range of 1–10 for each product. Hence, we collect expert scores for the six products from five websites[1,2,3,4,5,6] and use these scores as the golden standard for product rating comparison. We calculate the average product rating score of 30 selected reviews generated by both models. Since the product rating score on Amazon is in the range of 1–5, we first rescale the derived average product rating scores into the range of 1–10 in order to undertake the comparison.

Table 4 provides the comparison between the baseline model and our proposed method on product rating score for the six considered products, respectively. According to the results, we can see that the average product rating scores obtained by our proposed method are much closer to the golden standard compared to those of the baseline model for all products. This indicates that the review set produced by our approach can obtain more reliable opinion summarization results by providing a consistent product rating score with the one provided by the experts. We infer that the improvement is caused by the reviews of better quality in which more comprehensive information about the product has been provided in the selected reviews by our method.

5 Conclusions

In this paper, we introduced a review selection method to extract a set of reviews for representing the review corpus. Different from existing review quality prediction methods, we proposed to assess the quality of a review in an unsupervised manner by utilizing the structural relationships between product features to capture the review characteristic called comprehension which indicates how features have been covered in the review. The evaluation results on real world datasets have proven that our review selection method is effective and can be used to find reviews of good quality. In addition, the testing also shows that our proposed review selection method is able to optimize the selection results by providing a great balance between choosing the individual review of good quality and ensuring the extracted reviews as a whole to reflect the opinion distribution in the

---

[1] http://www.cnet.com/products/apple-macbook-pro-with-retina-display-13-3-inch/.
[2] https://www.canon.com.au/en-AU/Personal/Products/Cameras-and-Accessories/EOS-Digital-SLR-Cameras/EOS-6D.
[3] http://www.cameralabs.com/reviews/Canon_EOS_5D_Mark_III/.
[4] http://www.digitaltrends.com/digital-camera-reviews/nikon-d800-review/.
[5] http://www.dpreview.com/reviews/canoneos7d/30.
[6] http://www.dpreview.com/reviews/nikond7000/22.

review corpus. In the future, we plan to improve our method and to undertake the evaluation on more datasets of different categories.

References

1. Djuana, E., Xu, Y., Li, Y., Cox, C.: Personalization in tag ontology learning for recommendation making. In: Proceedings of the 14th International Conference on Information Integration and Web-based Applications and Services, pp. 368–377. ACM (2012)
2. Hai, Z., Cong, G., Chang, K., Liu, W., Cheng, P.: Coarse-to-fine review selection via supervised joint aspect and sentiment model. In: Proceedings of the 37th International ACM SIGIR Conference on Research and Development in Information Retrieval, pp. 617–626. ACM (2014)
3. Kim, S.-M., Pantel, P., Chklovski, T., Pennacchiotti, M.: Automatically assessing review helpfulness. In: Proceedings of the 2006 Conference on Empirical Methods in Natural Language Processing, pp. 423–430 (2006)
4. Lappas, T., Crovella, M., Terzi, E.: Selecting a characteristic set of reviews. In: Proceedings of the 18th ACM SIGKDD International Conference on Knowledge Discovery and Data Mining, pp. 832–840. ACM (2012)
5. Lau, R.Y.K., Song, D., Li, Y., Cheung, T.C.H., Hao, J.-X.: Toward a fuzzy domain ontology extraction method for adaptive e-Learning. IEEE Transactions on Knowledge and Data Engineering $21(6)$, 800–813 (2009)
6. Liu, J., Cao, Y., Lin, C. Y., Huang, Y., Zhou, M.: Low-quality product review detection in opinion summarization. In: Proceedings of the 2007 Joint Conference on Empirical Methods in Natural Language Processing and Computational Natural Language Learning (EMNLP-CoNLL), pp. 334–342 (2007)
7. Liu, Y., Huang, X., An, A., Yu, X.: Modeling and predicting the helpfulness of online reviews. In: IEEE Eighth International Conference on Data Mining ICDM 2008, pp. 443–452 (2008)
8. Lu, Y., Tsaparas, P., Ntoulas, A., Polanyi, L.: Exploiting social context for review quality prediction. In: Proceedings of the 19th International Conference on World Wide Web, pp. 691–700. ACM (2010)
9. O'Mahony, M. P., Smyth, B.: Using readability tests to predict helpful product reviews. In: Proceeding RIAO 2010 Adaptivity, Personalization and Fusion of Heterogeneous Information, pp. 164–167 (2010)
10. Tian, N., Xu, Y., Li, Y., Abdel-Hafez, A., Josang, A.: Product feature taxonomy learning based on user reviews. In: WEBIST 2014 10th International Conference on Web Information Systems and Technologies (2014)
11. Tsaparas, P., Ntoulas, A., Terzi, E.: Selecting a comprehensive set of reviews. In: Proceedings of the 17th ACM SIGKDD International Conference on Knowledge Discovery and Data Mining, pp. 168–176. ACM (2011)
12. Xu, N., Liu, H., Chen, J., He, J., Du, X.: Selecting a representative set of diverse quality reviews automatically. In: Proceedings of the 2014 SIAM International Conference on Data Mining, pp. 488–496 (2014)

13. Zhang, Z., Varadarajan, B.: Utility scoring of product reviews. In: Proceedings of the 15th ACM International Conference on Information and Knowledge Management, pp. 51–57. ACM (2006)
14. Zhuang, L., Jing, F., Zhu, X.-Y.: Movie review mining and summarization. In: Proceedings of the 15th ACM International Conference on Information and Knowledge Management, pp. 43–50. ACM (2006)
15. Yu, X., Liu, Y., Huang, X.J., An, A.: Mining online reviews for predicting sales performance: a case study in the movie domain. IEEE Transactions on Knowledge and Data Engineering 24(4), 720–734 (2012)

Recommendations

An Author Subject Topic Model for Expert Recommendation

Haikun Mou, Qian Geng, Jian Jin[✉], and Chong Chen

Department of Information Management,
Beijing Normal University, Beijing 100875, China
jinjian.jay@bnu.edu.cn

Abstract. A supervised hierarchical topic model, named the Author Subject Topic (AST) model, was introduced for expert recommendation in this study. The difference between the Author Topic (AT) model and the AST model is that the AST model introduces an additional supervised "Subject" layer. The additional supervised layer of AST allows subjects to be shared across authors and group documents under various topic distributions, rather than only grouping documents under a single author's topic distribution, which encourages to cluster documents and words with less noise. In considerations that interdisciplinary studies are a major trend in many research fields, a typical interdisciplinary, Information Management and Information System, is investigated and corresponding real data were gathered from WANFANG DATA (http://www.wanfangdata.com.cn/). Different comparative experiments were conducted, which demonstrates that the AST model outperforms the AT model on this dataset. It shows that the AST model is able to capture the subject class and distinguish the topics effectively for modeling the expert's research interests, which helps for expert recommendation.

Keywords: Topic model · Expert finding · Expert recommendation

1 Introduction

In the conventional process of periodical review, appointing experts is usually an artificial work. Many people criticize the use of only artificial selection, as it has many drawbacks. For example, an inefficient task is often conducted that editors are required to pre-read all papers. In addition, reviewers might be invited to review some papers that do not match their research interest. Also, only using the artificial work is somewhat subjective, which might potentially lead to the potential unfairness. Therefore, when recommending the expert as a reviewer, experts' research interests are expected to be modeled accurately. In view of the above issues, in this study, a novel topic model is proposed, named the Author Subject Topic (AST) model, to help editors select right reviewers.

To evaluate the effectiveness of the proposed model, a dataset regarding researchers and research papers in the field of Information Systems and Management was built. The motivation to choose such area is that it involves intensive interdisciplinary studies which characterizes modern trends in the research field. To obtain the corresponding

G. Zuccon et al. (Eds.): AIRS 2015, LNCS 9460, pp. 83–95, 2015.
DOI: 10.1007/978-3-319-28940-3_7

research data, scholars holding a research project that is funded by the National Natural Science Foundation of China[1] from 2004 to 2014 in this area are selected as seed experts (256 unique seed experts in total). Next, according to research papers of seed experts, co-experts are found in WANFANG DATA. Finally, 5,519 unique scholars and 75,880 papers abstracts are obtained.

Some interesting phenomena are observed when this dataset is explored. Scholars in Information Systems and Management are found to have a very broad range of research interests. These labels regarding authors' research subjects are listed clearly in WANFANG DATA. For example, the subjects of Feicheng Ma's research papers[2] in WANFANG DATA may belong to Education, Economy, Industrial Technology, General Social Science, Mathematical Science and Chemistry. The research interests of Yuying Jiao[3], another famous scholar in this research field, spans Education, Sports, Economy, Industrial Technology, Transportation and General Social Science. Also, many other similar cases can be easily observed in this research area. Considering this situation, one disadvantage of the conventional Author Topic (AT) model [1] is that it ignores the similarity of documents from different authors, and the topic distribution associated with each author is the only focused in terms of the author's own documents. Since the research field in this dataset is dispersed, only one single topic distribution to model an author's research interests has significant constraints. Similar to a document class layer in the Author Interest Topic (AIT) model [2], the inner similarity of documents is introduced into the proposed AST model to reduce topic noises. The differences between the AST model with the other author topic models, such as the AT model and the AIT model, are: first, the AST model introduces a supervised "Subject" layer for grouping documents [3]. Second, unlike the AIT model that gives each document a class label, the AST model gives each document a subject distribution, considering the fact that each document may cover several subjects.

The first contribution of this research is that a novel probabilistic model is introduced for expert recommendation, which takes an expert's interdisciplinary research interests into considerations. It is a critical and practical concern, which is neglected by many existing models. Another contribution of this study lies in that not only the text information but also other available metadata information is borrowed in the AST model. It aims to capture the latent connection between authors, subject labels and words.

The remainder of this paper is organized as follows: In Sect. 2, some relevant studies on expert recommendation and topic model are reviewed. In Sect. 3, the author subject topic model is proposed and the Gibbs Sampling approach is utilized for parameter estimation. Different experiments are conducted and comparative results are presented in Sect. 4. Finally, conclusions are drawn in Sect. 5.

---

[1] http://www.nsfc.gov.cn/.

[2] http://social.wanfangdata.com.cn/Auto/Achievement.aspx?Name=%E9%A9%AC%E8%B4%B9%E6%88%90&articleId=qbxb199902013.

[3] http://social.wanfangdata.com.cn/Auto/Achievement.aspx?Name=%E7%84%A6%E7%8E%89%E8%8B%B1&articleId=qbkx200412014.

2 Relate Work

2.1 Expert Recommendation

Based on the conventional information retrieval, expert recommendation is often regarded as an effective solution to ease the information overload by helping users to find matching experts from a large number of candidates. In this aspect, some scholars borrowed the metadata such as co-authors, h-index, document quantity, etc., which might be important to estimate experts' research interests. Y. Fang constructed a logistic regression classifier combining ten types of metadata value such as a score defined in the document language model, some document features, some basic association features, etc. [4]. Han took use full of the experts' social relationship when recommending Program Committee members [5]. Moreira used the method of data fusion, in which multiple sensors were built. Each sensor was responsible for getting specific information on experts and finally ranking the authors [6]. In addition, some scholars introduced the academic network into expert recommendation. The major concerns are to build the social networks of candidates and extract the candidates' correlation degree, such as co-authors and citations. These relationships are embodied in potential communications, and they can reflect experts' authority and influence [7–9].

However, these methods have their own disadvantages. First, metadata alone cannot exactly determine experts' research interests. Second, an expert who has superior authority in an academic social network does not necessarily mean that he/she is an expert in all of the related sub-fields. Therefore, using the academic social network also cannot fully determine the expert's research profile.

2.2 Topic Modeling

Recently, many generative topic models are applied for expert recommendation. An early model is the Author Topic (AT) model [1], which is an extension of LDA [10]. It is a generative model that represents each document with a mixture of topics. The AT model substitutes the author layer for document layer, allowing each author to be represented by a mixture of topics. Jie Tang extended the AT model by adding the publication venue, which means that each author's topic is correlated with the word and conference stamp; the application of the Author Conference Topic (ACT) model can then be used to retrieve the experts and conferences [11]. The Author Conference Topic Connection (ACTC) model [12] and the Author Citation Venue Topic (ACVT) model [13] extended the ACT model by adding the subject layer and citation information. The Author Persona Topic (APT) model [14] introduced a persona layer, which allows authors' documents to be divided into one or more clusters, with each cluster being related to that persona. The Author Interest Topic (AIT) model [2] and Latent Interest Topic (LIT) model [15] extended the APT model, which introduced the interest layer and author class layer that can be shared among different authors to reduce the noise.

3 The Author Subject Topic(AST)

3.1 The Probabilistic Generative Model

Intuitively, when people talk about a discipline system in a certain field, a hierarchical structure might be formed in their minds. The root is a macro concept, and the leaves are the specific pieces of research content related to the macro concept. The same system can be used to describe an expert's discipline system. An expert would have some "Subject" that he is interested in. Under these subjects, there is a distribution of topics that can explain the meaning of the subject.

Following that train of thought, in this research, a novel topic model, the Author Subject Topic (AST) model, is proposed, which is an extension of the AT model. In Fig. 1, the AT model and the AST model are compared.

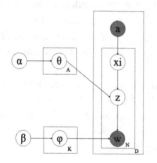   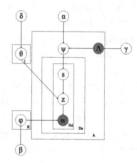

(a) The author topic model   (b) The author subject topic model

Fig. 1. The author topic model and the author subject topic model

The difference between the AT model and the AST model is that the latter introduces a supervised "Subject" layer, which can be shared among different authors. In contrast with the AST model, the AT model only learns the topic distribution with the authors' own documents and lacks the consideration on the similarity of documents, which might induce some noises. The "Subject" layer of the AST model is similar to the "Class" layer of the AIT model, which motivates to divide different documents into several groups. The difference between them is that the "Subject" layer is supervised. In this research, the subject categories of an expert can be collected in the WANFANG DATA, as these subject labels are given by the author. Accordingly, the "Subject" layer in the AST model is more accurate than the "Class" layer in the AIT model. Furthermore, in the AIT model, each document is given a class label from the class distribution, which can be regarded as a hard classification. Comparatively, in the AST model, each document is given a subject distribution from a "Subject" layer, which can be regarded as a soft classification. The soft classification is critical and practical since the selected Information Systems and Management is a cross-field domain and many papers span several of these subjects. For example, a paper entitled

"Study of Information Science Development and Education" falls into a category entitled "Intelligence, Information Science", although this paper also relates to other subjects such as Education. Despite the fact that it is convenient for a paper to be assigned to a category, it is inconvenient to mine an expert's real research interest.

Accordingly, a generative model is proposed, in which each paper is generated from words in that paper. Specifically, first, the author picks a subject from his/her own subject distribution. For this purpose, similar to [3], an author specific label projection matrix $L^{(a)}$ is introduced in this research, each row vector of which denotes one subject label of the author. For example, supposing four subjects are considered and one author's indicator vector $\Lambda^{(a)} = \{0,1,1,0\}$. It means that this author is interested in subject 2 and 3, and the corresponding $L^{(a)}$ can be denoted as $\begin{pmatrix} 0 & 1 & 0 & 0 \\ 0 & 0 & 1 & 0 \end{pmatrix}$.

Additionally, the distribution $\psi^{(a)}$ is to represent the author's research interests in the "Subject" layer. Next, each topic is chosen from the topic distribution related to this subject and each word is chosen from a word distribution related to this topic. Given several iterations, one paper is finally generated. The notations of this model are listed in Table 1 and the formal definition of the AST model are described as follows:

1. For each topic $k \in \{1,\ldots\ldots,K\}$:
2.   Generate $\varphi_k \sim Dir(\cdot|\beta)$
3. For each subject $j \in \{1,\ldots\ldots,J\}$:
4.   Generate $\theta_j \sim Dir(\cdot|\delta)$
5. For each experts $a \in \{1,\ldots\ldots,A\}$:
6.   For each subject $j \in \{1,\ldots\ldots,J\}$
7.   Generate $\Lambda_j^{(a} \in \{0,1\} \sim Bernoulli\ (\cdot|\gamma_j)$
8.   Generate $\alpha^{(a)} = L^{(a)} * \alpha$
9.   Generate $\psi^{(a)} \sim Dir\ (\cdot|\alpha^{(a)})$
10.   For each document $d \in \{1,\ldots\ldots,D\}$:
11.   For each word i in $\{1,\ldots\ldots,N\}$
12.   Generate $s_i \sim Mult\ (\cdot|\psi^{(d)})$
13.   Generate $z_i \sim Mult\ (\cdot|\theta_{si})$
14.   Generate $w_i \sim Mult\ (\cdot|\varphi_{zi})$

3.2   Parameter Estimation

The AST model employs Gibbs Sampling [16] to perform inference approximation. Gibbs Sampling is a popular method for parameter estimation in the topic model. In the AST model, what should be estimated are three sets of parameters: the probability that a subject is chosen by an expert is reflected by ψ. The probability that a topic is chosen by a subject is reflected by θ. The probability that a word chooses a topic is reflected by φ.

To simplify the operation, the parameters ψ, θ and φ should be integrated out first, so they do not appear in the joint probability. At the same time, because Λ is observed, γ is the "D-separation" from the rest of the model given that Λ and the hyper-parameter α are now restrained to the labeled subjects.

Table 1. Notations in the proposed topic model

SYMBOL	DESCRIPTION
α, β, δ	Hyper-parameters of Dirichlet distributions.
γ	Hyper-parameters of Bernoulli distribution.
Da	The total number of Expert A's documents.
Nd	The total number of words in document D.
s, z, w	s for subject, z for topic, w for word.
J	The total number of subjects.
K	The total number of topics.
A	The total number of authors.
θ	A $J*K$ matrix that indicates subject-topic distribution.
φ	A $K*V$ matrix that indicates topic-word distribution.
ψ	An $A*J$ matrix that indicates expert-subject distribution.
Λ	The indicator if the subject belongs to expert A.
$L^{(a)}$	The matrix to project α
x_i	Author x associated with i^{th} word in document D.

The joint probability can be formally derived in Eq. 1:

$$P(\mathbf{w}, \mathbf{s}, \mathbf{z} \mid \alpha, \beta, \delta, \Lambda, \gamma)$$
$$= \int P(s, \psi \mid \alpha, \Lambda, \gamma) d\psi P(z, \theta \mid s, \delta) d\theta P(w, \phi \mid z, \beta) \tag{1}$$

In Eq. (2), given w is observed, the conditional probability $P(s_{adi}|\mathbf{s}_{\neg adi}, \alpha, \Lambda, \gamma, z)$ and $P(z_{adi}|z_{\neg adi}, w, s, \beta, \delta)$ are expected. First, $P(s_{adi}|\mathbf{s}_{\neg adi}, \alpha, \Lambda, \gamma, z)$ is estimated. For simplicity, the hyper parameter is omitted. Now, supposing $s_{adi} = j$,

$$P(s_{adi} = j|s_{\neg adi}, z) \propto \frac{P(s, z)}{P(s_{\neg adi}, z_{\neg adi})}$$
$$= \frac{P(s)P(z|s)}{P(s_{\neg adi})P(z_{\neg adi}|s_{\neg adi})} \tag{2}$$
$$= \frac{n^j_{a, \neg adi} + \alpha}{\sum\limits_{j}^{J} \left(n^j_{a, \neg adi} + \alpha\right)} \cdot \frac{n^k_{j, \neg adi} + \delta}{\sum\limits_{k}^{K} \left(n^k_{j, \neg adi} + \delta\right)}$$

where s_{adi} represents the subject of token i in author a's dt_h paper. $n^j_{a, \neg adi}$ represents the number of subject j that author a assigned to, except $adi.n^j_{a, \neg adi}$ represents the number of topic k that subject j is assigned to, except adi. In addition, α is restrained with each author's Λ.

The second conditional probability is similar to the first one. Supposing $z_{adi} = k$,

$$P(z_{adi} = k|z_{\neg adi}, w, s = j) \propto \frac{P(w, z)}{P(w_{\neg adi}, z_{\neg adi})}$$

$$= \frac{P(z)P(w|z)}{P(z_{\neg adi})P(w_{\neg adi}|z_{\neg adi})}$$

$$= \frac{n_{k,\neg adi}^{v} + \beta}{\sum\limits_{v}^{V}\left(n_{k,\neg adi}^{v} + \beta\right)} \cdot \frac{n_{j,\neg adi}^{k} + \delta}{\sum\limits_{k}^{K}\left(n_{j,\neg adi}^{k} + \delta\right)} \quad (3)$$

where $n_{k,\neg adi}^{v}$ represents the number of word v that topic k assigned to, except $adi.n_{k,\neg adi}^{v}$ represents the number of topic k that subject j assigned to, except adi.

According to the Dirichlet Multinomial Conjugation, the parameters are estimated as follows:

$$\psi_{a,s} = \frac{n_a^{(s)} + \alpha}{\sum_{s=1}^{S} n_a^{(s)} + S\alpha} \quad \theta_{s,k} = \frac{n_s^{(k)} + \delta}{\sum_{k=1}^{K} n_s^{(k)} + K\delta} \quad \phi_{k,v} = \frac{n_k^{(v)}}{\sum_{v=1}^{V} n_k^{(v)} + V\beta} \quad (4)$$

4   Experimental Study and Discussion

4.1   Experiment Setup

In this experiment, 500 scholars with more than five subjects were randomly selected from the entire dataset. The NLPIR[4] was utilized for Chinese words segmentation and POS tagging on the abstracts. Only the noun, adjective and noun verb were used, in which the frequency of emergence is greater than or equal to five times for reducing the noise. Finally, 623 subjects and 15,533 unique words are left. Both the AT model and the AST model were also testified for benchmark.

4.2   Evaluating AST Model

4.2.1   Results of Perplexity

To evaluate the performance of the AST model, the perplexity was applied to estimate performance (lower perplexity is better). Perplexity is defined as the reciprocal geometric mean of the token likelihood in the given model as follows:

$$Perplexity = \exp\left\{ -\frac{\sum\limits_{d} \log(p(w_d|a_d))}{\sum\limits_{d} N_d} \right\} \quad (5)$$

---

[4] http://ictclas.nlpir.org/.

In this study, ten percent of words in each document were selected randomly as the testing dataset; the rest was used as the training dataset. The training dataset and the testing dataset are regarded to be drawn from an independent identical distribution, so the perplexity can be determined using the testing dataset of the AST model trained on the training dataset. For simplicity, the hyper parameters α, β, and δ of the AST model were set to $50/K$, 0.01, and 0.01, respectively. For performance comparison, the perplexity of the AT model is also estimated. The hyper parameters α and β of the AT model were also set to $50/K$ and 0.01 respectively. 1,000 iterations were conducted in each Gibbs Sampling procedure on a machine with Core i5 3.20 GHz processor. As each of the authors' "Subject" layers of the AST model are supervised, it was not needed to be set in advance. The "Topic" layer of the AST model and AT model were set from 50 to 500.

In Fig. 2, the perplexity values of both models are presented. As seen from this figure, by increasing the number of topics, the perplexity of AST model decreases at first and then converges to a relative stationary state when the topic is set to 450. Accordingly, 450 topics are applied in the following experiments. Also, it is found that the AST model presented a significantly lower perplexity value than that of the AT model, which means that the AST model is better as a topic model. The argument might be that the "Subject" layer of the AST model has a high degree of confidence and, at the same time, it catches the inter-similarity of documents to reduce some noises.

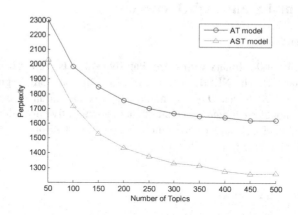

Fig. 2. The perplexity of AST model and AT model

4.2.2   Example of Topic Discovered

In this study, four subjects discovered by the AST model were presented as examples in Table 2. The four subjects are Computer technology, Informatics, Management science and Clinical medicine. Each subject is shown with top two topics and top six words. In Table 2, with regard to subject #358– Computer technology, these two topics are highly relevant, each of which describes a particular research field in this subject. For example, Topic #389 concerns about the model and algorithm, and Topic #222 concerns about the application of information system. Similar phenomena can be found in other three subjects.

Table 2. Example of subjects and topics discovered

SUBJECT #358 (Computer technology)		SUBJECT #574 (Informatics)	
topic	word	topic	word
#389	mathematical model constraint forecast matrix verification calculation method	#32	ontology text semantic information topic tracking metadata model
#222	information system application prospect forecasting method evaluating indicator constraint condition semantic	#289	experimental results SVM semantic clustering algorithm knowledge base extraction

SUBJECT #353 (Management science)		SUBJECT #12 (Clinical medicine)	
topic	word	topic	word
#147	supply chain enterprise efficiency supplier selection base expansion customer	#377	discrepancy age statistics clinical assessment patient
#222	information system application prospect forecasting method evaluating indicator constraint condition semantic	#165	infected serum transcription Chinese-Medicine specificity antigen

As seen from this table, some topics are shared by subjects. These topics always belong to a cross-field domain. For example, topic #222 has a high probability in the subject of Computer technology and Management science. Topic #222 concerns the application of Information system, which is a cross-domain. Computer technology and Management science all get involved in this area. Furthermore, it is also found that subjects may consist of both specific topics and cross-field topics [15]. For instance, in subject #353, topic #147 which concerns about the supply chain is one specific research field in Management science. Comparatively, topic #222, as mentioned above, is one cross-field topic. Some similar examples can be found within other subjects.

As reported from these results, the proposed AST model is found to be able to correlate topics and cluster words with a specific or cross-field topic effectively.

4.2.3 An Application of the AST Model

A typical application of the AST model is to recommend experts in a specific field for paper review of periodicals. A total of ten abstracts of papers were randomly selected from the dataset as testing data, which are built from WANFANG DATA. As mentioned in the previous section, the NLPIR was firstly used to pre-process the testing abstracts. The recommending function of the recommend system is defined as follows:

$$P(a|abs) = \prod_{wi} \sum_{s} \sum_{t} P(a|s)P(s|t)P(t|wi) \tag{6}$$

In this function, *abs* denotes the input testing abstract, which contains the words *wi*. The probability of $P(a|s)$, $P(s|t)$, $P(t|wi)$ are derived from the AST model respectively, which can be gained from a trained model. With this function, top@5, top@10 and top@20 authors are recommended as the experts. Subsequently, the recommended experts are invited for paper review of periodicals.

In considering the recommendation of review experts of periodicals, the difference between this research and the previous studies [11, 12, 14] is that, in this study, a group of experts are recommended whose research interests not only match the testing abstracts. In addition, all research interests of the recommended experts could cover the entire research domain of the testing abstract. To evaluate the performance on expert recommendations, the method of pooled relevance judgments [17] is often used. However, it is not applicable in this research. First, the pooled relevance judgments need manual evaluation. But the dataset in this research involves intensive interdisciplinary studies, which need a time-consuming process to identify the "ground-truth" research fields. Second, the pooled relevance judgments only lead to the evaluation on the precision. However, in this study, a group of experts are expected whose research domains are as far as possible to cover the research domains of the testing paper.

Accordingly, the topic coverage and the average symmetric KL-Divergence are utilized to measure the authors' topic distance. The topic coverage is defined as the percentage of topic aspects covered by reviewers [18]:

$$Coverage = \frac{N_a \cap N_{abs}}{N_{abs}} \tag{7}$$

N_a represents the number of topics that these N authors can cover. N_{abs} represents the number of topics that the abstract covers. In this experiment, top 100 topics of each author were selected as the whole topics of authors, and top 100 topics of each word were selected as the whole topics of the abstracts. The result of the coverage was displayed in Table 3. The testing abstracts are hardly trained in the AST model because of a lacking of information about authors; therefore, using the topic of words is an approximate way to represent the topics of testing abstracts. Overall, an encouraging result is obtained. For example, the coverage value reaches 0.55 if only five experts are recommended and that goes up to 0.87 if 20 experts are invited.

Table 3. Performances of coverage in AST model

	top@5	top@10	top@20
Min	0.46620046	0.52680652	0.67599067
Mean	0.50550913	0.65837041	0.78436371
Max	0.54950495	0.74504950	0.86881188

The KL-Divergence describes the distance between authors i and j with the topic distributions. The bigger the value of KL-Divergence, the more distinguished the two experts. The function of the KL-Divergence is defined as follows:

$$sKL(i,j) = \sum_{t=1}^{T} \left[\theta_{it} \log \frac{\theta_{it}}{\theta_{jt}} + \theta_{jt} \log \frac{\theta_{jt}}{\theta_{it}} \right] \tag{8}$$

θ_{it} represents the probability of expert i's topic t. θ_{jt} represents the probability of expert j's topic t. The comparative result of average symmetric KL-Divergence of the AT model and AST model is shown in Table 4.

Table 4. Performance of KL-Divergence in AT and AST model

	AST model			AT model		
	top@5	top@10	top@20	top@5	top@10	top@20
Min	0.19948	1.00034	3.94940	0.18276	1.02423	3.83440
Mean	0.37166	1.65791	6.24616	0.31914	1.39229	5.30737
Max	0.57940	2.58069	9.03237	0.48756	1.78555	6.85957

As seen from this table, it is found that authors' average symmetric KL-Divergence of the AST model at top@5, top@10 and top@20 is larger than that of the AT model, which implies that each author's topics in the AST model are more distinguished than those in the AT model. Since, in this research, a group of experts are recommended whose interest covers the testing abstract topics as much as possible, it is argued that the proposed AST model is better than the AT model in recommending a group of experts in a interdisciplinary research field. The reason about the KL-Divergence of the AST model is larger than that of the AT model might be that the AT model only infers authors' topics in the documents layer, which leads to an unavoidable decrease in the KL-Divergence over topics. However, the AST model allows the supervised "Subject" layer to be shared across each author and the documents can be clustered under the "Subject" layer, which will reduce some noises and increases the KL-Divergence.

5 Conclusions

In this study, a novel graph model of the Author Subject Topic model was proposed for expert recommendation. The supervised "Subject" layer is introduced in the AST model, which aims to be shared with authors and helps to cluster words and documents for noise reduction. Compared with the AT model, the AST model was proven to be a more effective topic model to model experts' research domains and more suitable for expert recommendation.

As there are increasingly more metadata is available, some potential research studies are suggested to extend the AST model with factors such as a time stamp, link structure, author authority such as H-index and citation times, etc. Additionally, some further studies may also include how to determine a group of experts accurately and comprehensively, which are believed to ease the burden of expert recommendation for periodical review.

Acknowledgments. This work was supported by the Fundamental Research Funds for the Central Universities (No. SKZZB2014037).

References

1. Rosen-Zvi, M., Griffiths, T., Steyvers, M., Smyth, P.: The author-topic model for authors and documents. In: UAI 2004, pp. 487–494. AUAI Press, Arlington (2004)
2. Kawamae, N.: Author interest topic model. In: SIGIR 2010, pp. 887–888. ACM, New York (2010)
3. Ramage, D., Hall, D., Nallapati, R., Manning, C.D.: Labeled LDA: a supervised topic model for credit attribution in multi-labeled corpora. In: EMNLP 2009, 248–256, vol. 1. Association for Computational Linguistics, Stroudsburg (2009)
4. Fang, Y., Si, L., Mathur, A.P.: Discriminative models of integrating document evidence and document-candidate associations for expert search. In: SIGIR 2010, pp. 683–690. ACM, New York (2010)
5. Han, S., Jiang, J., Yue, Z., He, D.: Recommending program committee candidates for academic conferences. In: CompSci 2013, pp. 1–6. ACM, New York (2013)
6. Moreira, C., Wichert, A.: Finding academic experts on a multisensor approach using Shannon's entropy. Expert Syst Appl. **40**(14), 5740–5754 (2013)
7. Datta, A., Yong, J.T.T., Ventresque, A.: T-RecS: team recommendation system through expertise and cohesiveness. In: WWW 2011, pp. 201–204. ACM, New York (2011)
8. Gollapalli, S.D., Mitra, P., Giles, C.L.: Ranking authors in digital libraries. In: JCDL 2011, pp. 251–254 ACM, New York (2011)
9. Wang, G.A., Jiao, J., Abrahams, A.S., et al.: ExpertRank: a topic-aware expert finding algorithm for online knowledge communities. Decis. Support Syst. **54**(3), 1442–1451 (2013)
10. Blei, D.M., Ng, A.Y., Jordan, M.I.: Latent dirichlet allocation. J. Mach. Learn. Res. **3**, 993–1022 (2003)
11. Tang, J., Zhang, J., Yao, L., Li, J., Zhang, L., Su, Z.: ArnetMiner: extraction and mining of academic social networks. In: KDD 2008, pp. 990–998. ACM, New York (2008)
12. Wang, J., Hu, X., Tu, X., He, T.: Author-conference topic-connection model for academic network search. In: CIKM 2012, pp. 2179–2183. ACM, New York (2012)
13. Yang, Z., Hong, L., Davison, B.D.: Academic network analysis: a joint topic modeling approach. In: ASONAM 2013, pp. 324–333. ACM, New York (2013)
14. Mimno, D., McCallum, A.: Expertise modeling for matching papers with reviewers. In: KDD 2007, pp. 500–509. ACM, New York (2007)
15. Kawamae, N.: Latent interest-topic model: finding the causal relationships behind dyadic data. In: CIKM 2010, pp. 649–658. ACM, New York (2010)

16. Griffiths, T.L., Steyvers, M.: Finding scientific topics. Proc. Nat. Acad. Sci. **101**(suppl 1), 5228–5235 (2004)
17. Buckley, C., Voorhees, E.M.: Retrieval evaluation with incomplete information. In: SIGIR 2004, pp. 25–32. ACM, New York (2004)
18. Karimzadehgan, M., Zhai, C., Belford, G.: Multi-aspect expertise matching for review assignment. In: CIKM 2008, pp. 1113–1122. ACM, New York (2008)

A Keyword Recommendation Method Using CorKeD Words and Its Application to Earth Science Data

Youichi Ishida[✉], Toshiyuki Shimizu, and Masatoshi Yoshikawa

Graduate School of Informatics, Kyoto University,
Yoshida-Honmachi, Sakyo-ku, Kyoto 606-8501, Japan
{yishida,tshimizu,yoshikawa}@db.soc.i.kyoto-u.ac.jp

Abstract. In various research domains, data providers themselves annotate their own data with keywords from a controlled vocabulary. However, since selecting keywords requires extensive knowledge of the domain and the controlled vocabulary, even data providers have difficulty in selecting appropriate keywords from the vocabulary. Therefore, we propose a method for recommending relevant keywords in a controlled vocabulary to data providers. We focus on a keyword definition, and calculate the similarity between an abstract text of data and the keyword definition. Moreover, considering that there are unnecessary words in the calculation, we extract CorKeD (Corpus-based Keyword Decisive) words from a target domain corpus so that we can measure the similarity appropriately. We conduct an experiment on earth science data, and verify the effectiveness of extracting the CorKeD words, which are the terms that better characterize the domain.

Keywords: CorKeD words · Domain corpus · Controlled vocabulary · Keyword definition · Abstract text · Earth science

1 Introduction

Due to the rapid advancement in information technologies and the remarkable dissemination of social media in recent years, diverse and vast amount of data has been generated. To classify those data accurately, and to obtain the right information quickly, it is effective to annotate them with metadata. Recently, people have annotated various data, such as user generated content(images, videos, web page bookmarks, and so on), academic research papers, earth science data.

As examples of metadata, there may be mentioned title, creation date, author, abstract text, keyword. We focus on keywords among these metadata. Annotation keywords are used to support search, browse and classification of various data. We consider that there are mainly two ways to add keywords to data. One way is that users themselves annotate various data with keywords [5,8,9], while the other is that data providers themselves add keywords to their own data in a research domain [2,3,6,7]. In the former case, since many general

© Springer International Publishing Switzerland 2015
G. Zuccon et al. (Eds.): AIRS 2015, LNCS 9460, pp. 96–108, 2015.
DOI: 10.1007/978-3-319-28940-3_8

users continuously add keywords to one data, there is an advantage that a set of keywords added to the data finally converges to useful one. Yet, in the latter case, since a data provider is the only person that annotates the data, the utilization value of the added keyword set depends on only the data provider. In addition, in many cases, they restrict keywords to add by using a controlled vocabulary of each domain. By this restriction, they can eliminate noise and omission in retrieval of data which are caused by changes in word form and orthographic variation. However, to select suitable keywords from a controlled vocabulary, it is required to gain extensive knowledge of the research domain and the large-scale controlled vocabulary which typically includes thousands of keywords. Therefore, even a data provider has difficulty in picking out keywords suitably from the vocabulary. In this paper, we focus on this latter case and propose a method for recommending suitable keywords in a controlled vocabulary on various research domains.

Abstract text of the dataset D8NDVI_J managed by DIAS-P
This dataset contains the daily value of the Normalize Difference Vegetation Index (NDVI) from 1982 to 2000 over the terrestrial areas of the Japan Islands that was derived from Pathfinder AVHRR Land (PAL) dataset. The horizontal resolution is 8 x 8 km. To reduce the cloud contamination, the original daily NDVI was temporally smoothed by Temporal Window Operation (TWO) method.

Keyword definition about ACID_RAIN in GCMD Science Keywords
Definition: Rain having a pH lower than 5.6, representing the pH of natural rainwater; the increased acidity is usually due to the presence of sulfuric acid and/or nitric acid, often attributed to anthropogenic sources.

Fig. 1. An example of an abstract text and a keyword definition

In this paper, we make use of an abstract text in metadata. In general, data providers annotate data with an abstract text describing the content of the data. For example, in the case of earth science data, the information of observation items, an observation method, usage of the data and so on are described in the abstract text. Researches on keyword recommendation [2,8] often propose a method for recommending keywords which are added to similar data to a target data in such text information. Yet, metadata quality is actually a pressing problem in the metadata portal called Europeana[1]. When, as in Europeana, the amount and quality of existing metadata set is insufficient in a metadata portal, their methods do not seem to be effective. In this paper, to propose a method which does not depend on the existing metadata set other than a target data, we utilize definition information given to each keyword itself as well as an abstract text given to the target data itself. In most cases, each keyword in

---

[1] http://pro.europeana.eu/publication/metadata-quality-task-force-report.

a controlled vocabulary has a keyword definition explaining the meaning of it. Fig. 1 shows an example of the keyword definition in the controlled vocabulary called GCMD(Global Change Master Directory) Science Keywords [1] and an abstract text of a dataset managed by the metadata portal called DIAS-P(Data Integration Analysis System Program)[2], which is managed in Japan.

As an initial attempt, we calculate the similarity between an abstract text of data and keyword definitions, and recommend keywords which have high degree of similarity. However, considering that not all the words in those documents contribute for deciding which keywords to recommend, we extract "CorKeD (Corpus-based Keyword Decisive) words" so that we can measure the similarity appropriately. We first consider that decisive words for keyword recommendation are domain specific words in a target domain, and extract the domain specific words by analyzing the occurrence tendency of each word between the target domain corpus and the other domain corpora. Secondly, by further refining useful words for recommendation from the domain specific word list, we restrict the words to use in the calculation. We call the restricted words "CorKeD words". Some researches have been conducted on extracting domain specific words [4, 11, 12], but we moreover extract from the domain specific word list the CorKeD words, which are the decisive words for keyword recommendation.

Our proposed method can be applied to various research domains, and this paper deals with earth science among such domains. Owing to the recent progress in earth observation technologies, the total amount of earth science data has explosively increased in various domains such as atmosphere, ocean, climate. Therefore, it is required to manage metadata portals so that those metadata can be properly handled. For instance, the metadata portal called GCMD[3] provides a search function for searching various metadata and manages the controlled vocabulary such as GCMD Science Keywords. As mentioned above, there is also a project called DIAS-P in Japan. DIAS-P is aiming to build a database which promotes the interoperability of heterogenous data collected from multiple fields, places, times.

A keyword of metadata in earth science is added to a dataset by selecting keywords relevant to the dataset from a controlled vocabulary. For example, a dataset on rainfall observations is most likely to be annotated with the keyword "PRECIPITATION AMOUNT". In DIAS-P, data providers themselves annotate their provided datasets manually with metadata such as keywords. Therefore, it is hard to select suitable keywords from a large-scale controlled vocabulary. As a result of investigating metadata in DIAS-P, there are actually many poorly annotated datasets. In this paper, we conduct an experiment on datasets managed by DIAS-P, and verify the effectiveness of our method.

Contributions. This study makes three contributions as follows:

1. Unlike the previous methods, we propose the method which does not depend on the quality of the other existing metadata set. We make use of not only an

---

[2] http://www.diasjp.net/.
[3] http://gcmd.nasa.gov/.

abstract text of metadata but also keyword definitions, which are associated with each keyword in a controlled vocabulary.
2. We restrict the words to use in the calculation by extracting domain specific words and moreover selecting the CorKeD words, which contribute for deciding suitable keywords. Some previous researches [4,11,12] only extract domain specific words.
3. We conduct an experiment on real datasets managed by DIAS-P and verify the effectiveness of extracting the CorKeD words.

2   Related Works

In recent years, researches on keyword recommendation based on the system of folksonomy have attracted attention [5,8,9]. However, most of those researches focus on personalized keyword recommendation utilizing a user's history. In such works, it is common for the users themselves to annotate multiple data arbitrarily with keywords without using a controlled vocabulary. On the other hand, for a highly specialized data such as a research data, it is not users but data providers that add keywords to those data with a controlled vocabulary. Since, in this case, sufficient information of their history is unavailable, content-based methods are considered to be useful. This section presents some related works which propose content-based methods for keyword recommendation.

We describe some researches on supporting social tagging with a content-based method [8,9]. In social bookmarking services such as Delicious[4], Lu et al. [8] propose a method of recommending suitable keywords for a webpage lacking tag information. Their approach calculates an assignment probability of each tag for a webpage, based on how much each tag is appearing in a set of tags added to the webpage and the similarity between the webpages. However, this work presupposes that multiple users annotate one webpage with the same tags as the other users do. Hence, this method cannot be applied to highly specialized research domains because in such domains only the data providers add keywords. They also calculate how trustworthy the webpage is, based on the total number of tags added to the webpage. Yet, in research domains, the number of keywords added to data has nothing to do with the reliability of the data. Belem et al. [9] propose a formula to calculate the relevance of each tag for a resource with learning-to-rank technologies, combining various indicators such as tag co-occurrence, descriptive power, term predictability. However, this work does not use a controlled vocabulary, and extracts recommended keywords from the whole terms of documents.

As a research of keyword recommendation for earth science data, Tuarob et al. [2] propose a method for recommending tags for data missing tag information from a controlled vocabulary. They create the feature vector of each dataset from the text information in the metadata, and recommend tags which are added to similar datasets by calculating the similarity between the feature vectors. Each

---

[4] https://delicious.com/.

document is represented with either a TF-IDF vector [13] or a probability distribution of LDA (Latent Dirichlet Allocation)[14]. However, when the amount and quality of existing metadata set is insufficient in a metadata portal, their method does not seem to be effective. In contrast, we propose a method which does not depend on an existing metadata set, and our proposed method can be applied to a new controlled vocabulary which has not been used much. Shimizu et al. [3] suggests the 14 keywords which represent categories of earth science with Labeled LDA [15]. They define the 14 keywords as labels, learning correspondence between an abstract text of a dataset and added keywords. Then, they recommend suitable keywords by applying the learning results to a target dataset. As in this study, when the number of the labels is small, Labeled LDA is useful for recommendation. Yet, it is very hard to prepare enough training data to define thousands of keywords as labels.

We also introduce some works on supporting an annotation for an academic research paper. Chernyak [6] propose a method for recommending topics from the controlled vocabulary called ACM Computing Classification System. Using self-learning methods such as TF-IDF, BM25, annotated suffix tree, they calculate the similarity between the topics and each paper's abstract. Santos et al. [7] address the problem of multi-label classification for research papers with machine learnings such as SVM, KNN, naive Bayes classification. Although the studies of annotations for research papers are different from earth science in that their studies can guess suitable keywords from reference information, they have much in common with our study in that both studies need a controlled vocabulary and in that keywords are added by a specific person such as an author. Our proposed method can be applicable to the annotation of research papers.

3   Proposed Method

In the following, we explain the case of applying our proposed method to earth science data. In this paper, we made use of an abstract text of a dataset in metadata. By viewing the abstract text, users can roughly comprehend the content of the dataset. We give an example of added keywords with GCMD Science Keywords in Fig. 2. As shown in Fig. 2, keywords are hierarchically managed in GCMD Science Keywords, but in this paper, we propose the method where we do not take the hierarchical structure into account so that our proposed method can be applied to a controlled vocabulary without hierarchical model.

At the beginning, as a method of simple string matching, we extracted keywords from an abstract text of a dataset in DIAS-P by matching each keyword in GCMD Science Keywords. However, as a result of applying the method, we could only recommend the average of about 2.7 keywords. Therefore, we decided to utilize implicit information such as a keyword definition as well as explicit information such as a keyword name. As an initial attempt, we considered that we recommend keywords which have high degree of similarity between an abstract text and the keyword definitions. Moreover, considering that there are unnecessary words in the calculation, we extracted the CorKeD words, which are the

- Atmosphere > Atmospheric Water Vapor > Humidity
- Atmosphere > Atmospheric Water Vapor > Water Vapor
- Atmosphere > Precipitation > Precipitation Amount
- Oceans > Oceans Temperature > Sea Surface Temperature
- Cryosphere > Snow/Ice > Snow Water Equivalent
- Land Surface > Soils > Soil Moisture/Water Content

Fig. 2. Keywords added to Aqua AMSR-E dataset managed by DIAS-P

decisive words for keyword recommendation. We first created the domain specific word list from the domain corpus of earth science, and then by further refining useful words for recommendation from the list, we extracting the CorKeD words to analyze in calculating the similarity. We preprocess the abstract texts and the keyword definitions by removing stopwords, and stemming each word.

3.1 Definition of a Domain Specific Word of Earth Science

As Kubo et al. [4] points out, a domain specific word in certain target domain is considered as a word which has a higher appearance frequency in the target domain than in the other domains. In other words, we can define a domain specific word of earth science as a word which appears at a higher frequency in a corpus of earth science. In this paper, as the other domains other than earth science, we used biology, chemistry and physics, which belong to the same natural science. The reason why we used those three domains is because we considered that we can extract the domain specific words of earth science more properly by comparing with those domains than with non-natural science domains such as the humanities or social science.

Corpus of Each Domain. To compare among the domains, we must construct a corpus of each domain. We created a corpus of earth science from the presentation summaries in 2013 Fall Meeting held by AGU(American Geophysical Union)[5], which is the organization of earth science. We obtained approximately 6 million words from 20028 summaries. As corpora of the other domains, we used summaries of papers published in journals of each domain[6]. In addition, we equalized a corpus size of each other domain at about 200 thousand words.

The Method of Creating the Domain Specific Word List. To construct the domain specific word list of earth science, we need to compare the relative

---

[5] http://sites.agu.org/.
[6] Chemistry : Journal of the American Chemical Society
Physics : The European physical journal
Biology : International journal of biological sciences, Journal of evolutionary biology.

frequency for each one word in the corpus of earth science between earth science and each other domain. This study utilized a formula called DP(the Difference between Population Proportions) that Kubo et al. [4] propose. This formula is based on 2-sample test for equality of proportions in statistics. It is described in detail below.

$$DP_d(t) = \frac{\dfrac{f_0(t)}{W_0} - \dfrac{f_d(t)}{W_d}}{\sqrt{\pi_d(t)(1 - \pi_d(t))\left(\dfrac{1}{W_0} + \dfrac{1}{W_d}\right)}}, \quad \pi_d(t) = \frac{f_0(t) + f_d(t)}{W_0 + W_d} \quad (1)$$

Let $f_0(t)$ and $f_d(t)$ be the appearance frequency of word t in the corpus of earth science and the other domain d, respectively. W_0 and W_d is the total number of words in the corpus of earth science and the other domain d, respectively. $\pi_d(t)$ is the ratio of the appearance frequency of word t in the both corpora, and the set D consists of {biology, chemistry, physics}. $DP_d(t)$ represents the relative frequency of word t in comparing between earth science and the other domain $d \in D$. This $DP_d(t)$ follows a normal distribution. Then, by Eq. 2, we calculated the average of the relative frequency obtained by comparing with each other domain. $|D|$ is the size of D, that is, $|D| = 3$.

$$w(t) = \frac{\sum\limits_{d \in D} DP_d(t)}{|D|} \quad (2)$$

When $w(t)$ was positive as calculation results, we defined the word t as a domain specific word. Table 1 represents the top 10 scores of $w(t)$. Certainly, all of the highly ranked words can be considered as domain specific words of earth science.

Table 1. The top 10 of $w(t)$'s score

	word t	$w(t)$
1	data	27.89
2	model	24.46
3	climat	24.26
4	water	20.18
5	region	20.03
6	soil	19.88
7	atmospher	19.00
8	fault	18.18
9	ic	18.15
10	event	18.07

However, it seems that ranking highly words such as "data","model", "region" have little information for deciding which keywords to recommend. Therefore, we furthermore discussed a method for refining decisive words for recommendation from the domain specific word list.

3.2   Whether a Word Contributes for Deciding Keywords

In earth science, there are further subdivided domains, such as atmosphere, agriculture, oceans. In the case of a word which contributes for deciding keywords, we considered that there is a bias in the appearance frequency for such words among the subdivided domains. On the other hand, in the case of a word which has little information for the decision, we considered that such words appear without depending on the subdivided domains. For instance, the word "climat" is likely to appear disproportionately in the subdivided domain "atmosphere", while the word "data" tends to appear at about the same frequency among the subdivided domains. Thus, by quantifying the bias of the frequency distribution for each word among the subdivided domains, we can judge whether the word contributes for deciding keywords or not.

In this paper, as the subdivided domains, we used the 49 categories taken as a classification axis of AGU index terms[7], which is a controlled vocabulary managed by AGU introduced in Sect. 3.1. Furthermore, we utilized χ square value, which is generally used as a method for quantifying a bias of a distribution. χ square value shows difference between an observed and an expected value. As the observed value, we calculated the document frequency(DF) of each word in the summaries of AGU. Besides, as the expected value, we calculated the DF of the word by assuming that the word appears at about the same proportion among the subdivided domains. It is described in detail below.

$$\chi^2(t) = \sum_{i=1}^{n} \frac{(O_i - E_i)^2}{E_i}, \quad E_i = S_i \times \frac{S_t}{S} \tag{3}$$

S represents the total number of the summaries in AGU 2013 Fall Meeting, which is 20028. n represents the number of the subdivided domains, that is, $n = 49$. Let O_i be the observed value and let E_i be the expected value in i^{th} subdivided domain. S_i is the number of the summaries in ith subdivided domain and S_t is the total number of the summaries containing word t. In addition, we considered that a word which has little information is highly likely to appear in any summaries, and calculated χ square value for each word contained in the top 0.5 % of DF values. Tables 2 and 3 show the part of the calculation results.

In Table 2, χ square value for each word which is likely to contribute for the decision shows a relatively large value. This indicates that these words appear disproportionately in some subdivided domains. Conversely, in Table 3, χ square value for each word which has little information shows a relatively small

---

[7] http://abstractsearch.agu.org/keywords.

Table 2. CorKeD words

word t	χ square value
climat	5735.9
water	3678.5
soil	3439.1
atmospher	4375.0
temperatur	1729.7

Table 3. Not CorKeD words

word t	χ square value
data	660.8
model	695.1
region	801.8
time	282.3
base	352.5

value. This shows that these words appear without depending on the subdivided domains. From the result of a preliminary investigation, we set a threshold 1700, and by eliminating words less than the threshold from the domain specific word list, we finally created a set of CorKeD words, which is used in the calculation.

3.3 How to Calculate the Similarity

Using only the CorKeD words as previously described, we calculated the similarity between an abstract text and the keyword definitions. We represented a query abstract A_i by a feature vector $\boldsymbol{DA}(A_i)$. Let Cs be the set of the CorKeD words. When a word is included in both the query abstract A_i and Cs, the word's element of $\boldsymbol{DA}(A_i)$ is 1. Conversely, when a word is not included in A_i or Cs, the word's element of $\boldsymbol{DA}(A_i)$ is 0.

$$t_{ij} = \begin{cases} 1 & (t_{ij} \in A_i \wedge t_{ij} \in Cs) \\ 0 & (otherwise) \end{cases} \tag{4}$$

$$\boldsymbol{DA}(A_i) = \{t_{i1}, t_{i2}, \cdots, t_{im}\} \tag{5}$$

We represented a keyword definition D_j by a feature vector $\boldsymbol{KD}(D_j, Cl)$, and each element of the feature vector is TF-IDF value for each word. On this occasion, we used LRTF(Length Regularized TF) introduced in [10] as TF(Term Frequency). These are described in detail below.

$$LRTF(t, D_j) = TF(t, D_j) \times \log_2\left(1 + \frac{ADL(Cl)}{len(D_j)}\right) \tag{6}$$

$$IDF(t, Cl) = \log_2\left(\frac{|Cl|}{DF(t, Cl)}\right) + 1 \tag{7}$$

$$KD(t, D_j, Cl) = LRTF(t, D_j) \times IDF(t, Cl) \tag{8}$$

$$\boldsymbol{KD}(D_j, Cl) = \{KD(t_1, D_j, Cl), \cdots, KD(t_n, D_j, Cl)\} \tag{9}$$

Let Cl be the keyword definitions set, and let $|Cl|$ be the number of the keywords. In addition, $len(D_j)$ is the length of the keyword definition D_j, $ADL(Cl)$ is the average of $len(D_j)$, and $TF(t, D_j)$ is the appearance frequency of word t in D_j. LRTF is a formula which normalizes TF value, considering the proportion between $len(D_j)$ and $ADL(Cl)$. We considered that LRTF is appropriate to this situation, where an abstract text is regarded as a query, because [10] says that LRTF is useful to a long query composed of more than 5 words. IDF (Inverse Document Frequency) value was calculated by the most standard formula, in which $|Cl|$ is divided by $DF(t, Cl)$. In this paper, we calculated the cosine similarity between the above two feature vectors using only the CorKeD words, and recommended keywords in descending order of cosine similarity values.

$$CosineSim(\boldsymbol{DA}(A_i), \boldsymbol{KD}(D_j, Cl)) = \frac{\boldsymbol{DA}(A_i) \cdot \boldsymbol{KD}(D_j, Cl)}{\|\boldsymbol{DA}(A_i)\| \times \|\boldsymbol{KD}(D_j, Cl)\|} \quad (10)$$

4   Evaluation

To verify the effectiveness of our proposed method, we conducted an experiment on 20 datasets managed by DIAS-P. We submitted the recommended keywords to each data provider to judge whether each recommended keyword is correct or not. We used GCMD Science Keywords as a controlled vocabulary, which includes 2017 keywords. To demonstrate effectiveness of creating the set of CorKeD words, we compare our approach with a method for calculating the similarity in using all words included in the keyword definitions and an abstract text.

4.1   Evaluation Metric

This experiment evaluated precision of top 10 keywords recommended by the two methods. In most cases, when precision is evaluated, recall and F-value are calculated at the same time. However, since it is hard to understand the whole keywords in the large-scale controlled vocabulary, even data providers have difficulty in obtaining perfectly the correct keywords set. Thus, we considered that accurate recall and F-value are difficult to calculate.

4.2   Results

Table 4 shows the average of precisions and each precision evaluated by the two methods. Table 4 indicates that our proposed method outperforms the comparative method. The reason is because we can calculate the similarity more properly by using only the CorKeD words. Table 5 describes an example of recommended keywords for the dataset called "GCOM_W1", whose precision is particularly improved. The correct keywords are shown in bold text. In Table 5, our proposed method can recommend many correct keywords which cannot be recommended by the comparative method. We give an example of the similarity between the dataset and the keyword "DEGREE DAYS". In this case, when we used the comparative method, the words to use in the calculation were "atmospher", "one", "temperatur", "day", "measur", "degre", whereas by applying our method, we could use only the CorKeD words such as "atmospher", "temperatur", which are useful words for deciding suitable keywords. The reason why the accuracy is

Table 4. The evaluation results of keyword recommendation for each dataset

Dataset ID	The Comparative Method	The Proposed Method
ALOS_AVNIR2	30 %	10 %
ALOS_PALSAR	10 %	10 %
ALOS_PRISM	10 %	10 %
AMY_HARIMAU_WPR_dataset	40 %	30 %
Aqua_AMSR_E	0 %	0 %
AVISO_SLA	20 %	10 %
CEOP_CAMP_Eastern_Siberian_Taiga	10 %	30 %
D8NDVI_J	40 %	40 %
D8NDVI_J	50 %	50 %
DIAS_ODAPv2.1	40 %	50 %
DIAS_ODAPv2.1	40 %	60 %
Fuji_Hokuroku_Flux	30 %	50 %
GCOM_W1	0 %	30 %
Global_map	20 %	10 %
Global_map	40 %	20 %
GPV	0 %	10 %
MAHAPGP	30 %	20 %
MIRAI_CTD	30 %	50 %
MOM_rNP	30 %	40 %
MSST	0 %	20 %
ODA_rNPhigh	30 %	40 %
ODA_rNPhigh	40 %	60 %
SSM_I	0 %	20 %
TRMM_PR	10 %	10 %
Average of precisions	**22.92 %**	**28.33 %**

(Note : When more than two data providers evaluates the same dataset, the precision evaluated by each data provider is described)

improved can be because our method could eliminate the words such as "one", "day", "measur", which have useless information for recommendation.

On the other hand, there are some datasets whose precision decrease. We give as an example the similarity between the dataset "ALOS_AVNIR2" and the keyword "LAND USE". In this case, when we used the comparative method, the words to use in the calculation were "earth", "land", "observ", "area", **"use"**. However, although the keyword is included in the abstract text, our method with the CorKeD words eliminated the word "use", which is in the part of the keyword name. In consequence, it was difficult for our method to recommend words related to "LAND USE", resulting in low recommendation accuracy. In addition, we can find some examples where, for the same reason, we cannot recommend keywords which are included in the abstract texts. These keywords can be extracted by processing in phrase units. Therefore, in the future task, we

Table 5. The result of recommended keywords for GCOM_W1 dataset

	The Comparative Method	The Proposed Method
1	PLANETARY BOUNDARY LAYER HEIGHT	PLANETARY BOUNDARY LAYER HEIGHT
2	SEA SURFACE HEIGHT	MOISTURE FLUX
3	DEGREE DAYS	**SEA SURFACE TEMPERATURE**
4	STRATOPAUSE	**SEA SURFACE TEMPERATURE INDICES**
5	TROPOSPHERIC/HIGH-LEVEL CLOUDS	SEA SURFACE HEIGHT
6	ALTITUDE	CLOUDS
7	ICE TEMPERATURE	INVERSION HEIGHT
8	DEW POINT TEMPERATURE	**ATMOSPHERIC WATER VAPOR**
9	SENSOR COUNTS	STRATOPAUSE
10	INVERSION HEIGHT	GEOPOTENTIAL HEIGHT

would like to consider the combination of our proposed method and processing in phrase units. For the dataset "Aqua_AMSR_E", the correct keywords are not recommended at all by either of the two methods. This is because this abstract text describes some advantages or features of an observational instrument, not explanation about the contents of the dataset. We consider that the information of the observational instrument is likely to help the keyword recommendation.

5 Conclusions and Future Works

To support keyword annotation for various data of research domains, we proposed the method for recommending keywords in a controlled vocabulary. We utilized each keyword definition itself as well as an abstract text of a target data, and proposed the method which does not depend on the existing metadata set other than a target data. Also, to calculate the similarity more properly, we refined the words by extracting domain specific words and moreover selecting the CorKeD words. In this paper, we conducted the experiment on real datasets managed by DIAS-P, and showed the effectiveness of extracting the CorKeD words.

In the future work, we need to compare our approach with the previous ones such as [2,8], and other recent approaches. In addition, we would like to compare DP[4] with the other measures for calculating the relative frequency of one word, such as self mutual information and log-likelihood ratio. Also, we are interested to use the other controlled vocabularies of earth science, and want to apply our approach to the other domains such as chemistry, biology.

References

1. Olsen, L.M., Major, G., Shein, K., Scialdone, J., Ritz, S., Stevens, T., Morahan, M., Aleman, A., Vogel, R., Leicester, S., Weir, H., Meaux, M., Grebas, S., Solomon, C., Holland, M., Northcutt, T., Restrepo, R.A., Bilodeau, R.: NASA/Global Change Master Directory (GCMD) Earth Science Keywords. Version 8.0.0.0.0 (2013)

2. Tuarob, S., Pouchard, L.C., Giles, C.L.: Automatic tag recommendation for metadata annotation using probabilistic topic modeling. In: JCDL, pp. 239–248 (2013)

3. Shimizu, T., Sueki, T., Yoshikawa, M.: Supporting keyword selection in generating earth science metadata. In: COMPSAC, pp. 603–604 (2013)

4. Kubo, J., Tsuji, K., Sugimoto, S.: Automatic term recognition using the corpora of the different academic areas (in Japanese). J. Jpn Soc. Inf. Knowl. **20**(1), 15–31 (2010)

5. Krestel, R., Fankhauser, P., Nejdl, W.: Latent dirichlet allocation for tag recommendation. In: RecSys, pp. 61–68 (2009)

6. Chernyak, E.: An approach to the problem of annotation of research publications. In: WSDM, pp. 429–434 (2015)

7. Santos, A.P., Rodrigues, F.: Multi-label hierarchical text classification using the acm taxonomy. In: EPIA, pp. 553–564 (2009)

8. Lu, Y.T., Yu, S.I., Chang, T.C., Hsu, J.Y.J.: A content-based method to enhance tag recommendation. In: IJCAI, pp. 2064–2069 (2009)

9. Belem, F., Martins, E., Pontes, T., Almeida, J., Goncalves, M.: Associative tag recommendation exploiting multiple textual features. In: SIGIR, pp. 1033–1042 (2011)

10. Paik, J.H.: A novel TF-IDF weighting scheme for effective ranking. In: SIGIR, pp. 343–352 (2013)

11. Utiyama, M., Chujo, K., Yamamoto, E., Isahara, H.: A comparison of measures for extracting domain-specific lexicons for english education (in Japanese). J. Nat. Lang. Process. **11**(3), 165–197 (2004)

12. Uchimoto, K., Sekine, S., Murata, M., Ozaku, H., Isahara, H.: Term recognition using corpora from different fields. Terminology **6**(2), 233–256 (2001)

13. Salton, G.: Automatic text processing: the transformation, analysis, and retrieval of information by computer. Addison-Wesley, Boston (1989)

14. Blei, D.M., Ng, A.Y., Jordan, M.I.: Latent Dirichlet allocation. J. Mach. Learn. Res. **3**, 993–1022 (2003)

15. Ramage, D., Hall, D., Nallapati, R., Manning, C.D.: Labeled LDA: a supervised topic model for credit attribution in multi-labeled corpora. In: EMNLP, pp. 248–256 (2009)

Incorporating Distinct Opinions in Content Recommender System

Grace E. Lee, Keejun Han, and Mun Y. Yi[✉]

Department of Knowledge Service Engineering,
KAIST, Daejeon, Republic of Korea
{leklek0816, keejun.han, munyi}@kaist.ac.kr

Abstract. As the media content industry is growing continuously, the content market has become very competitive. Various strategies such as advertising and Word-of-Mouth (WOM) have been used to draw people's attention. It is hard for users to be completely free of others' influences and thus to some extent their opinions become affected and biased. In the field of recommender systems, prior research on biased opinions has attempted to reduce and isolate the effects of external influences in recommendations. In this paper, we present a new measure to detect opinions that are distinct from the mainstream. This distinctness enables us to reduce biases formed by the majority and thus, to potentially increase the performance of recommendation results. To ensure robustness, we develop four new hybrid methods that are various mixtures of existing collaborative filtering (CF) methods and our new measure of *Distinctness*. In this way, the proposed methods can reflect the majority of opinions while considering distinct user opinions. We evaluate the methods using a real-life rating dataset with 5-fold cross validation. The experimental results clearly show that the proposed models outperform existing CF methods.

Keywords: Distinctness · Bias · Content · Recommender system · Collaborative filtering

1 Introduction

With the advancement of technology, the media content industry has been growing continuously, and an enormous amount of content is now generated on a daily basis. Various strategies such as advertising and Word-of-Mouth (WOM) have been implemented to draw people's attention [1]. Indeed, it has become very common that content providers hire celebrities or well-known bloggers to promote their content and strive to shape mainstream opinions because most users simply follow the majority opinions [2]. In recent years, these promotion strategies have become more unnoticeable and have been used to maneuver more people into biased choices for their purchases without their awareness [3, 4].

In this paper, we attempt to resolve the aforementioned problems by proposing a simple yet novel measure, called *Distinctness*, to estimate how unique a certain rating of a user is from the major trend of ratings. In the field of recommender systems, there have been some prior attempts to exclude the effects of possible biases in systems by

© Springer International Publishing Switzerland 2015
G. Zuccon et al. (Eds.): AIRS 2015, LNCS 9460, pp. 109–120, 2015.
DOI: 10.1007/978-3-319-28940-3_9

detecting users' biased opinions [5, 6]. To the best of our knowledge, however, none of those studies have explored distinct opinions when biases exist, nor did they utilize such opinions for recommender systems.

For instance, The Matrix, a well-known blockbuster movie, has 133, 229 ratings from individual users in our dataset (obtained from MovieLens[1]), the distribution of ratings for the movie is (52, 32, 10, 3, 1) in percentages; the numbers correspond to a rating scale of (5, 4, 3, 2, 1). It should be noted that the rating 5, representing 'strongly like', accounts for more than half of the total ratings and that the majority of ratings, 94 %, are positive ratings (above 3 out of 5). Let us assume that there are user A and user B, each of which rated the movie 5, and user C and user D, each of which rated the movie 1. In accordance with the rating distribution for the movie, it seems that the identical opinions made by users C and D are more distinct than the opinions made by users A and B: they are strongly against the movie while most users liked it. In other words, the two people who gave the movie a rating of 1 express distinct opinions (rating the movie 'strongly dislike'), in contrast to the majority of users, who gave positive ratings. These unique ratings are powerful evidence to explain the characteristics of users. In spite of the potential usefulness of this type of distinctness feature, common CF approaches overlook the feature by assigning the same similarity value for the two user groups because the relations within groups are treated equally [7].

Unlike the aforementioned CF approaches, we measure the relations of users while considering the degree of their differences from the majority. By doing so, we are able to identify distinct opinions that can be potentially highlighted to generate more accurate recommendation results. In our experiment, it is shown that, compared with the existing CF approaches, an approach that exploits the *Distinctness* feature can improve the accuracy for recommender systems. In addition, we introduce three hybrid collaborative filtering methods combined with *Distinctness*; the results show unanimous increases compared to the baselines. The experiment was conducted on a real-life movie dataset, MovieLens, which is well-known for containing reliable data that can verify the performance of recommender systems [8–10]. By choosing the common dataset, we expect that our work can be easily reproducible.

The remainder of the paper is followed by related work on bias in recommender systems in Sect. 2. We then present the details of the proposed measure, *Distinctness*, and four variations of the CF methods utilizing *Distinctness* in Sect. 3. In Sect. 4, we describe experiments and results. Finally, the conclusion and future work are given in Sect. 5.

2 Related Work

In this section, we discuss some of the state-of-the-art recommendation models dealing with possible influences within interactions between users and recommender systems, and examine the limitations of these systems.

---

[1] http://movielens.org.

When users interact with a recommender system, it is possible that a variety of biases are involved in the interaction. User tendencies for rating and item selection are identified as biases in [11] and one work [12] pointed out user background and personal interest as potential sources of biases. In addition to the biases from individual users, context information influences opinions and responses. Regarding context information, many approaches have been studied. Context information such as weather, time, and companions does matter in recommender systems that have a multidimensional perspective since it affects the way users react [14] and thus its influences correspond to biases. The interactive methods that are used to detect context information and adapt to changes of information are able to reduce possible biases [15].

Designs of recommender systems also lead to biases in interactions with users. For example, different kinds of interfaces of recommender systems are able to intentionally instruct users in certain ways of expressing opinions and lead users' behaviors without their perceptions [9, 16]. Visually effective readability drives presentation and temporal biases in online reviews and comments [9]. Different types of rating scales also guide users toward certain ways of interacting with recommender systems and a 1-to-5 star scale allows users to express extreme feelings with great ease [16].

In this paper, we specifically look into influences coming from outside of users and recommender systems. In terms of popularity biases, several models are suggested for improvement of recommendations. A recent work [10] analyzed the causes and effects of popularity bias and proposed an algorithm to weaken the phenomenon in which only popular items are frequently recommended to users regardless of users' preferences. In [17], it was found that a function to penalize popular items in item-based collaborative filtering was able to decrease the chances of popular items appearing as recommendations.

Further, WOM both offline and online has an impact on users' overall acceptance, purchases, and opinions. By continuously appearing on a front page, early written online reviews lead to sequential biases and finally influence others' opinions [6]. Also, a greater volume of WOM results in a higher box office performance in the movie industry and directly connects to revenue in the field [2].

The aforementioned research has mainly attempted to identify types of biases and to exclude the effects of biases in the system. However, in this paper we explore implicit and unbiased opinions and propose an efficient measure to build sophisticated relationships among users. To the best of our knowledge, unbiased and distinct opinions have never been exploited in recommender systems, though use of such opinions can potentially be effective in bolstering relations among users.

3   Proposed Methods

In this section, we will first introduce the concepts of popularity and entropy. Based on those concepts, we first describe a new measure called *Distinctness* to estimate distinct similarities between users and then move on to explain four CF methods based upon *Distinctness*.

Popularity. In content industries including movies and books, opinions about content are highly susceptible to advertising and WOM [1, 2]. This situation eventually leads to the Matthew Effect: "the rich get richer and the poor get poorer" [10]. As a result, once items become widespread and establish certain reputations, their images are accumulative and fossilized [5]. Therefore, when a major trend of opinion for an item is formed by the public, *Distinctness* of unique opinions increases. The total number of ratings can indicate the *Popularity* of a certain movie. The advantage of *Popularity* is that it is straightforward and easy to compute, but the possibility of prefix bias – popular items having enough ratings usually defeat unpopular items – is a disadvantage [18]. However, in this study, to lessen the influence of possible weakness, we take advantage of a property of the logarithm function, which can transform an exponential-like curve to a linear-like curve by compressing large values.

Entropy. *Entropy* of an item's ratings indicates the distribution of ratings. For instance, when ratings are evenly distributed on a 1-to-5 star rating scale, the value of *Entropy* is greater. In contrast, when the majority of ratings are positive ratings, 4 or 5, the *Entropy* value is smaller. As the *Entropy* value decreases, the meanings of distinct ratings increase in terms of *Distinctness*. *Entropy*, however, does not imply the total number of ratings for items. Two rating distributions, (1, 0, 0, 0, 5) and (100, 0, 0, 0, 500), have the same *Entropy* values, although the former distribution has fewer ratings. In other words, *Entropy* alone never enables us to completely represent the concept of *Distinctness* of ratings due to the total number of ratings. *Entropy* of item p's rating distribution is calculated as follows:

$$Entropy(p) = \sum_i P(i|p) \log_2 P(i|p) \tag{1}$$

where $P(i|p)$ denotes the relative frequency of rating i in an item p. In a 1-to-5 star rating scale, i can be 1, 2, 3, 4, or 5.

Distinctness. Both *Popularity* and *Entropy* are correlated with *Distinctness*. We estimate *Distinctness*, D_{pi} for each rating i in an item p, using Bayes' theorem [13] and as follows:

$$D_{pi} = \log N_p \times \frac{1}{Entropy(p)} \times \left(1 - \frac{N_{pi}}{N_p}\right) \tag{2}$$

$$\text{where } N_p = \sum_i N_{pi}$$

where N_p is the total number of given ratings to item p and N_{pi} is the number of rating i in an item p.

$$D_p = \sum_i D_{pi} \tag{3}$$

The total *Distinctness* value for item p, denoted by D_p, is derived from the sum of D_{pi}. N_p is the total number of given ratings to item p and N_{pi} is the number of rating i in an item p. In order to apply Bayes' theorem in *Distinctness*, *Popularity* and *Entropy* have less correlation [8, 18]. Unlike the previous studies applying the concepts of *Popularity* and *Entropy* [8, 10, 18], this research exploits the idea of *Entropy* in reverse. Skewed distributions of ratings represented in smaller values of *Entropy* play a crucial role in *Distinctness*.

Distinctness-Based Collaborative Filtering (DISTINCT). All the ratings i of every item p can be represented by *Distinctness* values, D_{pi}. We use the *Distinctness* values to calculate *Distinctness*-based user-user similarity, S_D as follows:

$$S_D(u, v) = \sum_{p \in P} \min\{D_{pr_{u,p}}, D_{pr_{v,p}}\} \tag{4}$$

where P is a set of items rated by both users, u and v, and p is each item included in P. $r_{u,p}$ and $r_{v,p}$ are user u's and v's ratings for p, respectively. Based on the similarity, S_D, the predicted rating of an unknown item i for the target user u is computed as follows:

$$\widehat{r_{u,i_D}} = \overline{r_u} + \frac{\sum_{v \in K} S_D(u, v) * (r_{v,i} - \overline{r_v})}{\sum_{v \in K} S_D(u, v)} \tag{5}$$

where $r_{v,i}$ is user v's rating for item i, $\overline{r_u}$ and $\overline{r_v}$ are user u's and v' average ratings, and K is a set of u's neighbors who rated the target item i, and at the same time satisfied with a parameter k from 0.1 to 1.0 which is detailed in Sect. 4.3.

Further, we present three hybrid methods combined with *Distinctness* and conventional rating-based collaborative filtering in recommender systems. Generally, in the conventional collaborative filtering, which directly uses ratings, user-user similarity using Pearson Correlation Coefficient is derived as follows:

$$S_R(u, v) = \frac{\sum_{p \in P} (r'_{u,p} - \overline{r'_u})(r'_{v,p} - \overline{r'_v})}{\sqrt{\sum_{p \in P} (r'_{u,p} - \overline{r'_u})^2} \sqrt{\sum_{p \in P} (r'_{v,p} - \overline{r'_v})^2}} \tag{6}$$

The predicted ratings are computed in the same way as shown in Eq. 5 except S_R replaces S_D. In the first two hybrid algorithms, the *Distinctness*-based CF (DISTINCT) and the conventional CF are linearly combined together.

Linearly Combined Similarities CF (LCS). The two similarities from each CF method are joined together in the first hybrid CF as follows:

$$a \times S_D(u, v) + (1 - \alpha) \times S_R(u, v) \rightarrow S'(u, v) \tag{7}$$

where $S_D(u, v)$ and $S_R(u, v)$ are similarity values from Eqs. 4 and 6, and α is simply set as 0.5 to balance the two similarity scores. The new combined similarity value S' is used to compute the predicted rating, $\widehat{r_{u,i}}$ as follows:

$$\widehat{r_{u,i}} = \overline{r_u} + \frac{\sum_{v \in K} S'(u, v) * (r_{v,i} - \overline{r_v})}{\sum_{v \in K} S'(u, v)} \tag{8}$$

Linearly Combined Ratings CF (LCR). The second hybrid method linearly combines two predicted ratings coming from the different CF strategies based on Eqs. 4 and 6. The combination of two predicted ratings of an unknown item i for the target user u is then calculated as:

$$a \times \widehat{r_{u,i}}_D + (1 - \alpha) \times \widehat{r_{u,i}}_R \rightarrow \overline{r_{u,i}'} \tag{9}$$

where $r'_{u,i}$ is the final predicted rating as a result of the second hybrid CF.

Distinctness Weighted CF (DWCF). In the last hybrid CF method, the *Distinctness* values are applied to S_R as weighting parameters to control the contributions of rating-based relationships as follows:

$$\frac{\sum_{p \in P} \min\{D_{pr_{u,p}}, D_{pr_{v,p}}\}}{\sum_{p \in P} D_p} \times S_R(u, v) \rightarrow S'(u, v) \tag{10}$$

In DWCF, $S'(u, v)$ from Eq. 10 is used to compute the final predicted rating in Eq. 8.

4 Experiment

4.1 Experiment Setup

To evaluate our approach, we carry out experiments on the MovieLens dataset, which consists of more than 10 million ratings given by 70,250 active users for approximately 10,000 movies. This dataset is commonly used to evaluate recommendation tasks, and thus, we expect that our work can be easily reproducible. The ratings are on a 1-to-5 star scale. In order to focus on the users having common rating behaviors, we randomly chose 1,000 users who rated individual movies between 1,000 and 5,000 times. For users having given a myriad of ratings, the rating behaviors show stricter standards in rating [12], and in the case of users who have rated movies only a few rating times, the cold start problem occurs [8], so such types of users are excluded from this study. We then follow 5-fold cross validation by categorizing 80 % of the data as training data which generates recommendations and the rest of the data as test data to evaluate the recommendation results.

To compare the performance of the proposed methods, Pearson Correlation Coefficient CF (PEARSON) and Cosine Similarity CF (COSINE) are used as baseline models. All experiments are evaluated for two types of accuracies. Mean Absolute Error (MAE) and Root-Mean-Square Error (RMSE) are used to evaluate the prediction accuracy and Normalized Discounted Cumulative Gain (NDCG) is used to assess the ranking accuracy. Due to the limits of space, details of the measures are referred to [7, 19].

4.2 Experiment Results

In this section, we analyze the performances of our approaches, which adopt *Distinctness*: DISTINCT, LCS, LCR, and DWCF compared with the baselines. Figure 1(a) shows MAE values from the different methods; and Fig. 1(b) displays RMSE values. Note that lower values indicate better performances for those metrics and all the improvements are significant with a confidence level of 0.01. From the two graphs, the baselines show lower performances than the methods using the *Distinctness* feature. DISTINCT using the *Distinctness* feature alone outperforms the COSINE and PEARSON in 5.5 and 4.56 %, respectively. This is due to the fact that the method contains not only the *Distinctness* feature that detects the unique characteristics of the users, but also the popularity feature that reflects the major trend of ratings. Furthermore, we can see that adopting the hybrid approaches boosts the overall performance. DWCF especially presents the highest performance among our approaches, as it indeed outperforms the baselines in 9.32 and 9.28 %, respectively.

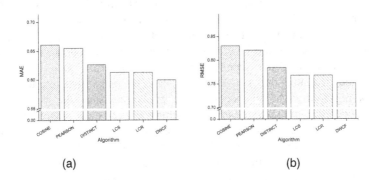

(a)   (b)

Fig. 1. Performance evaluation on (a) MAE and (b) RMSE

By looking at the NDCG scores, we also can observe which method consistently performs well in the higher ranking predicted accuracy. In detail, Fig. 2(a) represents NDCG score at N while N varies from 1 to 10; and it shows that distinctness methods including DWCF return higher NDCG score at every N compared to the baselines. Figure 2(b) shows that this tendency still resides while N increases from 10 to 40. A cursory look at the results is that utilizing the *Distinctness* feature reaps benefits in recommendations as the algorithms including the feature show better performances than the baselines. Especially, in both graphs, DWCF presents the highest performances in most cases of N compared to all the methods, implying that using the *Distinctness* feature as a weighting parameter mixed with the conventional CF methods promises the best recommendation results. Our proposed method DWCF outperforms baselines by approximately 1.48 % and 1.64 % on average for NDCG at 5 and 10, respectively. Note that the improvements are significant with a confidence level of 0.01.

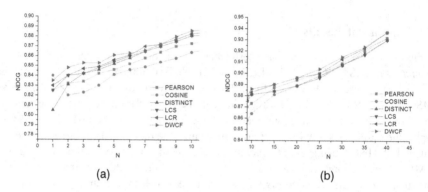

Fig. 2. Performance evaluation on NDCG at N varying from (a) 1 to 10 and (b) 10 to 40

Our assumption for *Distinctness* is that the relationship exhibited in a unique trend is stronger than the relationship in major trend. By focusing on the information on the minor rating on users, *Distinctness* enables us to successfully predict the minor ratings of users, although they have been hardly exploited for rating the movies. To verify the assumption and understand why the performance of the certain algorithm surpasses the other methods, we looked into how accurate the predicted ratings are in each rating scale. Let us denote HIT (is a counting measure) if an error value, the difference between a predicted rating and a target rating, lies in the range from zero to the MAE. HIT can be regarded as the number of correct predictions since the predicted rating values are quite close to the target answers within the setup error range; HIT Ratio is calculated that the number of the ratings is divided by the number of HITs. The higher the HIT Ratio is, the more precise prediction the method generates.

Table 1 shows a partial rating distribution for randomly chosen 40 users and HIT and HIT Ratio performances. In this random set, users frequently used 3 and 4 when rating the movies compared to 1, 2, and 5. The common ratings such as 3 and 4 have more opportunities to be predicted and obtain higher HITs. Nevertheless, if a method has higher HIT Ratio values for minor ratings like 1, 2, and 5, the method incorporates the information buried in ratings data including the distinct feature in efficient ways for recommender systems.

Table 1 demonstrates HIT and HIT Ratio of the methods for each rating scale as well as the rating distribution. In Table 1, a [a] indicate that the value is significantly higher within each method. From the table we can see that PEARSON has noticeable HIT accuracy when the target ratings to predict are 3 and 4 rather than 1 and 5. On the contrary, when the answer ratings are 1 and 5, the algorithm DISTINCT which uses

Table 1. Partial rating distribution and HIT and HIT Ratio performances

Rating scale		1	2	3	4	5	Total
# of ratings		68	244	639	745	285	1981
COSINE	HIT	1	62	466	329	50	908
	HIT Ratio	0.014	0.254	0.729	0.441	0.175	1.615
PEARSON	HIT	1	63	498	540	44	1146
	HIT Ratio	0.014	0.258	0.779[a]	0.724[a]	0.154	1.931
DISTINCT	HIT	48	62	480	540	64	1194
	HIT Ratio	0.706[a]	0.254	0.752	0.725	0.225[a]	2.665
LCS	HIT	23	62	498	555	64	1202
	HIT Ratio	0.338	0.254	0.779	0.745	0.224	2.341
LCR	HIT	22	66	498	555	73	1214
	HIT Ratio	0.324	0.270	0.779	0.745	0.256	2.374
DWCF	HIT	42	183	498	555	72	1350
	HIT Ratio	0.618[a]	0.750[a]	0.779[a]	0.745[a]	0.253[a]	3.145

only the *Distinctness* values has far higher HIT numbers than the baseline models (the HIT for 5 is reasonably better than PEARSON and COSINE).

DWCF is the combination of DISTINCT and one of the baseline models, PEARSON, since it exploits the *Distinctness* feature as weights on the conventional CF method. The performances of DWCF for 1 and 5 are as good as DISTINCT and for 2, 3, and 4 are more superb than PEARSON. This implies that DWCF mutually adopts both strengths from DISTINCT and PEARSON. A clear summarization of the above statements is given in Fig. 3 as it indicates how much each method generates HIT prediction for target ratings. In Fig. 3, the methods COSINE and PEARSON fail to predict the target rating 1 and 2, while DWCF yields the much better predictions in the target ratings. Again, as DWCF is the hybrid technique based upon PEARSON, it also shows a robust performance in predicting the major ratings such as 3 and 4 like PEARSON does.

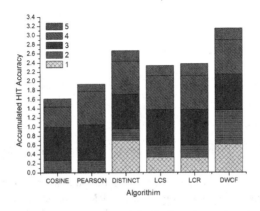

Fig. 3. Accumulated HIT accuracy graph

4.3   Impact of Parameter for Selecting Neighbors

Previously we introduced a parameter k in order to choose the size of similar neighbors for a certain user. It has been believed that the parameter plays a vital role to generate precise recommendations [11]. If the value of k is too small, some of the neighbors' ratings that match with the current user might be completely overlooked. On the other hand, if the value of k is too large, the ratings of users who have strongly the opposite tendency against the user might be potentially regarded for prediction. To evaluate the impact of the parameter, we observe the MAE, RMSE, NDCG at 5 and NDCG at 10 results by slowly increasing the parameter from 0.1 to 1.0. Note that the value of 1.0 indicates that all users are used as neighbors. Meanwhile, the value of 0.1 indicates that, after sorting all users in a descending order of each similarity method: COSINE, PEARSON, and DWCF, top 10 % of the users are chosen to be a set of neighbors with the target user.

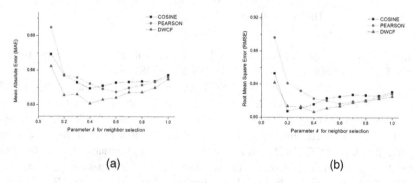

(a)   (b)

Fig. 4. Parameter k graph on (a) MAE and (b) RMSE

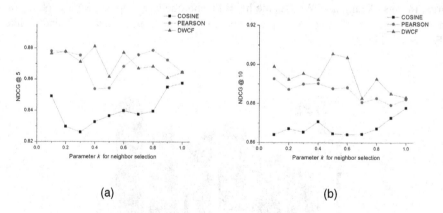

(a)   (b)

Fig. 5. Parameter k graph on NDCG at (a) 5 and (b) 10

Figures 4 and 5 present the effect of varying the parameter k. In specific, the MAE performance for each k is shown in Fig. 4(a), the RMSE performance is given in Fig. 4 (b), and the NDCG performance is given in Fig. 5. In terms of DWCF, we observe that the overall performance for each metric peaks approximately between 0.4 and 0.5, gradually decreasing afterwards. On the other hand, PEARSON and COSINE show lower performances than DWCF except for few cases. Although the performance varies depending on the value of k, we still can observe that the best value for DWCF presents the best performance compared to the best cases of the compared methods. Another observation is that each method peaks at different k and thus, we set the parameter to its best performing value in accordance with the method chosen for our experiment.

5   Conclusion and Future Work

The goal of this paper is to suggest a new collaborative filtering method for content recommender systems. To achieve this goal, we have presented a novel measure, *Distinctness*, to estimate unique and distinct ratings that do not follow major trends. Using *Distinctness*, we have developed four CF approaches: DISTINCT, LCS, LCR, and DWCF. Following the proposed approaches, similarity and predicted ratings have been computed while considering the degree of *Distinctness*. Throughout our experiment, we have showed that our models effectively utilize data and clearly outperform comparable models. Especially, by exploiting the concept of HIT and HIT Ratio, we have detailed an investigation of superior results from DWCF.

Our study needs further work. We will have to apply classification of users on the basis of analyzing rating patterns. We assume that there might be different types of users in terms of ways of reacting to external influences according to rating times and experiences. In our experiment, we have chosen only common users who have a certain range of rating times, to avoid cases of the cold-start problem and unusual rating behaviors. Additionally, in order to focus on discovering the distinctness information and validating this new feature for the first time, we have used easily computable algorithms as the baselines. Based on improvements from the distinctness feature proposed in this study, we believe that it is also worthwhile to incorporate sophisticated algorithms like Matrix Factorization approach into this feature in future. Despite the need for the further work, the proposed methods show the promising potential of *Distinctness* to improve the overall performance of recommendation results.

Acknowledgements. This work was supported by the Industrial Strategic Technology Development Program, 10052955, Experiential Knowledge Platform Development Research for the Acquisition and Utilization of Field Expert Knowledge, funded by the Ministry of Trade, Industry & Energy (MI, Korea).

References

1. Gopinath, S., Chintagunta, P.K., Venkataraman, S.: Blogs, advertising, and local-market movie box office performance. Manage. Sci. **59**(12), 2635–2654 (2013)
2. Duan, W., Gu, B., Whinston, A.B.: The dynamics of online word-of-mouth and product sales-an empirical investigation of the movie industry. J. Retail. **84**(2), 233–242 (2008)

3. Gal-Or, E., Geylani, T., Yildirim, T.P.: The impact of advertising on media bias. J. Mark. Res. **49**(1), 92–99 (2012)

4. Faber, R.J., O'Guinn, T.C.: Effect of media advertising and other sources on movie selection. Journalism Mass Commun. **61**(2), 371–377 (1984)

5. Liu, Y.: Word of mouth for movies: its dynamics and impact on box office revenue. J. Mark. **70**(3), 74–89 (2006)

6. Kapoor, G.: Sequential bias in online product reviews. J. Organ. Comput. Electron. Commer. **19**(2), 85–95 (2009)

7. Herlocker, J.L., Konstan, J.A., Terveen, L.G., Riedl, J.T.: Evaluating collaborative filtering recommender systems. ACM Trans. Inf. Syst. **22**(1), 5–53 (2004)

8. Rashid, A.M., Albert, I., Cosley, D., Lam, S.K., McNee, S.M., Konstan, J.A., Riedl, J.: Getting to know you: learning new user preferences in recommender systems. In: Proceedings of the 7th International Conference on Intelligent User Interfaces, pp. 127–134. ACM (2002)

9. Sparling, E.I, Sen, S.: Rating: How difficult is it? In: Proceedings of the Fifth ACM Conference on Recommender Systems, pp. 149–156 (2011)

10. Zhao, X., Chen, W., Niu, Z.: Opinion-based collaborative filtering to solve popularity bias in recommender systems. In: Decker, H., Lhotská, L., Link, S., Basl, J., Tjoa, A.M. (eds.) DEXA 2013, Part II. LNCS, vol. 8056, pp. 426–433. Springer, Heidelberg (2013)

11. Koren, Y.: Factorization meets the neighborhood: a multifaceted collaborative filtering model. In: Proceedings of the 14th ACM SIGKDD International Conference on Knowledge Discovery and Data Mining, pp. 426–434 (2008)

12. Tan, C., Chi, E.H., Huffaker, D., Kossinets, G., Smola, A.J.: Instant foodie: predicting expert ratings from grassroots. In: Proceedings of the 22nd ACM International Conference on Conference on Information & Knowledge Management, pp. 1127–1136 (2013)

13. Stoimenova, E.: Methodology in robust and nonparametric statistics. J. Appl. Stat. **40**(12), 2773–2777 (2013)

14. Oku, K., Nakajima, S., Miyazaki, J., Uemura, S.: Context-aware SVM for context-dependent information recommendation. In: Proceedings of the 7th International Conference on Mobile Data Management, p. 109 (2006)

15. Hariri, N., Mobasher, M., Burke, R.: Context adaptation in interactive recommender systems. In: Proceedings of the 8th ACM Conference on Recommender Systems, pp. 41–48 (2014)

16. Yeniterzi, R., Callan, J.: Analyzing bias in CQA-based expert finding test sets. In: Proceedings of the 37th International ACM SIGIR Conference on Research & Development in Information Retrieval, pp. 967–970 (2014)

17. Lai, S., Liu, Y., Gu, H., Xu, L., Liu, K., Xiang, S., Zhao, J., Diao, R., Xiang, L., Li, H., Wang, D.: Hybrid recommendation models for binary user preference prediction problem. In: JMLR: Workshop and Conference Proceedings 18, pp. 137–151 (2012)

18. Rashid, A.M., Karypis, G., Riedl, J.: Learning preferences of new users in recommender systems: an information theoretic approach. ACM SIGKDD Explor. Newsl. **10**(2), 90–100 (2008)

19. Yilmaz, E., Kanoulas, E., Aslam, J.A.: A simple and efficient sampling method for estimating AP and NDCG. In: Proceedings 31st SIGIR, pp. 603–610 (2008)

Twitter and Social Media

Detecting Automatically-Generated
Arabic Tweets

Hind Almerekhi[1,2](✉) and Tamer Elsayed[1]

[1] Computer Science and Engineering Department, College of Engineering,
Qatar University, Doha, Qatar
{200752261,telsayed}@qu.edu.qa
[2] Qatar Computing Research Institute, Doha, Qatar

Abstract. Recently, Twitter, one of the most widely-known social media platforms, got infiltrated by several automation programs, commonly known as "bots". Bots can be easily abused to spread spam and hinder information extraction applications by posting lots of automatically-generated tweets that occupy a good portion of the continuous stream of tweets. This problem heavily affects users in the Arab region due to the recent developing political events as automated tweets can disturb communication and waste time needed in filtering such tweets.

To mitigate this problem, this research work addresses the classification of Arabic tweets into automated or manual. We proposed four categories of features including formality, structural, tweet-specific, and temporal features. Our experimental evaluation over about 3.5 k randomly sampled Arabic tweets shows that classification based on individual categories of features outperform the baseline unigram-based classifier in terms of classification accuracy. Additionally, combining tweet-specific and unigram features improved classification accuracy to 92 %, which is a significant improvement over the baseline classifier, constituting a very strong reference baseline for future studies.

Keywords: Tweet classification · Arabic microblogs · Bots · Automated tweets · Crowdsourcing

1 Introduction

The growth of online microblogging services introduced new means of sharing opinions, news, and information such as Twitter. Users of Twitter can exchange messages up to 140 characters, commonly known as *tweets*. Statistics from 2014 show that the total number of tweets posted per day is about 500-million tweets[1]. Interestingly, as of March 2014, an average of 17-million *Arabic* tweets are posted daily[2]. Moreover, Twitter has no particular restriction on automation; thus,

---

[1] http://www.adweek.com.

[2] http://www.arabsocialmediareport.com.

© Springer International Publishing Switzerland 2015
G. Zuccon et al. (Eds.): AIRS 2015, LNCS 9460, pp. 123–134, 2015.
DOI: 10.1007/978-3-319-28940-3_10

automation programs can play a critical role in creating a huge volume of tweets. Such automation programs are formally known as *bots* [2]. The main purpose of bots is to mimic humans and post tweets (periodically in most cases) to the timelines of subscribed users. Tweets posted by bots, referred to as "automated" tweets, are sometimes partially-edited by a human, or completely automated (e.g., prayer times or temperature readings). In the Arab world, Arabic bots often use formal or Modern Standard Arabic (MSA) in their messages. Furthermore, tweets generated by bots are not personalized, as they discuss broad topics like news and famous quotes.

Examples of automated tweets written by bots are given in Table 1. The first tweet is a verse from Quran, the second is a supplication posted by an Arabic bot, the third tweet is a quote by the famous Greek philosopher Aristotle, and the fourth tweet is an advertisement for an Arabic words application. In terms of tweeting behavior, humans tend to use informal (dialectal) Arabic when tweeting. Moreover, they communicate with other users, sometimes misspell words, and often use abbreviations or emoticons in their tweets. The diversity of writing styles used by Arab users on Twitter makes human tweets unique by nature.

The bottom half of Table 1 shows examples of "manual" tweets written by humans. The first manual tweet is written in Egyptian dialect, while the second is written in Gulf dialect. The last two tweets are written informally in mix of different Arabic dialects.

Table 1. Examples of automated and manual Arabic tweets

Tweet Type	Example
Automated	1. [التغابن:٣] (خلق السماوات والأرض بالحق وصوركم فأحسن صوركم وإليه المصير)
	2. http://t.co/VF806XZIuE اللهم عافني في بَدَني اللهم عافني في سمعي اللهم عافني في بصري
	3. http://t.co/Jvz5ONdElw إذا أردت إن تكون سعيداً ينبغي أن يكون لديك اكتفاء ذاتي!. ارسطو
	4. http://t.co/tylxpStXTu عكنكم الآن الاشتراك في تطبيق كلمات من هنا
Manual	1. انا مبقتش أحاول أعمل أي حاجة في حياتي، انا قاعد بسمع مزّيكا وسرحان وسايب كل حاجة بتحصل لوحدها
	2. ترا نفهم الشيء اللي يكون من قلب و الشيء اللي ينعطي بمجامله و تسليك ممكن نتظاهر بس نكون فاهمين
	3. أجلس بمكان بارد وهادي!؟راحه عظيمه تمتلكني.
	4. @Mohmcd ـهـهـهـهـهـهـهـهـ والله انسحب عليك صحبة الذين كفروا ـهـهـهـهـ

Unfortunately, bots can be easily abused to spread spam such as advertisements and malicious hyperlinks [13]. Moreover, some bots interfere with information extraction applications, thus hindering their job. For example, trending topic detection systems can easily confuse actual trending topic tweets with automated tweets. Results of such confusion stem from the interference of automated tweets, which leads to biased result [7]. When users rely on Twitter to exchange valuable information, automated tweets might obstruct the communication between users, especially in cases that involve dangerous locations or survival needs. At times of need, users seek information from genuine tweets, not from automated sources. Therefore, it is important to distinguish between automated and manual tweets.

In this work, we tackle the problem of detecting automated Arabic tweets, since it noticeably affects users in the Arab region due to the evolving recent political events. To address the problem, we formulate it as a binary classification problem. Besides the typical unigram features (i.e., term occurrences in tweets) that constitute our baseline classifier, we experimented with four feature categories: formality, structural, tweet-specific [4], and temporal features [12,14]. Our classifier was trained using a set of manually-labeled tweets that were obtained through crowdsourcing. The classification performance was evaluated through a set of experiments that aim to answer the following research questions:

- *RQ1:* Would preprocessing of unigram features help improve the baseline unigram classifier?
- *RQ2:* Can any of the feature categories separately produce a classifier that outperforms the unigram classifier? Would temporal features specifically help classify the automated tweets?
- *RQ3:* What is the impact of combining the unigram features with each of the four feature categories? Will combining all feature categories improve the classification compared to individual categories?

The contribution of this study is three-fold:

- To our knowledge, this is the *first* study that focuses on automation behavior in *Arabic* microblogs. We conducted our experiments on a collection of 1.2-million tweets from 11 k different users that was obtained in 4 days.
- Our proposed classifier, achieving an accuracy of about 92 %, constitutes a very strong baseline classifier for future studies on the problem.
- Two lists of Arabic tweets were developed and made publicly available for further research[3]: the first one includes 1.2-million tweets (represented by their tweet ids) that were used in our experiments. The second contains a total of 3503 manually-labeled tweets, where 1944 were labeled as automated tweets and 1559 were labeled as manual tweets.

The remainder of the paper is organized as follows. Section 2 summarizes the related work. Section 3 introduces the proposed solution. Section 4 discusses our experimental evaluation in terms of setup and results. Section 5 concludes this work with some future research directions.

2   Related Work

In recent years, several researchers attempted to study the implications of automated tweets on the microblogging community. Unfortunately, the literature does not report any work on classifying tweets as either automated or manual. Therefore, all of the systems that will be discussed shortly tackle the problem of classifying Twitter accounts.

---

[3] http://faculty.qu.edu.qa/telsayed/datasets.aspx.

Chu et al. [2] proposed a methodology to classify Twitter users based on their posting activity. The classification system comprises of four main components: the spam detection component, the entropy component, the account properties component, and the decision making component. The system applies a set of measurements to classify a collection of over 500 K Twitter accounts. The goal of the proposed classification is to identify accounts that originate from humans, bots, or cyborgs. The study shows that the ratio of Twitter population given the human, cyborg, and bot categories is 5:4:1 [2]. In another research, Laboreiro et al. [7] introduced classes of features that aid in identifying automation. The work proposed classes of features that combine chronological and other content-based features to identify automation. The classes of features include chronological features and content-based features. With a collection of 72 K accounts, the authors show that out of the 26 K features that were generated through experiments, stylistic content-based features contribute the most to gaining high classification accuracy [7].

Since malicious bots are generally associated with spreading spam, there is an overlap between automation and spam detection. Hence, it was important to look at the work done on spam detection to understand how such systems identify these tweets.

The work done in [2] triggered the interest of researchers to further investigate the nature of automation. Zhang and Paxson [12] conducted a study that aims to identify spammers on Twitter. Their research relies on the timestamp associated with each tweet. The authors showed that their approach can analyze Twitter's landscape and identify the exhibited degree of automation. By crawling 19 K accounts from the public timeline, 16 % of the accounts were identified as automated. Moreover, verified Twitter users and accounts with many followers exhibit lower automation rates [12]. The work done by Lee et al. [8] studied the nature of content pollution on Twitter. By spending seven months on the study, the authors were able to create 60 honeypots and attract 36 K spamming users on Twitter. The unique features that were used in this study include user friendship network, user demographic, user content, and user history [8].

Another work by Yang et al. [11] aimed at discovering features that identify spam accounts. By looking at the relationship between spammer nodes and their neighbors, the study looked at three graph-based features. In addition to other new features that consider the posting time and whether tweets are automated or not. This suggests spam detection systems also consider automated and non-automated tweets to be possible sources of spam. Similarly, the work done by Zhu et al. [13] aimed at modeling the social network in twitter and identify spammers based on their account associations. By deploying a supervised Matrix Factorization technique, the proposed method relied highly on relations between users and actions performed by users. On the same note, the study conducted by Ghosh et al. [3] focused on identifying the nature of link farming in Twitter. The authors show that complex features that rely on mutual linking between spammers and non-spammers can aid in the spread of spam. Similarly, the work done in [5, 10] studying the communication between users on Twitter to identify spam bots and prevent spread of spam.

In contrast to the work in [3,8,11,13] which focuses on classifying spam accounts. Martinez-Romo and Araujo [9] introduced a spam classification system that detects spam within trending topics by analyzing tweets rather than user accounts. The method introduces two approaches that uniquely distinguish spam. The first approach studies spam tweets without considering their source user. The second approach relies on statistical analysis of language to detect spam. The proposed online content-filtering technique relies on language as a detection tool. Amidst the spam detection research, a study conducted by Aggarwal et al. [1] tackled the problem of detecting phishers in Twitter. The methodology relies on a combination of Twitter based features and URL features to classify tweets. A recent study by Hu et al. [6] followed an approach similar to the work proposed in [1] but with a focus on optimizing the process of spam detection.

This research is different from all the work discussed earlier because it tackles the problem of classifying tweets rather than Twitter accounts. Additionally, this work focuses on automated and manual Arabic tweets as opposed to the all the work discussed earlier which only considers English tweets and spam content.

3   Proposed Solution

We formally define our problem as follows. Given a collection of Arabic tweets, the goal is to classify each of them into one of two main classes: *automated* tweets and *manual* tweets. This section describes in detail our proposed solution to the problem which involves an overview of the system architecture and the categories of features extracted from the data collection to help classify the tweets.

3.1   System Architecture

The architecture of the proposed system is illustrated in Fig. 1. It was inspired by Martinez-Romo and Araujo's system in [9] which was used for filtering spam tweets. The model in Fig. 1 shows the four major components of the automated tweet detection system. The preprocessing component performs punctuation and stop-word removal to process the raw tweets, then several feature categories and unigram features are extracted from the preprocessed tweets. The last component performs the classification and evaluation of tweets using the extracted feature vectors and the labels obtained through crowdsourcing.

3.2   Feature Extraction

Since features play a critical role in the classification process, it is essential to detail the list of features that will be used to detect automation in Arabic tweets. The focus of this section will be on describing the four main feature categories used in the study, which are: *formality features, structural features, tweet-specific features,* and *temporal features.*

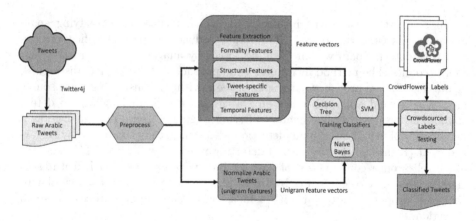

Fig. 1. The architecture of the proposed system for detection of automatically-generated Arabic tweets.

Formality Features. Formality features are features that measure how formal a tweet is. In this category, three features are involved: *emoticons*, which is the total number of emoticons (like smiley faces (:-)) found in Arabic tweets. *Diacritics* feature checks if a tweet contains a decorating diacritic or not. The third feature is *elongation*, which checks if tweet words contain more than four consecutive characters, such as " ـﻮﻮﻮﻟ ", which is equivalent to "oool" [4].

Structural Features. Studies show that the structure of a tweet can provide good features for tweet classification [4]. Such features include the *length* of the tweet in terms of total number of characters, the total number of *question marks*, and the total number of *exclamation marks*.

Tweet-Specific Features. Observing the data associated with tweets can provide some useful features for classification. Fortunately, a lot of research was done in this area to identify such features [4]. For classifying automated and manual Arabic tweets, a total of six tweet-specific features were used: *retweet*, which checks if the tweet was a retweet or not, *reply* also checks if the tweet is a reply or not, *hashtags* computes the total number of hashtags in the tweet, and *URLs* checks if a tweet contains a hyperlink or not.

A tweet object (obtained through Twitter API) embeds several fields that provide some information about the tweet. For instance, the "source" field can indicate the type of client application used in posting the tweet. A user can post a tweet through different devices, like the Web service, a dedicated mobile application, a third party application, or an API [2]. For bots, some devices are easier to access than others due to authentication requirements. Moreover, humans can combine different clients to post tweets, such as the Web client and the mobile client. The *source* feature checks if the tweet contains a source field or not. Checking just the existence of the field (instead of using the source type) was sufficient for this study because most of the crawled tweets did not contain any value in the source field.

Temporal Features. Temporal features constitute an essential part of this study because they allow us to study the characteristics of tweets in terms of posting nature. For that purpose, the following temporal features were proposed:

- *Activity period on Twitter*: Some Twitter accounts exhibit constant usage over time, such as bot accounts. While other accounts focus on a particular time period to be active on Twitter, like human accounts. Therefore, studying the posting time of tweets during a particular time period can be a strong indication of automation. This feature, which was proposed by Zhang et al. [7] assigns the activity period to be a full day. However, in this study, the period was set to five hours due to the limited amount of tweets per user. For each hour, we count the number of tweets posted in that hour by the author of the tweet, resulting in five features, one per hour.

- *Spread Velocity*: This is the sixth temporal feature, adopted from the research done by Zubiaga et al. [14] on trending topic classification. The reason behind choosing this feature is because it captures temporal characteristics based on the differences in seconds between the posting time of tweets per user. To compute this temporal feature, Eq. 1 was used, where Δt is the time period in seconds between the first and last posted tweet of the author, and $|T|$ is the total number of tweets posted by the author in that period.

$$AM(sv)_t = \frac{|T|}{\Delta t} \tag{1}$$

4   Experimental Evaluation

4.1   Experimental Setup

Dataset. To build the data collection, tweets were first randomly sampled through Twitter streaming API[4] over a period of four days. The query used to collect the random sample is "lang:ar" using Twitter4j java library. The language query ensures that the crawled tweets were in Arabic. To avoid the case where the collected set of tweets are biased towards specific users, a post-processing step was applied to keep just one tweet for each unique user. The result of this step was a collection of 11,764 tweets from unique users. The second step involved collecting more tweets for each user to measure temporal features. For each user, the most recent 120 tweets were obtained through the REST API. However, some users tend to tweet in English and Arabic simultaneously. Therefore, the most recent 120 for each unique user were filtered again with the query "lang:ar". The final collection was processed to remove tweets with duplicate content. The result of this step is a collection of 1,202,815 tweets posted by 11,764 different users.

---

[4] https://dev.twitter.com/rest/public.

Manual Labeling. For any classification problem, it is essential to obtain labeled data that identify each class. For this study, there was no labeled data that is publicly available for identifying automated and manual Arabic tweets. To solve this problem, the crowdsourcing platform CrowdFlower was used. Annotators were given the task of labeling Arabic tweets as either automated or manual. Annotators were presented with clear Arabic instructions and a total of 18 test questions that qualified them to take part in the labeling process. This measure ensures the accuracy of the aggregated labels. Due to the difficulty of the task and time constraints, annotators were able to label a total of 3503 tweets out of the 11764 collected tweets, where each tweet was judged by at least three annotators. Out of the 3503 labeled tweets, 1944 (55 %) were labelled as automated tweets and 1559 (45 %) were labelled as manual tweets. Annotator agreement of the submitted labeling job was around 93 %, with a Fleiss kappa value of 0.61, which is a good value considering the ambiguity of some tweets to the annotators.

Classification. We experimented with three classification algorithms: Support Vector Machines (SVM), Naïve Bayes, and Decision Trees; we used their implementations in Weka[5], the machine learning library. These algorithms were reported as extremely effective for binary classification problems in the literature [13]. For each machine learning algorithm, a 10-fold cross validation approach was used for training and testing purposes. As for evaluation measures, this study reports the classification accuracy, precision, recall, and F1 measures.

Baseline. To evaluate the proposed solution, it was essential to identify and construct a baseline classifier to compare against. Finding a baseline system was a real challenge because there was no previous study in the literature that addresses the same problem. Hence, we selected the basic unigram model to comprise our baseline classifier. This choice was made due to the simplicity of the unigram model and its effectiveness in many classification problems [14].

4.2   Experimental Results

RQ1: **Baseline Performance.** Before conducting experiments on any of the four feature categories, it is important to think about the proper settings of the baseline unigram model. The unigram model does not take into consideration any features other than the frequency of words in tweets. To construct the term frequency feature, terms must be tokenized and preprocessed, then the frequency of each word is computed for each tweet. The problem with this model is that the number of dimensions is not fixed as it changes depending on the number of unique words in the collection. Since Arabic tweets are rich with all types of decorations and symbols, there are many options when it comes to preprocessing. We tried the following three scenarios:

---

[5] http://www.cs.waikato.ac.nz/ml/weka/.

- **uni(raw):** Nothing is removed from the unigrams, thus they are raw.
- **uni(emoti):** Emoticon symbols are removed from the unigrams.
- **uni(stop):** Standard MSA Arabic stop words are removed from the unigrams.

The first experiment looks at the classification accuracy of different variations of the unigram model. This experiment aims to achieve two goals. The first goal is to get some insights on the impact of stop words and emoticons on the quality of unigrams. The second crucial goal is to select a candidate baseline system that will be compared with in the rest of the experiments.

The results shown in the left section of Table 2 depict the classification accuracy of the three proposed unigram models. The accuracy values answer *RQ1* because the unigrams(stop) clearly outperforms the two other unigram models, which indicates that the best baseline model is the unigrams with stop words removal.

Table 2. Classification accuracy of different categories of features: U(raw), U(emoticon), U(stop): Unigram, F: Formality, S: Structural, TS: Tweet-Specific, Temp: Temporal. Starred results indicate statistically-significant improvement over the best baseline using two-tailed paired t-test, with $\alpha = 0.05$.

Model	U(raw)	U(emoti)	U(stop)	F	S	TS	Temp
SVM	0.600	0.598	**0.663**	0.620	0.618	**0.892***	0.666
J48	0.613	0.611	**0.672**	0.679	0.679	**0.886***	0.740
NB	0.612	0.610	**0.672**	0.589	0.683	**0.861***	0.649

RQ2: **Performance of Individual Feature Categories.** Along with unigrams(stop) as a baseline model, the remaining models that were used in the following experiments are the categories of features that were described in Sect. 3.2. To answer *RQ2*, the right side of Table 2 shows the classification accuracy of each category of features. The results highlighted in bold are the highest average classification accuracy of each model. Results clearly show that the tweet-specific features model outperforms all the models including the baseline. Furthermore, the SVM and Decision Tree classifiers show higher accuracy values when compared to the Naïve Bayes classifier.

Focusing more on temporal features, Table 3 shows that they outperform the baseline model in the classification accuracy of SVM and Decision Trees algorithms only. The results show that for the temporal features, the precision drops in the Naïve Bayes classifier for the automated class. As for the manual class, both the recall and F1 drop when compared to the baseline. We suspect that the limited number of tweets per user (only 120 tweets) were not enough to compute effective temporal features, and hence the limited classification performance. Nonetheless, the temporal features are on par with the baseline model, even though they did not outperform it in one classification algorithm.

Table 3. Precision, Recall, and F1 measures for the baseline unigrams(stop) model (U) and temporal features model (Temp).

Algorithm	Automated			Manual		
	Precision	Recall	F1	Precision	Recall	F1
U(SVM)	0.500	0.001	0.001	0.663	1.000	0.798
U(J48)	0.812	0.060	0.112	0.676	0.993	0.804
U(NB)	0.855	0.031	0.059	0.670	0.997	0.801
Temp(SVM)	0.646	0.879	0.745	0.726	0.399	0.515
Temp(J48)	0.810	0.695	0.748	0.677	0.796	0.732
Temp(NB)	0.633	0.877	0.735	0.705	0.365	0.481

RQ3: **Performance of Combined Features.** Results in Table 4 addresses *RQ3*. The left side of the table shows the impact of adding the unigram features to each of the other feature categories. No or slight improvements in accuracy are shown when compared to the results in Table 2. The right side shows the results of combining all feature categories. The column "All" indicates the combination of all feature categories in addition to the unigram features. The highest classification accuracy values are highlighted in bold for each classification model. The decision tree classifier seems to have the highest classification accuracy at 91.97 % when the unigram and tweet-specific features are combined, while the Naïve Bayes shows the lowest classification accuracy overall. Hence, combining all feature categories with the baseline unigram features did not improve classification accuracy.

Table 4. Classification accuracy of different combinations of feature categories along with unigram features.

Model	U+F	U+S	U+TS	U+Temp	F+S+TS+Temp	All
SVM	0.620	0.620	**0.892**	0.667	0.892	0.892
J48	0.796	0.844	**0.920**	0.873	0.865	0.920
NB	0.491	0.667	**0.861**	0.651	0.828	0.828

5   Conclusion and Future Work

This work showed that different feature categories can be leveraged to classify Arabic tweets as either automated or manual. With the aid of conventional classification techniques, it is possible to classify Arabic tweets at a high accuracy of 92 %. In fact, this research shows that combining tweet-specific and unigram features outperforms all other experimented combinations. Moreover, crowdsourcing labelers identified a total of about 2 k automated Arabic tweets and 1.5 k manual tweets in a random sample of tweets used in this study; this indicates that automation plays a huge role in the Arabic space of Twitter.

Although the achieved classification accuracy was pretty high, there is still some room for improvement. Since this work presented a preliminary study, future work will focus on task-specific features that target different kinds of automated tweets (e.g. propaganda and spam tweets). The temporal features in this work did not manage to fully outperform the baseline model due to the limited number of features and tweets per user. Hence, as future work, it would be interesting to investigate additional temporal features, like Pearson's X^2 test, which considers minutes-of-the-hour and hours-of the-minute as features [12]. We also plan to sample more tweets per user to allow for more accurate temporal feature values. As for the tweet-specific features, we plan to extend the source field to account for different types of sources. Moreover, dialect detection and bigram (or generally n-gram) features can be leveraged to extend the list of feature categories. Another direction of future work can focus more on Arabic-specific preprocessing of the tweets, e.g., character normalization, and compare the classification results with an English corpus.

References

1. Aggarwal, A., Rajadesingan, A., Kumaraguru, P.: Phishari: automatic realtime phishing detection on twitter. In: IEEE eCrime Researchers Summit (eCrime), pp. 1–12. IEEE (2012)
2. Chu, Z., Gianvecchio, S., Wang, H., Jajodia, S.: Detecting automation of twitter accounts: are you a human, bot, or cyborg? IEEE Trans. Dependable Secure Comput. 9(6), 811–824 (2012)
3. Ghosh, S., Viswanath, B., Kooti, F., Sharma, N.K., Korlam, G., Benevenuto, F., Ganguly, N., Gummadi, K.P.: Understanding and combating link farming in the twitter social network. In: Proceedings of the 21st International Conference on World Wide Web (WWW), pp. 61–70. ACM (2012)
4. Hasanain, M., Elsayed, T., Magdy, W.: Identification of answer-seeking questions in arabic microblogs. In: Proceedings of the 23rd ACM International Conference on Information and Knowledge Management (CIKM), pp. 1839–1842. ACM (2014)
5. Hentschel, M., Alonso, O., Counts, S., Kandylas, V.: Finding users we trust: scaling up verified twitter users using their communication patterns. In: Eighth International AAAI Conference on Web and Social Media (ICWSM) (2014)
6. Hu, X., Tang, J., Liu, H.: Online social spammer detection. In: Twenty-Eighth AAAI Conference on Artificial Intelligence (AAAI) (2014)
7. Laboreiro, G., Sarmento, L., Oliveira, E.: Identifying automatic posting systems in microblogs. In: Antunes, L., Pinto, H.S. (eds.) EPIA 2011. LNCS, vol. 7026, pp. 634–648. Springer, Heidelberg (2011)
8. Lee, K., Eoff, B.D., Caverlee, J.: Seven months with the devils: a long-term study of content polluters on twitter. In: Fifth International AAAI Conference on Web and Social Media (ICWSM). Citeseer (2011)
9. Martinez-Romo, J., Araujo, L.: Detecting malicious tweets in trending topics using a statistical analysis of language. Expert Syst. Appl. 40(8), 2992–3000 (2013)

10. Wald, R., Khoshgoftaar, T.M., Napolitano, A., Sumner, C.: Predicting susceptibility to social bots on twitter. In: IEEE 14th International Conference on Information Reuse and Integration (IRI), pp. 6–13. IEEE (2013)
11. Yang, C., Harkreader, R., Zhang, J., Shin, S., Gu, G.: Analyzing spammers' social networks for fun and profit: a case study of cyber criminal ecosystem on twitter. In: Proceedings of the 21st International Conference on World Wide Web (WWW), pp. 71–80. ACM (2012)
12. Zhang, C.M., Paxson, V.: Detecting and analyzing automated activity on twitter. In: Spring, N., Riley, G.F. (eds.) PAM 2011. LNCS, vol. 6579, pp. 102–111. Springer, Heidelberg (2011)
13. Zhu, Y., Wang, X., Zhong, E., Liu, N.N., Li, H., Yang, Q.: Discovering spammers in social networks. In: Twenty-Sixth AAAI Conference on Artificial Intelligence (AAAI) (2012)
14. Zubiaga, A., Spina, D., Martínez, R., Fresno, V.: Real-time classification of twitter trends. J. Assoc. Inf. Sci. Technol. 66(3), 462–473 (2014)

Improving Tweet Timeline Generation
by Predicting Optimal Retrieval Depth

Maram Hasanain[1]([✉]), Tamer Elsayed[1], and Walid Magdy[2]

[1] Computer Science and Engineering Department, College of Engineering,
Qatar University, Doha, Qatar
{maram.hasanain,telsayed}@qu.edu.qa
[2] Qatar Computing Research Institute, HBKU, Doha, Qatar
wmagdy@qf.org.qa

Abstract. Tweet Timeline Generation (TTG) systems provide users
with informative and concise summaries of topics, as they developed
over time, in a retrospective manner. In order to produce a tweet timeline that constitutes a summary of a given topic, a TTG system typically
retrieves a list of potentially-relevant tweets over which the timeline is
eventually generated. In such design, dependency of the performance of
the timeline generation step on that of the retrieval step is inevitable.

In this work, we aim at improving the performance of a given timeline
generation system by controlling the depth of the ranked list of retrieved
tweets considered in generating the timeline. We propose a supervised
approach in which we *predict* the optimal depth of the ranked tweet list
for a given topic by combining estimates of list quality computed at different depths.

We conducted our experiments on a recent TREC TTG test collection of 243 M tweets and 55 topics. We experimented with 14 different
retrieval models (used to retrieve the initial ranked list of tweets) and 3
different TTG models (used to generate the final timeline). Our results
demonstrate the effectiveness of the proposed approach; it managed to
improve TTG performance over a strong baseline in 76 % of the cases,
out of which 31 % were statistically significant, with no single significant
degradation observed.

Keywords: Tweet summarization · Microblogs · Dynamic retrieval cutoff · Query difficulty · Query performance prediction · Regression

1   Introduction

Coping with a flood of user-generated content about ongoing events through the
online social media is getting more challenging over time. With several trending
topics of interest that are active simultaneously, losing track of some of them is
sometimes inevitable due to the large amount of posts compared to the limited
time. One potential solution is to have the ability to get a retrospective timeline
of posts that cover trending topics or events; Tweet Timeline Generation (TTG)
systems aim at addressing this problem [14].

© Springer International Publishing Switzerland 2015
G. Zuccon et al. (Eds.): AIRS 2015, LNCS 9460, pp. 135–146, 2015.
DOI: 10.1007/978-3-319-28940-3_11

TTG task is typically query-oriented. The user provides a query representing the topic of interest and requires the TTG system to provide a list of tweets that were posted prior to query time and that are both *relevant* to the topic and *non-redundant*. This construction of the problem suggests a natural design of a TTG system that consists of two consecutive steps. First, it *retrieves* a ranked list of tweets that are potentially-relevant to the topic (called the *retrieval* step) and then *generates a timeline* of non-redundant tweets out of that list (called the *timeline generation* (TG) step). This design, in turn, imposes a natural dependency of the quality of the generated timeline on the quality of the retrieved list of tweets. Additionally, an important decision that a TTG system usually makes is how many retrieved tweets (or in other words, which depth of the retrieved ranked list) to start the timeline generation step with. Out of **13** teams participated in the first offering of the TTG task at TREC-2014, at least **10** teams[1] have used this design and **7** of them have used a static (i.e., fixed) depth (or rank cutoff) of the retrieved tweets over all queries [14].

Figures 1 and 2 show how the performance of an example clustering-based TTG system is sensitive to the depth of the retrieved list of tweets. Performance is measured using weighted F_1 (denoted by wF_1) used as the official evalaution measure of the TTG task at TREC-2014 [14].

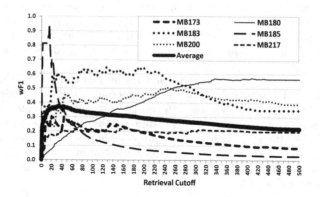

Fig. 1. TTG performance using different cutoffs over 6 TREC-2014 queries.

Figure 1 demonstrates this for 6 different TREC-2014 queries and for the average performance over all queries as well. It shows that different queries behave differently in terms of the effect of changing retrieval depth on TTG performance.

Given 55 TREC-2014 queries, Fig. 2 shows how an optimal per-query rank cutoff can improve the performance over a static one by comparing the performance of two oracle TTG systems: the first used the best global static cutoff (i.e., a fixed cutoff over all queries that maximizes the average performance, which

---

[1] No published work on the system design of the remaining 3 teams.

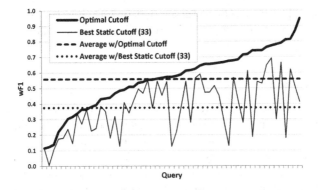

Fig. 2. Effect of using Static vs. dynamic cutoffs on a TTG system performance.

happened to be at depth 33), while the other used an optimal cutoff per query (i.e., a different cutoff per query that maximizes the performance of each query separately). The figure indicates that it is possible to achieve large improvements, reaching 50 %, by just dynamically selecting the right retrieval cutoff per query, without any changes to neither the retrieval nor the TG components of the system.

Motivated by the above observations, we address the problem of improving the performance of a *given TTG system* by controlling the depth of the retrieved list of tweets. We propose to tackle the problem by learning a regression model that predicts optimal list depth needed to optimize the TTG performance. The model is learned over features that estimate the retrieval quality at different depths of the list. The problem of estimating the performance of a retrieval system given a query, called query performance prediction (QPP), has been studied extensively [3] and had recently showed promising results in microblog search [10,18]. In this work, we leverage QPP techniques to predict a suitable retrieval cutoff for a TTG query. To our knowledge, this is the first study that leverages QPP for *improving* TTG or (more generally) tweet summarization systems.

Our contribution is two-fold. First, we showed that the performance of TTG systems is highly sensitive to the depth (and thus the quality) of the retrieved list of tweets over which the timeline is generated. Second, we proposed a learning framework that leverages QPP to improve the overall performance of any typical TTG system that starts with the retrieval step.

The rest of the paper is organized as follows. In Sect. 2, we summarize the related work. We define the problem in Sect. 3. The proposed approach is introduced in Sect. 4. The experimental results are presented and discussed in Sect. 5 before we conclude and give some directions of future work in Sect. 6.

2   Related Work

Several research studies have targeted the TTG problem and the more general tweet summarization problem; many were part of the TTG task in TREC 2014 microblog track. There were also studies that investigated the performance prediction of summarization systems. We touch upon those related directions in this section.

2.1 Tweet Timeline Generation

Lv et al. [16] designed a TTG system that first applies hierarchical clustering on the top k tweets from a ranked list retrieved for a query. The timeline is then composed of the highest-scoring tweet from each cluster. Value of k was determined differently per query using a *static* retrieval score threshold. Our method allows both different depth and score cutoffs across queries.

Xu et al. [22] applied set-based novelty detection over the top k tweets retrieved. A tweet is added to the novel set (and timeline) if its similarity with any of the tweets in a sequentially-updated novel tweet set is below a threshold. Magdy et al. [17] used 1NN clustering of the top k tweets to generate a timeline. Their results show that the choice of the value of k can affect the TTG performance. Similarly, Xu et al. [22] tuned the parameter k for their TTG system, indicating that it had an effect on the performance.

Xiaohui et al. [4] also used the idea of clustering in TTG, but looking at tweets in a different way. They compute what they call a sequential pattern over each tweet in an initially retrieved list of tweet. The pattern captures term co-occurrence and semantics in the tweet. Once patterns are computed, the system clusters tweets with similar patterns together and select the tweet with highest retrieval score from each cluster to be added to the timeline.

2.2 Tweet Summarization

Shou et al. [19] employed an online incremental clustering algorithm with data structures designed to maintain important cluster information as the tweet stream evolves. Their system allowed for creating summarizes in two modes: online where summaries are created based on current clusters in memory, and historical which creates summaries based on history of clusters maintained in a data structure called Pyramidal Time Frame. In both modes, the system uses an algorithm that constructs a cosine similarity graph between tweets in all clusters then applies the LexRank method [8] to select most novel and central tweets to add to the summary.

In a more recent work, Chen et al. [5] worked with tweets retrieved by a search model which is similar to TTG but for the problem of tweet summarization. In their system, they classify tweets in the full list into genres and proceed to summarize tweets in each genre. They re-enforce the per-genre list of tweets by retrieving documents from the Web using a search query composed of terms selected from that list of tweets. Each terms in the tweet list is weighted using a measure that focuses on the authority of authors of tweets in which this term appeared. Once those Web documents are retrieved, they are split into sentences and added to the list of tweets (i.e., creating artificial tweets). A graph-based summarizer is then applied and top-scoring sentences/tweets are selected to create the summary.

2.3 Predicting Summarization Performance

In another direction, Louis and Nenkova investigated the prediction of summarization performance in the context of classical summarization tasks. They attempted to use predictors computed on input documents to predict the performance of summarization systems [15]. Some of these predictors are usually used in predicting query performance in adhoc search tasks. They ran experiments using single- and multi-document summarization and evaluated their approach over a large set of training/testing datasets. Their results showed promising correlation between predicted and actual summarization system performance. This encouraged us to consider query performance prediction in the context of TTG, not to predict TTG performance but to improve it by predicting the optimal cutoff of a retrieved ranked list of tweets to start with.

Another line of studies investigated finding the optimal cutoff for the ranked list of results, but for the purpose of effective re-ranking techniques [2,12]. However, that work focused on finding the optimal *global* cutoff over all queries. In our case, we predict the optimal cutoff *per* query.

3 Problem Definition

Given a query q posted at time t_q and a tweet collection C, TTG aims at generating a timeline T of *non-redundant* tweets that are *relevant* to q and posted prior to t_q. A TTG framework is usually composed of a retrieval component (represented by a *retrieval model*) that provides a ranked list R of tweets that are potentially-relevant to q, and a timeline generation (TG) component (represented by a *TTG model*) that accepts the list R and generates the tweet timeline T extracted from the top k tweets in R. We address the problem of improving the quality of T generated by *a given TTG system* by optimizing the cutoff value k.

4 Approach

We formulate the problem as a learning problem. Given a query q and, a ranked list R of tweets retrieved by a retrieval model, and a TTG model, we aim to learn a *regression* model that estimates (or *predicts*) a cutoff value k applied to R that is needed to optimize the TTG performance for q. k determines depth of R to be used in generating the timeline.

4.1 Features

To train the regression model, we propose to leverage the idea of query performance prediction (QPP). We compute a predictor for the query at m different cutoff (i.e., depth) values applied to R, resulting in m predicted values that together constitute the feature set. Each predicted value (i.e., feature) is an estimation of the retrieval quality for q at the corresponding cutoff. Similarly, a feature vector can be generated for each query in a query set Q, yielding a set of

Q feature vectors generated using the same retrieval model. Moreover, since different retrieval models can theoretically retrieve different ranked lists of tweets given the same query, the regression model can be trained using feature vectors generated using different retrieval models, which results in a larger set of feature vectors.

Query performance predictors are usually computed using a list of documents retrieved in response to the query using a retrieval model [3]. Several predictors were proposed in microblog search context [18] in addition to those proposed in other domains like news and Web search [6,7,20].

We experimented with **10 predictors** including the most effective ones in microblog search, in addition to effective predictors typically used with non-microblog collections. We selected microblog-specific predictors computed based on the following two measures of the topical-focus of R_k, the list composed of the top k tweets in R. Each measure is computed *per tweet* in R_k [18]:

1. Query Terms Coverage (QTC): QTC computes the *coverage* of query terms in the tweet, i.e., the number of query terms appeared in the tweet, and normalize it by the length of the query.
2. Top Terms Coverage (TTC): Given the n most frequent terms in R_k, TTC measures the coverage of these terms in the tweet normalized by n.

Once a measure is computed over each tweet in R_k, we compute mean, median, lower percentile, and upper percentile of QTC/TTC values over all tweets in R_k. Each one of these statistics represent a predictor. We also experimented with variants to these predictors using inverse document frequencies (IDF) of terms when computing the coverage.

As for typical predictors, we used the normalized query commitment (NQC) due to its reported effectiveness over different test collections [20].

4.2   Retrieval Models

We used retrieval approaches covering a large spectrum of effective techniques usually used in microblog search. We group these approaches into the following main groups:

- Standard query-likelihood (QL).
- Query Expansion (QE) models based on Pseudo Relevance Feedback (PRF).
- QE that benefits from web resources to select the expansion terms (QEW).
- Learning-to-rank (L2R)-based models. L2R models can also be combined with other models such as QE.
- Temporal (TEMP) models that emphasize temporality of the data (tweets) and the adhoc search task when performing retrieval.

In total, we used **14** retrieval models. We acquired ranked results (i.e., tweets) retrieved by these models using the 55 queries and dataset provided by the TREC-2014 TTG task (further details in Sect. 5) from 3 participated teams [9,17,22]. We evaluated the performance of the models using mean average precision (MAP), which is the commonly-used measure to evaluate microblog adhoc search [14]. MAP was computed over the top 500 tweets per query. We summarize the models used in Table 1.

Table 1. Summary of retrieval models used

Group	ID	MAP	Group	ID	MAP
QL	QL1 [22]	0.385	QE+L2R	QEL [17]	0.470
	QL2 [9]	0.398		HQEL [17]	0.482
QE	QE1 [17]	0.464		WQEL [22]	0.497
	QE2 [9]	0.466		WQETL [22]	0.571
	QE3 [9]	0.456	TEMP	TDC [9,13]	0.406
	QE4 [9]	0.490		TRM [9,11]	0.445
QEW	HQE [17]	0.477			
	WQE [22]	0.485			

4.3 TTG Models

We worked with TTG models selected from existing literature based on their reported effectiveness, while attempting to diversify the types of considered models. We considered models based on two main concepts:

– Topical Clustering. TTG and tweets summarization systems based on topical clustering were effective in related studies [9,17,19]. These models create topical clusters for the input tweets assuming that each cluster reflects a sub-topic of the main topic. Some tweets from created clusters are selected to form the timeline based on several factors including: (a) which clusters to represent in the timeline, (b) number of tweets to select from each cluster and (c) a selection criterion that minimizes redundancy in the timeline.
– Temporal clustering [1]. It groups tweets of the input set (viewed as a stream) considering temporal signals, usually extracted based on posting time of tweets. The timeline is generated based on selecting tweets from these clusters considering the same factors as in topical clustering.

We implemented and experimented with the following effective, existing TTG models. Specifically, we implemented *1NN* and *Centroid* which were the top **second** and **fourth** TTG systems (respectively) in TREC-2014 TTG task of the microblog track. Additionally, we implement a temporal TTG system.

– *Centroid* [9]: This model incrementally clusters the input tweet list by computing similarity between the tweet and centroids of clusters updated with each new tweet. The tweet with the maximum retrieval score is used as the centroid of a cluster and it is included in the final timeline.

- *1NN* [17]: Similar to *Centroid*, tweets are incrementally clustered, but model only adds a tweet to a cluster if its similarity to any tweet in a cluster exceeds a threshold, earliest tweet in each cluster is added to the timeline.
- *Z-Score* [1]: This is a model that considers temporal signals in tweets. The model creates fixed-length time buckets of tweets given the ranked list sorted chronologically. Each term in each bucket is scored using the Z-Score–a measure designed to help detect spiking terms in a bucket. A tweet in a bucket is scored by summing the Z-Scores of all of its terms and the tweet with the maximum score is included in the timeline.

4.4 Regression Models

We combine predictors computed at different cutoffs using Weka's[2] implementation of two regression models: typical linear regression and the M5P algorithm that is based on generating model trees [21]. For the M5P model, we experimented with both pruned and un-pruned trees.

5 Experimental Evaluation

5.1 Experimental Setup

Dataset. In our experiments, we used TREC-2014 microblog track test collection, which includes access to Tweets2013 of 243 Million tweets and 55 queries [14]. For simplicity, we assume that we experiment with H adhoc retrieval models, T TTG systems, and Q queries. Though the dataset used is relatively small, we increase the size of training examples in our ground truth by using a large set of adhoc models as discussed next.

Generating Ground Truth. To generate our ground truth, we pre-identified the optimal retrieval cutoff for each query for each retrieval model by changing the cutoff from 1 to 500 (with step 1 with cutoffs 1–100 and step 10 with 110–500[3]), and identifying the one maximizing TTG performance. We repeat that for each TTG system. The optimal cutoff values represent the target function that the regression model is learning. Overall, the ground truth includes $T * H * Q$ samples (i.e., feature vectors).

We adopted weighted F_1 (denoted by wF_1) performance evaluation measure that was the official measure in TREC-2014 [14]. wF_1 combines precision and recall of the TTG system over a set of semantic clusters of relevant tweets to query q. The retrieved clusters are weighted by the number of tweets in each cluster and the tweets themselves are weighted by their relevance grade where "relevant" tweets get a weight of one and "highly-relevant" tweets get a weight of two. Additionally, we report overall system performance by averaging per topic wF_1 over all queries to compute average wF_1, which is a more accurate way than the one used to compute reported TTG results in [14].

---

[2] http://www.cs.waikato.ac.nz/ml/weka/.

[3] We observed that TTG performance is less sensitive to change in list depth with large cutoffs.

Training and Testing Data. For training and testing our regression model, we adopted *leave-one-query-out* cross validation. For each TTG system, we train our model on $Q - 1$ queries and test it on the remaining unseen one query. Since we have H retrieval models, for each query, we train a regression model using $(Q - 1) * H$ training samples and test it over $1 * H$ testing samples. The trained model is used to *predict* the cutoff value for each unseen sample, which is eventually used by the TTG system to generate a corresponding timeline. We repeat that process Q times.

Baseline. We compare the performance of our proposed approach with a baseline that applies one optimal *static* cutoff to all queries. This baseline follows a similar approach used by the best team at TREC TTG task to select the list depth by learning a static *score cutoff* over training queries [16]. Our baseline system learns the optimal static cutoff using a leave-one-query-out approach that is similar to the one described above for the regression model. It is trained over the same training set (i.e., $Q - 1$ queries) used to train the regression model by changing the cutoff in the same way as above and picking the cutoff that maximizes the average TTG performance while using it for all queries in the training set. We then apply the learned cutoff to the remaining testing query. This process is followed independently on each retrieval model and on each TTG system. We believe this is a strong baseline as it benefits from the training data to learn an optimal, but static, retrieval cutoff.

Statistical-significance testing in our experiments was performed using two-tailed paired t-test, with $\alpha = 0.05$.

5.2   Results and Discussion

We studied 10 sets of features (one for each predictor) and two different regression models. Additionally, using both regression models, we also attempted to combine sets of features in an attempt to combine predictors used, but that resulted in poorer performance with some TTG systems compared to using single predictors. We suspect this is because of the small training/testing dataset we have, impeding learning a regression model over such large set of features.

Due to space limitation, we only report the results of the best performing setup using pruned **M5P** model trees and the feature set based on the IDF-variant of lower percentile of TTC values described in Sect. 4.

Averaging percent-improvement that our method achieved over the baseline on all queries and retrieval models, our proposed approach improved for all of the 3 TTG models. The overall percent-improvement ranges from 3.2 % with the *Z-Score* model to 6.8 % with *1NN* model; surprisingly, those models were the worst and best performing respectively, according to the baseline results, among the models we used.

Table 2 shows improved wF_1 (denoted by wF_1^*) and the percent-improvement over baseline for each retrieval model used with each of the TTG models. In 32 out of 42 cases, our proposed approach improved over the baseline, reaching

Table 2. Baseline wF_1 for each TTG model with each retrieval model vs. improved wF_1 (wF_1^*) along with the percent-improvement over baseline. Bold improvements are statistically significant.

Retrieval model	Centroid		1NN		Z-score	
	wF_1	$wF_1^*(\%)$	wF_1	$wF_1^*(\%)$	wF_1	$wF_1^*(\%)$
QL1	0.3372	0.3783(**+12.2**)	0.3701	0.3697(−0.1)	0.3450	0.3204(−7.1)
QL2	0.3757	0.3809(+1.4)	0.3854	0.4102(+6.4)	0.3002	0.3373(**+12.4**)
QE1	0.3844	0.3898(+1.4)	0.3847	0.4117(+7.0)	0.3540	0.3754(+6.0)
QE2	0.3374	0.3779(**+12.0**)	0.3464	0.3814(+10.1)	0.2973	0.3405(**+14.5**)
QE3	0.3486	0.3659(+5.0)	0.3634	0.3673(+1.1)	0.3450	0.3204(−3.2)
QE4	0.3684	0.3774(+2.4)	0.3644	0.4057(+11.3)	0.3137	0.3587(**+14.3**)
HQE	0.3730	0.4069(**+9.1**)	0.3917	0.4282(**+9.3**)	0.3315	0.3752(**+13.2**)
WQE	0.3764	0.3748(−0.4)	0.3417	0.3774(+10.5)	0.3646	0.3621(−0.7)
QEL	0.3927	0.3928(+0.0)	0.3979	0.4105(+3.2)	0.3240	0.3801(**+17.3**)
HQEL	0.3962	0.4193(+5.8)	0.4089	0.4368(+6.8)	0.3709	0.3825(+3.1)
WQEL	0.3711	0.3672(−1.1)	0.3776	0.3954(+4.7)	0.3458	0.3602(+4.2)
WQETL	0.4068	0.3979(−2.2)	0.4011	0.4149(+3.4)	0.3965	0.3529(−11.0)
TDC	0.3777	0.3789(+0.3)	0.3845	0.4160(+8.2)	0.3561	0.3311(−7.0)
TRM	0.3326	0.3685(+10.8)	0.3219	0.3710(**+15.2**)	0.3129	0.2999(−4.2)

up to 17 % increase in wF_1; 10 of those cases were statistically significant, while none of the cases where the performance dropped was statistically significant. Furthermore, the results demonstrate the strength of our method as it managed to improve both systems with low and high TTG performance.

Finally, we went a step further and studied the relationship between the difference in performance (between the proposed approach and the baseline) and the optimal cutoff value per query. We compared the average optimal cutoff over H retrieval runs and the average percent-improvement over the baseline over the same H runs for different queries and per TTG system. The results showed that almost all queries of which performance was degraded have an average optimal cutoff value that is ≤ 100. Moreover, at least 9 of the top 10 degraded queries (i.e., the ones with largest degradation) have an average optimal cutoff value ≤ 50 in all TTG systems. This is logical as the error in low cutoffs has a larger effect on performance than in large cutoffs, because tweets at higher ranks are potentially more relevent and therefore missing them becomes more costly. Another possible reason is the general sensitivity of query performance predictors to the depth of the retrieved list. Since the predictors (used as features) were not tuned per retrieval model, it might produce poor results at shallow lists in some cases. This indicates that more attention should be given to features at the first 100 cutoffs and possibly to a regression model that penalizes errors in queries of low optimal cutoffs than in those of higher cutoffs. We also notice that almost all queries of high optimal cutoffs ($\geq$150) were improved in all TTG systems.

6   Conclusion and Future Work

In this work, we used query performance predictors to predict the optimal depth of retrieval ranked list to use with a TTG system. Our results showed the effectiveness of this method in improving performance of 3 sample different TTG systems across 14 different retrieval approaches. Out of 42 different cases, 32 were improved with 10 of them had significant improvement, while in only 10 cases TTG effectiveness was degraded but insignificantly.

For future work, more analysis of failure instances is needed especially for instances where optimal cut-offs are low. Other performance predictors can also be tried and we plan to experiment with more regression models. Another evident direction is to study how good the predictors are in predicting actual retrieval performance and how is that related to their performance in predicting optimal cutoff for TTG. With larger test collction (more importantly larger set of queries), extensive experiments can be conducted for more concrete results.

Acknowledgments. This work was made possible by NPRP grant# NPRP 6-1377-1-257 from the Qatar National Research Fund (a member of Qatar Foundation). The statements made herein are solely the responsibility of the authors.

References

1. Alonso, O., Shiells, K.: Timelines as summaries of popular scheduled events. In: Proceedings of the 22nd International Conference on World Wide Web Companion, pp. 1037–1044. WWW 2013 Companion (2013)
2. Arampatzis, A., Kamps, J., Robertson, S.: Where to stop reading a ranked list?: threshold optimization using truncated score distributions. In: Proceedings of the 32nd International ACM SIGIR Conference on Research and Development in Information Retrieval, pp. 524–531. SIGIR 2009 (2009)
3. Carmel, D., Yom-Tov, E.: Estimating the Query Difficulty for Information Retrieval. Synthesis Lectures on Information Concepts, Retrieval, and Services **2**(1), 1–89 (2010)
4. Chen, X., Tang, B., Chen, G.: BUPT_pris at TREC 2014 microblog track. In: TREC 2014 (2014)
5. Chen, Y., Zhang, X., Li, Z., Ng, J.P.: Search engine reinforced semi-supervised classification and graph-based summarization of microblogs. Neurocomputing **152**, 274–286 (2015)
6. Cronen-Townsend, S., Zhou, Y., Croft, W.B.: Predicting query performance. In: Proceedings of the 25th Annual International ACM SIGIR Conference on Research and Development in Information Retrieval, SIGIR 2002, pp. 299–306 (2002)
7. Cummins, R.: Predicting query performance directly from score distributions. In: Salem, M.V.M., Shaalan, K., Oroumchian, F., Shakery, A., Khelalfa, H. (eds.) AIRS 2011. LNCS, vol. 7097, pp. 315–326. Springer, Heidelberg (2011)
8. Erkan, G., Radev, D.R.: Lexrank: graph-based lexical centrality as salience in text summarization. J. Artif. Intell. Res. **22**(1), 457–479 (2004)
9. Hasanain, M., Elsayed, T.: QU at TREC-2014: online clustering with temporal and topical expansion for tweet timeline generation. In: TREC 2014 (2014)

10. Hasanain, M., Malhas, R., Elsayed, T.: Query performance prediction for microblog search: a preliminary study. In: Proceedings of the First International Workshop on Social Media Retrieval and Analysis, SoMeRA 2014, pp. 1–6 (2014)
11. Keikha, M., Gerani, S., Crestani, F.: Time-based relevance models. In: Proceedings of the 34th International ACM SIGIR Conference on Research and Development in Information Retrieval, SIGIR 2011, pp. 1087–1088 (2011)
12. Lan, Y., Niu, S., Guo, J., Cheng, X.: Is top-k sufficient for ranking? In: Proceedings of the 22nd ACM International Conference on Conference on Information and Knowledge Management, CIKM 2013, pp. 1261–1270 (2013)
13. Li, X., Croft, W.B.: Time-based language models. In: Proceedings of the Twelfth International Conference on Information and Knowledge Management, CIKM 2003, pp. 469–475 (2003)
14. Lin, J., Efron, M., Wang, Y., Garrick, S.: Overview of the TREC-2014 microblog track (notebook draft). In: TREC 2014 (2014)
15. Louis, A., Nenkova, A.: Performance confidence estimation for automatic summarization. In: Proceedings of the 12th Conference of the European Chapter of the Association for Computational Linguistics, EACL 2009, pp. 541–548 (2009)
16. Lv, C., Fan, F., Qiang, R., Fei, Y., Yang, J.: PKUICST at TREC 2014 microblog track: feature extraction for effective microblog search and adaptive clustering algorithms for TTG. In: TREC 2014 (2014)
17. Magdy, W., Gao, W., Elganainy, T., Zhongyu, W.: QCRI at TREC 2014:applying the KISS principle for the TTG task in the microblog track. In: TREC 2014 (2014)
18. Rodriguez Perez, J.A., Jose, J.M.: Predicting query performance in microblog retrieval. In: Proceedings of the 37th International ACM SIGIR Conference on Research and Development in Information Retrieval, SIGIR 2014, pp. 1183–1186 (2014)
19. Shou, L., Wang, Z., Chen, K., Chen, G.: Sumblr: Continuous summarization of evolving tweet streams. In: Proceedings of the 36th International ACM SIGIR Conference on Research and Development in Information Retrieval, SIGIR 2013, pp. 533–542 (2013)
20. Shtok, A., Kurland, O., Carmel, D., Raiber, F., Markovits, G.: Predicting query performance by query-drift estimation. ACM Trans. Inf. Syst. $30(2)$, 11:1–11:35 (2012)
21. Wang, Y., Witten, I.H.: Induction of model trees for predicting continuous classes. In: Proceedings of the 9th European Conference on Machine Learning Poster Papers, ECML 1997 (1997)
22. Xu, T., McNamee, P., Oard, D.W.: HLTCOE at TREC 2014: microblog and clinical decision support. In: TREC 2014 (2014)

Abstract Venue Concept Detection from Location-Based Social Networks

Yi Liao[1]([⊠]), Shoaib Jameel[2], Wai Lam[1], and Xing Xie[3]

[1] Department of Systems Engineering and Engineering Management,
The Chinese University of Hong Kong, Hong Kong, China
{yliao,wlam}@se.cuhk.edu.hk
[2] School of Computer Science and Informatics, Cardiff University, Cardiff, Wales
jameels1@cardiff.ac.uk
[3] Microsoft Research, Beijing, China
xingx@microsoft.com

Abstract. We investigate a new graphical model that can generate latent abstract concepts of venues, or Point of Interest (POI) by exploiting text data in venue profiles obtained from location-based social networks (LBSNs). Our model offers tailor-made modeling for two different types of text data that commonly appears in venue profiles, namely, tags and comments. Such modeling can effectively exploit their different characteristics. Meanwhile, the modeling of these two parts are tied with each other in a coordinated manner. Experimental results show that our model can generate better abstract venue concepts than comparative models.

Keywords: Location-based social networks · Abstract venue concept · Graphical model

1 Introduction

With the advent of online social networks such as Facebook, Foursquare, etc., geo-tagging has become a popular activity online where people broadcast their location [1,2]. Such geo-tagged information can be useful to advertisers who want to recommend a venue or a product based on the user's past movement patterns or construct more interesting lifestyle patterns as proposed in [3], where rather than estimating lifestyles from the check-in data directly, we can first convert each check-in to an abstract data, in which the specific name of the venue can be replaced by an abstract name. This will help mitigate sparsity problem to a large

The work described in this paper is substantially supported by grants from the Research Grant Council of the Hong Kong Special Administrative Region, China (Project Codes: 413510 and 14203414) and the Microsoft Research Asia Urban Informatics Grant FY14-RES-Sponsor-057. This work is also affiliated with the CUHK MoE-Microsoft Key Laboratory of Human-centric Computing and Interface Technologies.

© Springer International Publishing Switzerland 2015
G. Zuccon et al. (Eds.): AIRS 2015, LNCS 9460, pp. 147–157, 2015.
DOI: 10.1007/978-3-319-28940-3_12

extent. We present a novel generative probabilistic model that exploits textual information obtained from location-based social networks (LBSNs) to uncover meaningful latent concepts related to venues, or Point of Interest (POI). We call such automatically discovered concepts "abstract venue concepts". For example, a concept representing "upscale hotels" may be discovered and it may contain representative terms such as "five-star", "luxury", "expensive", etc. Abstract venue concepts enable semantic characterization of venues, facilitating a better understanding of venues for both users and service providers, which could potentially benefit services such as venue recommendation [4]. While we could use the categories provided by Foursquare and similar services, the taxonomies which are used by these LBSNs are not always sufficiently fine-grained. For example, by investigating the category tree[1] of Foursquare, we can easily observe that the LBSN assigns all types of hotels the to same category *hotel*, rather than distinguishing finer properties of the hotel, such as "upscale hotel". Moreover, these taxonomies are LBSN-specific, which causes problems when we want to integrate check-in data from different LBSNs.

Text data obtained from LBSNs has also been used in geographical discovery such as [5–7]. Kim et al. [8] recently applied Latent Dirichlet Allocation (LDA) [9] to elicit the semantic concepts of venues with aggregated text data from venue profiles. However, we observe that the text data in venue profiles originates from two different sources. The first source is *tags*, which is a set of discrete terms describing the intrinsic properties of the venue, e.g. "hotels", "shopping mall", etc. Tags are usually drawn from a relatively fixed word lexicon. The second source is *comments*, which are sentences written by users expressing their opinions about the venue. The linguistic property of comments is rather different from tags in that it consists of natural expressions which can be grammatical or ungrammatical, written by any users. Table 1 shows the venue document of the Ritz-Carlton, an upscale hotel located in Hong Kong. It contains the whole tag set and an example of comments.

One novelty of our model is that we consider tags and comments separately, and our proposed model offers tailor-made modeling for these two kinds of text data, exploiting their different characteristics. Meanwhile, the modeling of these two parts are tied to each other in a coordinated manner which makes our approach considerably different from existing approaches. Experimental results obtained by our model are more superior than other comparative models.

Table 1. Venue document of the Ritz-Carlton, Hong Kong

Types	Value
tags	hotel, five-star, icc, international commerce centre, luxury
comments	"Amazing stay. Gorgeous design in every detail. Guests enjoy panoramic view of Hong Kong from all corners. Stunning views + over the top design. Great staffs!"

[1] https://developer.foursquare.com/categorytree.

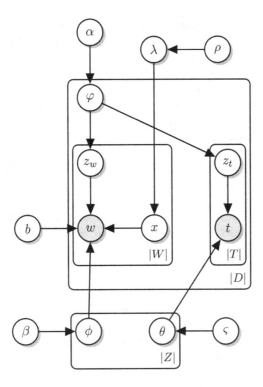

Fig. 1. The graphical model for abstract venue detection.

2   Abstract Venue Concept Detection

For each particular venue on a LBSN such as Foursquare, textual information, namely tags and user comments, are aggregated into a single document, called venue profile document. The category names of the given venue are also included as tags in our model. As a result, a venue profile document is composed of a bag of *tags* and a bag of *words* extracted from user comments.

The graphical model for our proposed abstract venue concept detection is depicted in Fig. 1. As mentioned in Sect. 1, one characteristic of our model is that it exploits different characteristics of words and tags, and offers tailor-made modeling for each of them. At the same time, the modeling of words and the modeling of tags are tied to each other in a coordinated manner. Precisely, an abstract venue concept is modeled as a probability distribution of tags, denoted by θ and a probability distribution of words, denoted by ϕ. Tags and words may have different vocabularies. The variable $|Z|$ denotes the number of abstract venue concepts.

Let D denote the set of venue profile documents. The outermost big plate in our graphical model represents a venue profile document, which contains a set of words, denoted by W and a set of tags, denoted by T. The number of words and tags is denoted by $|W|$ and $|T|$ respectively. Each venue profile document is associated with a distribution of abstract concepts, denoted as φ. φ is assumed to be drawn from a Dirichlet distribution with a hyper-parameter α. φ has two components, namely tag concept assignment denoted as z_t, and word concept assignment denoted as z_w. Each tag t is associated with z_t. θ captures the distribution of tags for the concept represented by z_t.

Words in user comments are modeled in a different manner due to the different characteristics of user comments compared with tags. It is common that user comments may contain some unrelated content which has no relationship with the abstract venue concept at all. We employ a background distribution to model the general words in user comments, denoted by the variable b, which shares some resemblances with the modeling paradigm in [10]. The dimension of the variable b is the total number of words in the word vocabulary. User comments are treated as mixture of words in the background and words related to the abstract venue concept. Thus each word w is associated with either z_w or b, which is governed by a binary variable x. x is associated with a Bernoulli distribution λ with parameter ρ. The generative process of our model can be written as:

1. Draw φ from **Dirichlet**(α) and λ from **Beta**(ρ)
2. For each abstract venue concept
    i. Draw global tag distribution θ from **Dirichlet**(ς)
    ii. Draw global word distribution ϕ from **Dirichlet**(β)
3. For each venue profile document
    i. For each word $w \in W$ in the aggregated user comment
        a. Draw the word concept assignment z_w from **Multinomial**(φ)
        b. Draw switch x from **Bernoulli**(λ)
        c. Draw w from ϕ_{z_w} if $x = 1$, otherwise draw from b
    ii. For each tag $t \in T$
        a. Draw the tag concept assignment variable z_t from **Multinomial**(φ)
        b. Draw the tag t from θ_{z_t}

We use Gibbs sampling to compute the approximate posterior in our model. Let $|R_w|$ denote the number of tokens in the vocabulary built from user comments. Let $|R_t|$ denote the number of tokens in the vocabulary built from venue tags. Let $|B|$ denote the number of tokens in the background corpus. Let β_w denote an element in the hyper-parameter vector related to the word w. Let $n_{z_w w}$ denote number of times a word w in the user comment has been sampled from the abstract venue concept z_w. Similarly, let ς_t denote an element in a vector ς. Let $n_{z_t t}$ denote number of times a tag t in the tag vocabulary has been sampled from the abstract venue concept z_t. Let α_z represent accessing an element in the hyper-parameter vector α. Let q_{z_t} and q_{z_w} denote the number of times a global abstract venue concept has been sampled in a venue profile document. Note that when we have excluded the counts of the current case in our sampling equations. Let $\Theta = \{\boldsymbol{w}, \boldsymbol{t}, \beta, \alpha, \rho, \varsigma\}$. The complete likelihood of the model is denoted in Eq. 1.

$$P(z_w, z_t, w, x, t | \alpha, \beta, \varsigma, \rho) = \int P(z_w | \varphi) P(\varphi | \alpha) d\varphi \cdot \int P(z_t | \varphi) P(\varphi | \alpha) d\varphi. \quad (1)$$

$$\int P(x | \lambda) P(\lambda | \rho) d\lambda \cdot P(w | x, z_w, \beta).$$

$$\int P(t | z_t, \theta) P(\theta | \varsigma) d\theta$$

Where

$$P(w | x = 1, z_w, \beta) = \int P(w | z_w, \phi) P(\phi | \beta) d\phi \quad (2)$$

$$P(w | x = 0, z_w, \beta) = b_w \quad (3)$$

Eqs. 4, 5 and 6 depict the formulations used in our Gibbs sampler.

$$P(z_t | \Theta) \propto \frac{\alpha_{z_t} + q_{z_t}}{\sum_{k=1}^{|Z|} (\alpha_k + q_k) - 1} \cdot \frac{\varsigma_t + n_{z_t t}}{\sum_{r=1}^{|R_t|} (\varsigma_r + n_{z_t r})} \quad (4)$$

$$P(z_w | x = 1, \Theta) \propto \frac{\alpha_{z_w} + q_{z_w}}{\sum_{k=1}^{|Z|} (\alpha_k + q_k) - 1} \cdot \frac{\beta_w + n_{z_w w}}{\sum_{r=1}^{|R_w|} (\beta_r + n_{z_w r})} \quad (5)$$

$$P(z_w | x = 0, \Theta) = \frac{b_w}{\sum_{k=1}^{|B|} b_k} \quad (6)$$

After sampling sufficient number of times, the parameters θ and ϕ are calculated with Eqs. 7 and 8.

$$\theta = \frac{\varsigma_t + n_{z_t t}}{\sum_{r=1}^{|R_t|} (\varsigma_r + n_{z_t r})} \quad (7) \qquad \phi = \frac{\beta_w + n_{z_w w}}{\sum_{r=1}^{|R_w|} (\beta_r + n_{z_w r})} \quad (8)$$

3 Labeling Abstract Venue Concepts

After the abstract venue concepts are detected, the next component is to automatically select one label to semantically describe the meaning of each concept.

For a discovered concept, the output from our model are a ranked list of tags and a ranked list of words from comments, obtained from matrices θ and ϕ respectively. The terms in the list coherently describe one venue concept. However, due to the intrinsic difference of tags and comments as shown in Sect. 1, these two lists generally contain some similar as well as different terms. For example, consider an abstract venue concept representing colleges, the corresponding abstract venue concept distribution for tags may consist of terms such as "library", "electronics", "college", "bookstore". Whereas the distribution of words in the abstract venue concept from the user comments may consist of "nice", "library", "excellent", "awesome", etc. Our objective is to automatically select representative tokens, such as "college" to serve as labels for that concept.

We adopt a technique based on the average Pointwise Mutual Information (PMI) described in [11], which also uses the same technique for finding topic labels. The value of PMI for a pair of words w_i and w_j is calculated with Eq. 9, where $P(w_i, w_j)$ denotes the probability of observing both w_i and w_j in the same list. $P(w_i)$ and $P(w_j)$ are the overall probability of token w_i and w_j respectively. Then the average PMI is calculated by averaging over all the tokens in the list, denoted by Eq. 10. PMI measures the association between one event to other events using information theory and statistics. In our case, intuitively, tokens that has more co-occurrence with other tokens will get higher PMI. For each discovered concept, we basically choose two concept labels that have the highest average PMI from the tag list and word list. We select the word with the highest avgPMI from the two ranked lists discovered by our model.

$$PMI = \log \frac{P(w_i, w_j)}{P(w_i) \cdot P(w_j)} \tag{9}$$

$$avgPMI = \frac{1}{N} \sum_{j}^{N} PMI(w_i, w_j) \tag{10}$$

4   Experiments and Results

4.1   Datasets

We used the official Foursquare API to crawl text data related to tags and user comments corresponding to that venue. We crawled data from the following countries and in brackets we list the number of venue profile documents: (1) Australia (30,880), (2) Canada (50,063), (3) Hong Kong (5,282), (4) India (12,277), (5) Indonesia (302,725), (6) Singapore (18,082), and (7) USA (879,476). We selected those venues which had text content in both tags and user comments. Each venue document obtained from Foursquare contains several tags and up to 20 comments.

4.2   Comparative Models

We choose a range of comparative models including some state-of-the-art topic models. Specifically, we compare our proposed model denoted as "Our Model" with (1) Latent Dirichlet Allocation (LDA) model [9]. We compare with both variational inference [9] and collapsed Gibbs sampling based algorithms [12] denoted by vLDA and cLDA respectively. (2) Topical N-gram (TNG) [13,14] model which is a phrase discovery topic model, (3) Hierarchical Dirichlet Processes (HDP) topic model [15,16], which is a nonparametric extension to the LDA model, (4) Biterm topic model (BTM) [17,18], which is a topic model suited for short texts as most of the documents in our collection are short. We use publicly available source codes of all these models. We used the same parameter settings of these models as described in their respective works. We use fixed symmetric Dirichlet distributions in our model in which we set $\alpha = 0.5$, $\beta = 0.01$, $\varsigma = 0.01$. In addition, we fixed $\rho = 0.01$ in our model. All models are run for 1000 iterations. We combine user comments and tags in one document for the comparative methods as used in [8].

4.3   Concept Coherence Evaluation

The first evaluation measures the quality of concepts generated by the models. To enable large scale evaluation, we evaluate topical coherence using an automated technique called observed coherence model discussed in [19]. The idea is to automatically find out whether the list of tokens in each concept are semantically related, which in turn leads to better concept interpretability.

In all models, we varied the number of concepts from 10 to 200 in steps of 10 except the HDP model which automatically finds out the number of latent concepts. We run the topic models for five times due to randomization as adopted in [20]. Therefore, for each concept, each model was run for five times and the average coherence score was computed in each run. Then the macro-average coherence score was computed for all five runs. We then computed the average across different number of concepts from 10 to 200.

Table 2. Average coherence scores obtained for different models in different datasets. The higher the average coherence score, the better is the model.

	vLDA	cLDA	TNG	HDP	BTM	Our Model
Australia	0.120	0.150	0.110	0.090	0.002	**0.220**
Canada	0.020	0.040	0.006	0.001	0.005	**0.120**
Hong Kong	0.013	0.300	0.120	0.090	0.002	**0.210**
India	0.010	0.020	0.090	0.002	0.006	**0.200**
Indonesia	0.011	0.013	0.008	0.008	0.003	**0.320**
Singapore	0.009	0.020	0.020	0.005	0.003	**0.140**
USA	0.003	0.040	0.010	0.009	0.003	**0.110**

We present the results obtained from different models in Table 2. We see from the table that our model has obtained the best average coherence score with improvements that is statistically significant according to two-tailed test with $p < 0.01$ against each of the comparative models. One may argue that comparative models may perform better if we separately model user comments and venue tags as separate documents. We found that results obtained from such strategy are even worse due to the sparsity problem. Our model jointly models words from user comments along with other useful information from venues, leading to more coherent concepts which mitigates the sparsity issue. In addition, introducing the background distribution helps us get rid of many irrelevant words which were dominant in many comparative models. Presenting only the average performance for all topics hides the per-topic performance of a model, but it must be noted that our model performs consistently better than the comparative models at different number of topics.

4.4   Concept Label Evaluation

We also evaluate the quality of the concept labeling task. We hired five human annotators to give ratings to concept labels. In our annotation task, each concept was presented in the form of its top-20 words ranked with decreasing probability value, followed by suggested label for the concept. Since our model generates two word distributions each of which will have a label of its own. In order to reduce the cognitive load on the human annotators, we only gave them the list of five concepts from each model, i.e. $|Z| = 5$. For HDP, five concepts were randomly selected. We gave the ordinal scale rating questions to the annotators as described in [21]. The ratings range from 0 to 3. We considered the scores as voting given by the human annotators and computed the average score from all annotators for each model.

Table 3. Average ratings given by annotators. The higher the average score, the better is the model.

	vLDA	cLDA	TNG	HDP	BTM	Our Model
Australia	1.56	1.32	1.12	0.28	0.45	**2.80**
Canada	1.68	1.04	0.96	0.40	0.20	**2.60**
Hong Kong	1.04	1.20	1.04	0.40	0.20	**2.40**
India	1.16	0.76	1.36	0.16	0.30	**2.40**
Indonesia	1.40	1.04	0.72	1.08	0.42	**2.15**
Singapore	0.92	0.80	1.20	0.28	0.16	**2.75**
USA	0.76	1.00	0.92	0.60	0.22	**2.40**

We present the results in Table 3. We see from the results that our model has obtained the highest value compared with other models.

Specially, the standard LDA methods, i.e. vLDA and cLDA, which aggregate tags and comments and do not model them differently, perform worse than our model. This observation shows the advantage of our tailor-made modeling, which exploits different characteristics of tags and comments. In addition, the relatively worse performances of HDP, compared with LDA, show that the venue profile documents do not have distinct hierarchical properties. The BTM, which is suitable for short texts, fails to get better results, although most of the documents in our collection are short. TNG shows comparable performance with standard LDA.

4.5   Sample Concept Case Study

We present top 20 terms from some abstract venue concepts from our model discovered from the "Australia" dataset in Table 4. We have merged the top ten words from two lists output by our model, and arranged the words in decreasing order of their probability values in the list presented below. There are two terms,

Table 4. Top 20 terms for some concepts. Labels are in Bold font

Top 20 terms and labels
asian, **restaurant**, indonesian, noodles, ramen, chinese, seafood, pizza, food,caf, diner, japanese, arcade, bbq, soup, breakfast, steakhouse, sushi, italian
university, video, games, music, **library**, electronics, building, school, store, **college**, office, education, bookstore, books, academic, tv, dvd, bowling, camera
theater, movie, theatre, apparel, cineplex, music, arts, movies, concert, performing, bowling,**entertainment**, hall, art, winery, popcorn, venue, gallery, alley
park, **playground**, **outdoors**, golf, hotel, field, beach, baseball, dog pool, lake,trail, apartments, museum, run, boat, scenic, entertainment, lookout

one from each distribution, selected as labels which are in bold font. We see from the sample terms that our model has generated meaningful and coherent terms. For example, the second concept, which is labeled with "college" and "library", apparently represents a college. Most of the words are related to college where university students play video games in the residential buildings. They go to the library to read books, listen to music, watch television or DVD, etc.

Our model has generated more superior results than the comparative models because of the following reasons. First, the background distribution in our model helps get rid of many general words from the distributions. This helps focus on only relevant content words in the user comments. Comparative models such as LDA, TNG, etc. are not designed to handle this. Although we could adopt aggressive pre-processing methods and then input the pre-processed text to the comparative models, it involves manual labour to select such general terms and removing them. Automatically removing the general words from the corpus can also be adopted, for example, using term-frequency and inverse document frequency score and removing those words which have low scores. But this involves selecting an appropriate threshold value, and we need to expend some computing time too. We could have considered a background distribution in the comparative models by modifying the models slightly and their sampling algorithms, however, those models still lack the ability to separately model user comments and tags in order to generate high quality abstract venue concepts.

5   Conclusion

We have proposed a new model to generate abstract venue concepts from LBSNs venue profiles. Our model jointly models user comment text and tags. Meanwhile, the model offers tailor-made modeling for these two kinds of text data, exploiting their different characteristics.We conducted extensive experiments and found our model to be superior compared with comparative models in both the coherence of concept and the quality of labels.

References

1. Yuan, Q., Cong, G., Sun, A.: Graph-based point-of-interest recommendation with geographical and temporal influences. In: Proceedings of the 23rd ACM International Conference on Information and Knowledge Management, pp. 659–668 (2014)
2. Zhong, Y., Yuan, N.J., Zhong, W., Zhang, F., Xie, X.: You are where you go: inferring demographic attributes from location check-ins. In: Proceedings of the Eighth ACM International Conference on Web Search and Data Mining, pp. 295–304 (2015)
3. Yuan, N.J., Zhang, F., Lian, D., Zheng, K., Yu, S., Xie, X.: We know how you live: exploring the spectrum of urban lifestyles. In: Proceedings of the First ACM Conference on Online Social Networks, pp. 3–14 (2013)
4. Liu, B., Xiong, H.: Point-of-interest recommendation in location based social networks with topic and location awareness. In: Proceedings of the 2013 SIAM International Conference on Data Mining, pp. 396–404 (2013)
5. Wang, C., Wang, J., Xie, X., Ma, W.Y.: Mining geographic knowledge using location aware topic model. In: Proceedings of the 4th ACM Workshop on Geographical Information Retrieval, pp. 65–70 (2007)
6. Wang, X., Zhao, Y.L., Nie, L., Gao, Y., Nie, W., Zha, Z.J., Chua, T.S.: Semantic-based location recommendation with multimodal venue semantics. IEEE Trans. Multimedia **17**(3), 409–419 (2015)
7. Hong, L., Ahmed, A., Gurumurthy, S., Smola, A.J., Tsioutsiouliklis, K.: Discovering geographical topics in the twitter stream. In: Proceedings of the 21st International Conference on World Wide Web, pp. 769–778 (2012)
8. Kim, E., Ihm, H., Myaeng, S.H.: Topic-based place semantics discovered from microblogging text messages. In: Proceedings of the Companion Publication of the 23rd International Conference on World Wide Web Companion, pp. 561–562 (2014)
9. Blei, D.M., Ng, A.Y., Jordan, M.I.: Latent Dirichlet allocation. J. Mach. Learn. Res. **3**, 993–1022 (2003)
10. Chemudugunta, C., Steyvers, P.S.M.: Modeling general and specific aspects of documents with a probabilistic topic model. In: Proceedings of the 2006 Conference in Neural Information Processing Systems 19, vol. 19, p. 241 (2007)
11. Lau, J.H., Newman, D., Karimi, S., Baldwin, T.: Best topic word selection for topic labelling. In: Proceedings of the 23rd International Conference on Computational Linguistics: Posters, pp. 605–613 (2010)
12. Porteous, I., Newman, D., Ihler, A., Asuncion, A., Smyth, P., Welling, M.: Fast collapsed gibbs sampling for latent Dirichlet allocation. In: Proceedings of the 14th ACM SIGKDD International Conference on Knowledge Discovery and Data Mining, pp. 569–577 (2008)
13. Wang, X., McCallum, A., Wei, X.: Topical n-grams: Phrase and topic discovery, with an application to information retrieval. In: Seventh IEEE International Conference on Data Mining, pp. 697–702. IEEE (2007)
14. Wang, X., McCallum, A.: A note on topical n-grams. Technical report, DTIC Document (2005)
15. Teh, Y.W., Jordan, M.I., Beal, M.J., Blei, D.M.: Hierarchical Dirichlet processes. J. Am. Stat. Assoc. **101**(476), 1566–1581 (2006)
16. Teh, Y.W., Kurihara, K., Welling, M.: Collapsed variational inference for HDP. In: Advances in Neural Information Processing Systems, pp. 1481–1488 (2007)

17. Yan, X., Guo, J., Lan, Y., Cheng, X.: A biterm topic model for short texts. In: Proceedings of the 22nd International Conference on World Wide Web, pp. 1445–1456 (2013)
18. Cheng, X., Yan, X., Lan, Y., Guo, J.: BTM: topic modeling over short texts. IEEE Trans. Knowl. Data Eng. **26**(12), 2928–2941 (2014)
19. Lau, J.H., Newman, D., Baldwin, T.: Machine reading tea leaves: automatically evaluating topic coherence and topic model quality. In: Proceedings of the 14th Conference of the European Chapter of the Association for Computational Linguistics, p. 530 (2014)
20. Zhu, J., Ahmed, A., Xing, E.P.: MedLDA: maximum margin supervised topic models. J. Mach. Learn. Res. **13**(1), 2237–2278 (2012)
21. Lau, J.H., Grieser, K., Newman, D., Baldwin, T.: Automatic labelling of topic models. In: Proceedings of the 49th Annual Meeting of the Association for Computational Linguistics: Human Language Technologies, vol. 1, pp. 1536–1545 (2011)

Web Search

Snapboard: A Shared Space of Visual Snippets - A Study in Individual and Asynchronous Collaborative Web Search

Teerapong Leelanupab[✉], Hannarin Kruajirayu, and Nont Kanungsukkasem

Faculty of Information Technology, King Mongkuts Institute of Technology
Ladkrabang (KMITL), Bangkok 10520, Thailand
teerapong@it.kmitl.ac.th, hannarin_k@hotmail.com, k.nont@outlook.com

Abstract. People often engage in many search tasks that be collaborative, where two or more individuals work together with the joint information needs. We introduced and built *CoZpace*, a web-based application that enables a group of users to collaborate on searching the web. We also presented the main feature of CoZpace, named *Snapboard*, which is a shared board for a collection of group-created visual snippets. The visual snippet is a snapshot of focused and salient information captured by a user. It acts as a visual summarization of web pages, which allows any user to quickly recognize information and to revisit web pages. This paper describes example usage scenarios and initially investigates the ways Snapboard facilitates users in individual and asynchronous collaborative search. We then analyze users' interactions and discuss how Snapboard supports search collaboration among study participants.

Keywords: Collaborative web search · Visual snippet · Search user interface · Exploratory search · Collaborative information behavior

1 Introduction

Although technology behind search engines such as Google, Yahoo! Search has advanced over the years, the designs of their web search interfaces have still largely remained for solitary information seeking; they are typically created for one person to use independently. Relatively, little support is available for a group of people, who share the same information need, to collaborate on search tasks. Examples of such tasks include planning a holiday with family members, organizing a social event with friends, or working on an annual report with colleagues. Additionally, some search tasks are too complex or difficult to be completed by a sole person, such as finding possible areas where the MH370 crashed or listing past and ongoing missions of New Horizons space probe to Pluto. These tasks require input and expertise from others. In fact, information seeking is considered collaborative [16]; multiple users can carry out search by working together. Golovchinsky et al. [7] defined four dimensions of the collaboration model, i.e., intent (explicit vs. implicit), depth of mediation (from user-interface

© Springer International Publishing Switzerland 2015
G. Zuccon et al. (Eds.): AIRS 2015, LNCS 9460, pp. 161–173, 2015.
DOI: 10.1007/978-3-319-28940-3_13

only to deeper algorithmic mediation), concurrency (synchronous vs. asynchronous) and location (co-located vs. remote).

We created a web-based application, CoZpace, for collaborative web searching. It enhances users' awareness to understand the activities of others engaging in a cooperative effort through instant messaging, result assessment, activity history and so forth. Users are aware of the degree to success in their search task, benefit from the work done previously or concurrently by others, and can plan subsequent activities for querying the current set of retrieved web pages. Importantly, we also addresses a key investigated feature of CoZpace, namely Snapboard, which helps a group of search partners to identify relevant information and re-finding the right web pages. Snapboard collects all visual snippets captured by users during either asynchronous or synchronous search sessions. By putting the visual snippets all together within the Snapboard, we allow users to review important web contents in one place without losing their search context. Users can use the visual snippets to evaluate the relevance of web pages together if they each cannot conclude the relevance alone. Focused information can be promptly recognized and its source can be easily revisited.

The type of collaboration, supported by CoZpace, is explicit in that the intention of the users is to work together on search, and the mediation is mainly supported by features in the interface but not algorithms[1]. Although CoZpace can also support both asynchronous or synchronous collaboration, for the evaluations that we describe here we focus on asynchronous collaboration among users and intentionally leave other types and dimensions of search such as co-located and synchronous collaboration for future investigation. Also, we examine the individual use of Snapboard done by a single user. Built as a web application, CoZpace supports both co-located and remote search. Even though location is not our interest here, we assume in our study that users remotely collaborate where they are not allowed to talk with each other during the experiment. We conducted a laboratory-based user study to evaluate the supportiveness of *Snapboard* in individual and asynchronous collaborative search.

2 Related Works

2.1 "Active" Collaborative Search Systems

Much research in the area of collaborative search focuses on *active* forms of collaboration (e.g., searching together by being aware of search partners' activities, by being able to plan search tactics, and by working as a team to achieve a shared goal.) An example from this line of work includes a very early system, SearchTogether, which allows remote users to synchronously and asynchronously collaborate on search [13]. It keeps a common history of search queries entered by any group members. Each user is provided with her split search result tab

---

[1] Our previous work experimented an algorithm that generates query suggestions extracted from terms present in visual snippets [10].

and allowed to communicate with other group members via integrated messaging. Similar to SearchTogether, CollabSearch [20] supports both implicit and explicit communication. It has an additional workspace where group members can share the saved whole web pages and textual snippets generated by Google API. CoSearch [1] leverages additional available devices such as extra mice and mobile phones for co-located collaborative web search. It assists users in control and division of labor while jointly searching the web by a single computer.

WeSearch [14] is an interactive tabletop display for face-to-face collaboration. It is designed to support co-located collaborative web search, browsing, and sense-making among a group of up to four people. Perez et al. [15] present a CoFox system, which shows a live video stream of a remote user's search screen to a local user to increase awareness of synchronous search collaboration. In multimedia retrieval domain, a grouping interface for video search, ViGOR [8], is built to assist asynchronous collaboration between users. The main feature of ViGOR is the provision of a workspace for creating and organizing groups of related videos. Each group can have multiple annotations and be used as a starting point for further search queries.

2.2   Studies of Search Habits

Also referred to exploratory search [19], *informational* search is identified as a common web search activity that would benefit from collaboration [3]. In this class of searches[2] [5], a user aims to seek some information on one or more web pages. To achieve this, informational search potentially involves multiple refinements of query terms and often spans many search sessions, and those are what CoZpace is designed for. Yue et al. [20] analyzed transitions of user search actions under three different conditions by varying two search factors (i.e., collaborative vs. individual and with vs. without explicit communication.)

2.3   Summarization of Web Pages

Presenting search results as a summarization of each web page is considered useful for information search, where users can quickly judge which ones of these results are of their interest [18]. For re-finding information, users can also use this similar summarization to access previously visited web pages, usually saved in the bookmarks of a web browser [6]. For simplicity and compactness, the summarization of web pages is typically represented as a textual snippet, consisting of its page title, URL and short text summary of web contents. Nevertheless, reading textual summarization is often time-consuming and difficult to comprehend its information if the textual snippet is very short.

As the classic quote states "A picture is worth a thousand words", this is simply because images convey information that words cannot capture. As such,

---

[2] The remaining two classes of searches are *navigational* and *transactional*, where users aim to find a single specific website or to perform some web-mediated activity, respectively.

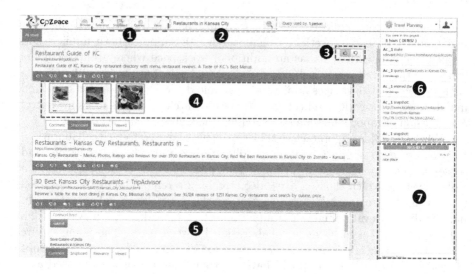

Fig. 1. Interface of CoZpace.

search and re-finding tasks can be facilitated if the web pages and, in particular, focused information are visually summarized since users can get a quick understanding by seeing an image than reading text. Visual summarization has been studied in several research communities [2,9,17] for search, revisitation and bookmarking of web pages.

3   Snapboard in Collaborative Search

3.1   CoZpace: A Collaborative Search System

CoZpace is a web-based application, which allows a group of remote users to create a collaborative search task. Within the created task, a user can invite and communicate with other members in the group. Figure 1 illustrates the interface of CoZpace that returns search results after a user submitted an example query, "Restaurants in Kansas City". Seven parts of the screen are described as follows:

1. The task summarization tab that includes four clickable image icons for: (i) relevance awareness, presenting three categories of judged web search results as indicated by any users in a group, i.e., relevant, non-relevant, and not sure[3], (ii) Task Snapboard collecting all group-captured visual snippets, (iii) query awareness, showing a history of all used queries, and (iv) view list, displaying all web search results clicked to view by any users, respectively;
2. The search bar for entering search query, which also suggests alternative queries and remind for the queries[4] that have already been used in the task;

---

[3] Web pages are marked as "not sure" if there is no consensus on relevance judgments/voting by group members.

[4] Already used queries are highlighted in yellow as part of query suggestion.

3. The relevant buttons for marking search results considered relevant (thumbs-up) or non-relevant (thumbs-down);
4. *Per-document Snapboard* showing multiple visual snippets captured from a single document/web page;
5. The comment bar for a user to leave comments about a website;
6. The project timeline showing real time stream of activities; and
7. The instant messaging for real-time discussion among the collaborators.

Note that all features in CoZapce are interactive and real-time; all interactions of group members are recorded and instantly shown according to their types of summarization. Furthermore, any search results that had been clicked to view by any users in a group will be highlighted in yellow. Our aim of providing the summarizations of different aspects is to facilitate awareness of a shared search task being pursued by other group members, so that unnecessary redundancy of effort can be avoided. For further details about the complete design and features of our CoZpace, we refer interested readers to [11].

3.2 Snapboard

Our Snapboard feature applies visual snippet which is an attractive representation and summarization of a web page, used in Human-Computer Interaction (HCI) to share a relevant part of a web page. However, our visual snippet is different from visual snippets in HCI that are a part of a web page captured by system which might not be relevant to user's information need. Our visual snippet allows a user, in a self-managed manner, to capture a part of focused or salient information in a web page by herself. Moreover, as Aula et al. [2] indicated that a thumbnail which shows the whole web page in a small picture can make users underestimate the relevance of the page and only textual summaries can make users overestimate, it makes us concern that using only a picture in visual snippet might face with the same problem as using thumbnail. Therefore, we allow to include both textual and visual information in our visual snippet.

Figure 2 displays an example of taking a snapshot of focused information as a visual snippet (1) and its visual snippet (2) generated from a template given a salient snapshot and other metadata of its source/web page, i.e., title, common textual snippet from a search engine[5], and URL of the page. Hovering a mouse over the upper part of the visual snippet (2A) will show a preview of its corresponding snapshot while clicking the lower part (2B) will open the link to the source of the visual snippet in a new web browsing tab within CoZpace.

In CoZpace, visual snippets are shown in a board, called *Snapboard*, which shares them among all members in the same group. There are two types of Snapboard, i.e., for the whole task and per-document. Task Snapboard demonstrates all visual snippets in a single tab (See Fig. 3). Per-document Snapboard displays all visual snippets created from an individual web page (See Fig. 1(4)), implemented as part of search results where a user can expand or hide it.

---

[5] In this study, we use Bing API for search and textual snippet generation. Other open source search engines can alternatively be used, such as Lemur Indri, Lucene and Terrier toolkits.

Fig. 2. An example of creating a visual snippet.

Fig. 3. Task Snapboard: a shared space of all visual snippets in a group.

4   Experimental Design

In this section, we describe our experimental methodology based on the model of interactive information retrieval evaluation as suggested by Borlund [4]. This user study is conducted to answer the following three research questions:

RQ1: Does our proposed Snapboard feature support exploratory search tasks that require collaboration of two or more users to search together?

RQ2: For an individual user, does the Snapboard also support such search tasks to search alone?

RQ3: Do visual snippets support users in collaboration in a search task when they asynchronously search the web?

4.1   Search Systems

In this experiment, two variations of CoZpace use Bing Search API[6] to provide search and query suggestion functionalities. We restrict the search results in

---

[6] http://www.bing.com/dev/.

CoZpace by showing only web pages in English. The two systems are: *(i)* CoZpace with a Snapboard function (S1) as an experimental condition; and *(ii)* CoZpace without the Snapboard (S2) as a control condition.

4.2 Search Tasks

All exploratory search tasks used in this study are selected from the top four search tasks on which people tend to cooperate as surveyed by Morris [12].

(T1) Travel Planning: All participants have to imagine that they are planning a trip to the Kansas City, USA. They want to search for information about how they will spend their vacation in USA. Their goal is to find where they will stay, what they will do, and how will they get there, etc.

(T2) General Shopping: All participants have to imagine that they are given USD 30,000 to buy a car. Their goal is to find the technical specifications of cars, brands of cars, and stores which sell cars, etc.

(T3) Literature Search: All participants have to imagine that they are assigned to write an article about the US civil war. Their task is to find causes of the civil war, economic causes, consequences of the civil war, civil war effects in the present, and weapons used during civil war, etc.

(T4) Technical Information: All participants have to imagine that they want to reduce the use of air conditioner in their house. Their task is to find the best material to use if the purpose is cooling down the roof so that the house temperature remains low, and other solutions regarding roof coating, etc.

4.3 Participants

All participants were 32 volunteers (12 males and 20 females); 12 participants were high school students in an English program from Nakhonnayok Wittayakom school, 10 participants were undergraduate students and the rest were graduate students from the Faculty of Information Technology, King Mongkut's Institute of Technology Ladkrabang (KMITL). All of them are highly proficient in English and web searching. According to the entry questionnaire, they often search more than twice a day. Most of them used to collaborate with other people in web searching by using social networks. Besides, they searched together in a group of three people, on average. However, they have never used a search system that is specifically designed for collaborative search before. All participants are then paired into 16 groups that will asynchronously collaborate on given search tasks. In each pair, we call "P1" a participant who firstly performs search tasks with fresh sessions that do not contain any summarizations and search histories, and call "P2" the other participant who pursues search tasks formerly done by P1.

4.4   Experimental Scenarios

We study the effectiveness of Snapboard in two search scenarios as follows:

(IN) Individual Search: In this scenario, a participant (P1) is assigned to search individually. He/She needs to find and mark as many relevant documents as possible, and then gives a reason why the documents are marked as relevant by taking a snapshot (in S1) or commenting (in S2).

(AC) Asynchronous Collaborative Search: Similar to IN, a participant (P1) in this scenario begin by the first session to perform a search task. Afterwards, the next participant (P2) continues to perform the same search task as P1. Review session starts immediately after finishing each search task.

Review session is a session that participants review the results from a previous session of the same task. It is designed to let everyone use our Snapboard. For review session, in S1, P1 is assigned to review only the visual snippets (Snapboard) of the web pages that marked as relevant by P2 but not marked by P1. In S2, P1 is assigned to review the comments of the web pages instead. Then, P1 has to judge whether the web pages are relevant or not. Afterwards, P2 is also assigned to review and judge the web pages that marked as relevant by P1 but not marked by P2.

4.5   Experimental Procedure

In the experiment, a pair of study participants are asked to collaboratively perform, in asynchronous manner[7], four simulated exploratory search tasks using the two collaborative search systems. To neutralize the effect of human learning behavior in our experiment, we applied a Graeco-Latin Square design to control and rotate the sequence of blocking factors (i.e., systems and search tasks).

In total, our experiment lasted around three hours. The experiment started with an individual introductory session, where participants were given an information sheet, asked to fill in an entry questionnaire, and demonstrated how to use the two search systems. This introduction took approximately five minutes, and was followed by a training session, where each participant was allowed up to ten minutes of interaction and familiarization with the systems. After training, they were asked to perform four exploratory search tasks, as described in Sect. 4.2. For each task, they had a maximum of fifteen minutes to carry it out, followed by a five-minutes review session. After finishing each task, they have to answer a post-task questionnaire. Also, after finishing all tasks, they have to complete an exit questionnaire. After two tasks, a five minute break was given to the subjects, as required by the ethical regulations at KMITL.

4.6   Data Collections

In order to achieve and answer our research questions, we collect both qualitative and quantitative data. Two methods are used to collect the qualitative data, i.e.,

---

[7] Any search session in which one participant performs first is considered the IN scenario and we call such a participant "P1".

(i) open-ended questionnaires, and *(ii)* focused group discussions. We collect all logs from participants' interactions and answers from close-ended questionnaires, and then analyze them to be quantitative data, such as, the number of submitted queries, the number of viewed web pages, the number of websites that are marked as relevant, the number of comments, the number of snapshots and the rating scores of user feedback in different aspects.

5 Results and Discussion

5.1 The Effectiveness of Snapshot in Individual and Asynchronous Collaborative Search

Table 1 presents a comparison of the performance between two systems in CoZpace, i.e., S1 and S2, operated in individual search (IN) and asynchronous collaborative search (AC) conditions (Cond). The results show that the average number of visual snippets (Vsnip) made in S1 is obviously higher than that of comments (Comt) in S2 in both IN and AC. It can be interpreted that the participants prefer taking the snapshots as visual snippets rather than commenting to summarize the web pages. The average number of queries used in S1 is higher than that in S2 in both IN and AC. However, it is not necessary to be inferred that participants make more effort to formulate queries in S1 than in S2. The reason behind this outcome is that the feedback from our participants in Table 2 shows that all tasks are easy to formulate query.

Furthermore, Table 1 shows that S1 outperforms S2 in term of the average number of web pages marked as relevant (All Rel). In AC, the Rel Web are counted from only the web pages that are marked by both participants P1 and P2 in the same team to be more sure about the relevance of the websites. In addition, the difference between the average number of relevant unique web pages (Uniq Rel) and the All Rel, which is really high in IN and higher in AC. This result shows that many relevant web pages are marked by more than one participants in IN or more than one groups of participants in AC. As a result, we can be more confident that marking the relevant web pages by participants does not happen by chance.

We also check how many percent of all unique web pages clicked to view by participants (Uniq View) are unique relevant web pages (Uniq Rel). The "% Uniq Rel per Uniq View" in S1 is higher than that in S2 in both IN and AC, indicating that S1 is more effective than S2 in the support of users to find relevant web pages from the returned and clicked web pages. From the above results, we can answer RQ2 and RQ3 that the Snapboard support users in both individual and asynchronous collaborative search.

We analyzed the statistical significant differences (using a one-way ANOVA) in individual search between two systems (i.e., S1 and S2) and among activities (i.e., Vsnip vs. Comt, Query, All Rel, and All View), at $p < 0.1$. The results showed better performance for S1 than that for S2 on average in Vsnip vs. Comt, All Rel, Query, and All View with statistical significance at $p = 0.035^*$, but not statistical significance at $p = 0.634$, $p = 0.188$, and $p = 0.966$, respectively.

Table 1. User interaction statistics – mean (averaged over 16 participants/groups) and standard deviation (in bracket) in Individual and Asynchronous collaborative search for all search tasks.

Cond	Sys	Vsnip	Comt	Query	Rel Web		Viewed Web		% Uniq Rel
					All Rel	Uniq Rel[†]	All View	Uniq View[†]	per Uniq View
IN	S1	**20.81** (4.25)	-	**13.69** (3.57)	**12.50** (3.33)	**8.44**	**23.19** (5.15)	14.38	**58.70**
	S2	-	15.88 (5.23)	11.31 (2.72)	11.75 (4.14)	7.88	23.06 (7.99)	**14.44**	54.54
AC	S1	**33.50** (4.37)	-	**22.38** (3.80)	**25.75** (3.61)	**13.88**	**53.50** (6.18)	**26.69**	**52.26**
	S2	-	28.25 (4.64)	20.44 (2.78)	18.69 (3.46)	10.44	49.31 (7.42)	25.31	41.25

[†] Note that we do not analyze standard deviation of *unique* relevant and viewed web page as they are summed over participants/groups before being averaged.

A one-way ANOVA was also conducted to determine the statistical significant differences in asynchronous collaborative search between two systems (i.e., S1 and S2) and among activities (i.e., Vsnip with Comt, Query, All Rel, and All View), at $p < 0.1$. The results demonstrate better performance for S1 than that for S2 on average in Vsnip vs. Comt, All Rel, Query, and All View with statistical significance at $p = 0.095^*$ and $p = 0.004^*$, but not statistical significance at $p = 0.395$ and $p = 0.337$, respectively.

Table 2. Average feedback from a post-task questionnaire using Five-Point-Likert scale (closer to 1 = disagree, closer to 5 = agree)

Question	System	Mean				Mean	S.D.
		T1	T2	T3	T4		
Q1. I feel that the search task was easy after finished it.	S1	**3.19**	**3.19**	2.13	**2.88**	**2.84**	1.12
	S2	2.63	2.31	**2.25**	2.69	2.47	1.42
Q2. I was relaxed while carrying out the search task.	S1	**3.94**	**3.81**	2.63	**3.52**	**3.52**	1.04
	S2	3.31	3.25	**3.31**	3.13	3.25	1.08
Q3. It was easy to formulate initial queries on these topics.	S1	**4.06**	**4.06**	**4.19**	**4.00**	**4.08**	0.55
	S2	3.63	3.88	3.94	3.88	3.83	0.72
Q4. I had enough time to do an effective search task.	S1	**4.06**	**4.19**	3.19	**3.88**	**3.83**	1.10
	S2	3.56	3.44	**3.63**	3.44	3.52	1.14
Q5. I have succeeded in my performance of the search task.	S1	**4.13**	**4.00**	3.38	**4.38**	**3.97**	0.91
	S2	3.13	3.38	**3.94**	3.75	3.55	0.87

5.2   User Feedbacks

Table 2 illustrates the average of user feedbacks in a post-task questionnaire. The participant feedbacks includes their satisfaction, opinion and their experience when using CoZpace with or without Snapboard. We analyzed the statistical significant differences (using a two-way ANOVA) between two systems (i.e., S1 and S2) and among search tasks (i.e., T1, T2, T3, and T4), at $p < 0.1$. The analysis shows better feedbacks for S1 than that for S2 on average in Q1, Q2, Q3, Q4, and Q5 with statistical significance at $p = 0.0568^*$, $p = 0.0856^*$, $p = 0.0829^*$, $p = 0.0934^*$, and $p = 0.0053^*$, respectively.

Moreover, as T1, T2, and T4 are ordinal interaction effect, the main effect can be considered. The main effect of system shows that the user's satisfaction on average in Q1, Q2, Q3, Q4, and Q5 are better for S1 than for S2 in T1, T2, and T4. There is no main effect of task among T1, T2, and T4 but there is a little bit effect from T3. We also tried to determine the statistical significant differences (using a two-way ANOVA) between two systems (i.e., S1 and S2) and among questionnaires (i.e., Q1, Q2, Q3, Q4, and Q5), at $p < 0.1$. The results show better feedbacks for S1 than that for S2 on average in T1, T2, T3 and T4 with statistical significance at $p = 0.0042^*$, $p = 0.0015^*$, $p = 0.0634^*$, and not statistical significance at $p = 0.1315$, respectively. As the tasks T1 and T2 basically require users to find information better presented in a visual form (i.e., pictures better convey the content toward collaborators than text), the results seem to be pertinent to our scenario of using Snapboard feature that it is appropriate for any task that needs to share pictures, such as sharing attractions for travel planning.

The difference in T3 is due to the fact that the literature search mostly related to finding text results might not take benefits from the Snapboard feature. As the difference in T4 is extremely not significant, the better feedbacks for S1 should not be claimed. However, it might be interpreted that the better feedbacks are not significant because T4 is a task that might not need to share pictures. The average feedbacks from Q1 in Table 2 can infer that all the four search tasks are quite difficult when using S2 but become easier when using S1 in T1, T2 and T4. This inference answers our RQ1 that Snapboard feature supports exploratory search tasks.

Table 3 shows quantitative data gathered from an exit questionnaire. Users were asked to rate their satisfaction to the interface of CoZpace as well as its three main features, i.e., retrieval, comment and snapshot functions. The satisfaction of Snapshot function on average is 4.41 with a low standard deviation showing a very high level of agreement. On the other hand, the comment feature got the lowest average score, 3.34. For the CoZpace interface, the results are very positive with a mean value of 4.03.

Upon the completion of study, we arranged a discussion session with participants. They provided open comments with respect to the feature of Snapboard. Some examples include: "Visual snippet helps me to easily share needed information with collaborators" and "Snapboard increases the awareness of how our tasks progress, and I can use it to identify relevant information in web pages or revisit them later". Most of the participants agree that those benefits can help them to collaboratively complete the search tasks easier.

Table 3. Average feedback about user satisfaction to our CoZpace system from an exit questionnaire using Five-Point-Likert scale (closer to 1 = disagree, closer to 5 = agree)

Question	Mean	S.D.
1. The search interface was easy to use	4.03	0.82
2. Retrieval function was helpful in exploring web	4.13	0.71
3. Comment function in S1 was helpful in exploring web	3.34	1.24
4. Snapshot function in S2 was helpful in exploring web	**4.41**	0.71

6   Conclusion

In this paper, we conducted a user study to evaluate the supportiveness of Snapborad, i.e., visual snippets, in carrying out exploratory search tasks in two scenarios of individual and asynchronous collaborative search. We experimented on two variations of CoZpace with pairs of participants by collecting all activities logs and their feedbacks on the usage of the systems, in particular those related to Snapboard. The evaluation showed that Snapboard helped the participants in the collaborative search tasks, especially Travel planning and General shopping, in both scenarios. The reason is that most traveling and shopping web pages consist of not only texts but also several images which are easier to recognize and to describe in picture than in text format. Furthermore, according to the evaluation, using Snapboard can generate more relevant web pages. The number of relevant web pages in S1 is higher than that in S2 in both individual search and asynchronous collaborative search. Thus, it means that the Snapboard support and improve in both searching conditions.

For future work, we plan to conduct a user study to evaluate the effectiveness in synchronous collaborative search. In addition, we will compare the proposed our visual snippets with other kinds of summarization (e.g. thumbnails.)

References

1. Amershi, S., Morris, M.R.: Cosearch: a system for co-located collaborative web search. In: Proceedings of CHI 2008, Florence, Italy, pp. 1647–1656 (2008)
2. Aula, A., Khan, R.M., Guan, Z., Fontes, P., Hong, P.: A comparison of visual and textual page previews in judging the helpfulness of web pages. In: Proceedings of WWW 2010, Raleigh, USA, pp. 51–60 (2010)
3. Bates, M.J.: The design of browsing and berrypicking techniques for the online search interface. Online Inf. Rev. **13**(5), 407–424 (1989)
4. Borlund, P.: User-centered evaluation of information retrieval systems. In: Information Retrieval: Searching in the 21st Century, pp. 21–37 (2009)
5. Broder, A.: A taxonomy of web search. SIGIR Forum **36**(2), 3–10 (2002)
6. Bruce, H., Jones, W., Dumais, S.: Keeping and re-finding information on the web: what do people do and what do they need? In: ASIST 2004 (2004)

7. Golovchinsky, G., Pickens, J., Back, M.: A taxonomy of collaboration in online information seeking. In: Proceedings of the 1st International Workshop on Collaborative Information Retrieval, JCDL 2008, Pittsburgh, USA (2008)

8. Halvey, M., Vallet, D., Hannah, D., Feng, Y., Jose, J.M.: An asynchronous collaborative search system for online video search. Inf. Process. Manage. **46**(6), 733–748 (2010)

9. Jiao, B., Yang, L., Xu, J., Wu, F.: Visual summarization of web pages. In: Proceedings of SIGIR 2010, Geneva, Switzerland, pp. 499–506 (2010)

10. Kruajirayu, H., Leelanupab, T.: Extracting visual snippets for query suggestion in collaborative web search. In: Proceedings of the 20th International Society on Artificial Life and Robotics, AROB 2015, Beppu, Japan (2015)

11. Kruajirayu, H., Tangsomboon, A., Leelanupab, T.: Cozpace: a proposal for collaborative web search for sharing search records and interactions. In: Student Project Conference (ICT-ISPC), 2014 Third ICT International, pp. 165–168. IEEE (2014)

12. Morris, M.R.: A survey of collaborative web search practices. In: Proceedings of CHI 2008, Florence, Italy, pp. 1657–1660 (2008)

13. Morris, M.R., Horvitz, E.: Searchtogether: an interface for collaborative web search. In: Proceedings of UIST 2007, Newport, USA, pp. 3–12 (2007)

14. Morris, M.R., Lombardo, J., Wigdor, D.: Wesearch: supporting collaborative search and sensemaking on a tabletop display. In: Proceedings of CSCW 2010, Savannah, USA, pp. 401–410 (2010)

15. Rodriguez Perez, J., Leelanupab, T., Jose, J.M.: Cofox: a synchronous collaborative browser. In: Proceedings of AIRS 2012, Tianjin, China, pp. 262–274 (2012)

16. Shah, C., Marchionini, G.: Awareness in collaborative information seeking. JASIST **61**(10), 1970–1986 (2010)

17. Teevan, J., Cutrell, E., Fisher, D., Drucker, S.M., Ramos, G., André, P., Hu, C.: Visual snippets: summarizing web pages for search and revisitation. In: Proceedings of CHI 2009, Boston, USA, pp. 2023–2032 (2009)

18. Turpin, A., Tsegay, Y., Hawking, D., Williams, H.E.: Fast generation of result snippets in web search. In: Proceedings of SIGIR 2007, Amsterdam, The Netherlands, pp. 127–134 (2007)

19. White, R.W., Roth, R.A.: Exploratory search: beyond the query-response paradigm. Synth. Lect. Inf. Concepts, Retrieval, Serv. **1**, 1–98 (2009)

20. Yue, Z., Han, S., He, D.: A comparison of action transitions in individual and collaborative exploratory web search. In: Proceedings of AIRS 2012, Tianjin, China, pp. 52–63 (2012)

Improving Ranking and Robustness of Search Systems by Exploiting the Popularity of Documents

Ashraf Bah[(✉)] and Ben Carterette

Department of Computer Sciences, University of Delaware, Newark, DE, USA
{ashraf, carteret}@udel.edu

Abstract. In building Information Retrieval systems, much of research is geared towards optimizing a specific aspect of the system. Consequently, there are a lot of systems that improve effectiveness of search results by striving to outperform a baseline system. Other systems, however, focus on improving the robustness of the system by minimizing the risk of obtaining, for any topic, a result subpar with that of the baseline system. Both tasks have been organized by TREC Web tracks 2013 and 2014, and have been undertaken by the track participants. Our work herein, proposes two re-ranking approaches – based on exploiting the popularity of documents with respect to a general topic – that improve the effectiveness while improving the robustness of the baseline systems. We used each of the runs submitted to TREC Web tracks 2013 – 14 as baseline, and empirically show that our algorithms improve the effectiveness as well as the robustness of the systems in an overwhelming number of cases, even though the systems used to produce them employ a variety of retrieval models.

1 Introduction

During the last two years of its running in 2013 and 2014, the TREC Web track focused not only on measuring ad-hoc effectiveness, but also on measuring the risk involved when systems try to improve results over a baseline. In regard to measuring ad-hoc effectiveness, as in previous years, the track proceeded by using traditional effectiveness measures such as ERR and nDCG, as well as diversity measures such as ERR-IA and α-nDCG. As for measuring the robustness or risk-sensitivity of the systems, the organizers of the track introduced a risk-sensitive utility measure that determines, for a set of queries, the average of the differences between the retrieval effectiveness of a given system and that of a baseline system. The goal is to determine and quantify the ability of each system to minimize the risk of providing results that are less effective than the baseline's results, for any given query.

Most research in Information Retrieval (IR) has focused on improving the average effectiveness of systems, using measures like the ones mentioned above. However, it is very often the case that the improved systems fare worse than the baseline on many of the queries, even though the average effectiveness score is higher than the baseline's. The concept of robust ranking appears therefore to be key, when it comes to remedying those cases. In this present work, robustness refers to the ability of the ranker to reduce

G. Zuccon et al. (Eds.): AIRS 2015, LNCS 9460, pp. 174–187, 2015.
DOI: 10.1007/978-3-319-28940-3_14

and mitigate poor performance on individual queries while striving to improve the overall performance as well.

In this paper, we present two re-ranking techniques – based on exploiting the popularity of documents with respect to a general topic – that, given a baseline or a state-of-the-art ranking, improves the average effectiveness of the ranking while improving the robustness of the ranking. Both methods merely re-rank the documents that were retrieved by the baseline that is being considered, without ever adding any new document to the set of retrieved documents.

2 Related Work

Ad-hoc Information Retrieval (IR) has long focused on improving average overall effectiveness, without worrying much about reducing the probability of getting poorer results for individual queries. Popular state-of-the-art retrieval models that have been used in ad-hoc IR include language modeling [19] and Okapi BM25 [21]. The Markov Random Field model for term dependencies has been proposed by Metzler and Croft [18]. And so were other term proximity models such as those described by Buttcher et al. [4] and Tao and Zhai [23]. In recent years, learning to rank algorithms have been adopted [3, 16], as well as learning to re-rank algorithms [15].

Another aspect of ad-hoc retrieval is focused on improving diversity and novelty search and retrieval. Many of the diversity ranking approaches are inspired by the MMR algorithm introduced by Carbonnell and Goldstein [5]. The fundamental idea of MMR is to optimize for both document relevance and coverage of intents and/or aspects. Differences in implementations lie in how similarities are computed: Carbonnell and Goldstein suggest using any similarity function such as cosine similarity. Zhai et al. advocate for modifying the language modeling framework to incorporate a model of relevance and redundancy [26]. Other researchers utilized the correlation between documents as a measure of their similarity in the pursuit of diversification and risk minimization in document ranking [24]. Carterette and Chandar [6] introduced a greedy result set pruning wherein there are 2 steps: in the first step, they rank documents in decreasing order of their similarity to the query; and in the second step, they proceed to iteratively prune documents whose similarity to any previously selected document is greater than a certain threshold. They also use a set-based probabilistic model to maximize the likelihood of covering all the aspects [6]. Radlinski and Dumais exploited a commercial search engine to obtain aspects of queries, and proceeded to diversify the ranking using query-query reformulations [20]. Santos et al. utilize a query-driven approach wherein they explicitly account for aspects by using sub-queries to represent a query. They then estimate the relevance of each retrieved document to every identified sub-query, and the importance of each sub-query [50].

Research on risk-sensitive ranking is still in its infancy. Perhaps the work most related to this paper is Wang et al.'s effort [25] to address robustness by proposing a principled learning to rank framework that optimizes for both effectiveness and robustness of a retrieval system. Essentially the authors proceeded by proposing a learning method that optimizes for both reward and risk with respect to a given baseline. While their approach consists in learning to control the tradeoff between effectiveness and robustness, our approach is a general re-ranking method that can be used on top of any retrieval model (e.g. language model, query expansion, learning to rank, data fusion, or Markov Random Field retrieval models) and that consists in re-arranging documents in the original ranking, in order to get a more effective and robust final ranking.

Another major contribution in the literature is that of Dincer et al. [13] that focused on uncovering biases inherent to the way TREC Web track evaluated systems prior to 2014. The authors argued that, given there were several various retrieval methods used by the track participants in building the systems (query expansion, learning to rank, etc.), comparing robustness using one single baseline that was created with a specific retrieval model creates inherent biases. That is, systems that build on top of retrieval models similar to the baseline will have an advantage over others, and systems that build on top of retrieval models very different from the baseline will be at great disadvantage. The authors proposed several ways to mitigate that issue including the use of mean within-topic system effectiveness as a baseline. In the present work, our focus is not to show that we can improve robustness with respect to one single specific baseline. Rather we focus on showing that, given any ranking obtained using a certain retrieval model, we can apply our method to improve the robustness as well as the overall effectiveness of the system. For that purpose, we used each and every run from TREC Web 2013 and 2014 – as baselines – to empirically show that.

Wang and Zhu [24] proposed a risk-aware ad-hoc retrieval model that utilized the correlation between documents as a measure of their similarity in the pursuit of diversification and risk minimization in document ranking. In a similar effort to establish a risk-aware framework for retrieval, Zhu et al. [27] proposed to model uncertainty and utilized a one-parameter loss function to model the level of risk acceptable by a user. Their loss function was applied to a language modeling framework. Our approach, however, is a general re-ranking approach that we apply to the ranking of any retrieval model, and that we show to improve average overall effectiveness as well as risk-sensitive measures. There are other efforts for ad-hoc risk-aware retrieval methods that focus on query expansion cases [12] and pseudo-relevance feedback cases [17].

It is to be noted however that the term robustness as used in this paper differs from the sense it is given in other work like in Battacharjee's work wherein robust ranking means ranking algorithm that is sensitive (or less vulnerable) to spams and noise in the training set [2].

3   Methodology

In order to show that we can improve the ranked list for a given query by exploiting a pre-fetched list of documents sorted by decreasing probability of retrievability of a document, we proceed as follows:

a. For obtaining pre-fetched lists of documents ranked by decreasing probability of retrievability, we propose to use the method described in Sect. 3.1.

b. In order to obtain a baseline ranked-list for each given query, we use every ranked list submitted at TREC Web tracks 2013 and 2014 as baseline. That allows us to show that our re-ranking works on a wide-range of systems.

c. Then at query run-time, we propose to use one of the two algorithms proposed in Sect. 3.2 for obtaining the final ranked list.

Finally, we evaluate the effectiveness of each method on each baseline using diversity measures and the robustness using risk-sensitive measures.

3.1   Estimating Document Retrievability

Given a query, we want to estimate how likely a document is to be retrieved. That is, we want to estimate its retrievability with respect to the topic. We treat this as the popularity of the document with respect to that general topic.

Suppose q is the observed query, that is, an actual user query. Let $\mathcal{Q}$ represent the space of possible queries from which q is one sample. In our experiments, we obtain the sample of possible queries by using Bing and Yahoo! Suggestions. We submit a query to each service through their APIs and we obtain a list of suggested queries.

Now let us say a document is *retrieved* if it appears in a top-k ranking of documents for some query q. In our experiments, we use k = 100.

We cannot observe that space fully, but we assume that the probability that a document is retrieved for a query in that space is approximately the same as the probability that a document is retrieved for q:

$$P(ret \mid q' \in Q, D) \sim P(ret \mid q \in Q, D)$$

Then let us define the probability that a document is retrievable for the possible query space as:

$$P(ret \mid \mathcal{Q}, D) = \sum_{q' \in \mathcal{Q}} P(ret \mid q', D) P(q' \mid \mathcal{Q}, D)$$

If we decide to use uniform weights for $P(q' \mid \mathcal{Q}, D)$ and if we assume that $P(ret \mid q', D) = 1$ if document D appears in top-k ranking for q' and $P(ret \mid q', D) = 0$ otherwise, then this equation is proportional to a data fusion method so-called Comb-CAT [1], which merges retrieval results based on the total number of times a document appears. We can think of – and use – this probability of retrieval of a document as a type of popularity score for the document with respect to the sample space.

3.2   Re-Ranking

We propose two different methods for re-ranking the baseline.

Method 1. Suppose a user provides a query q to our system. Given a pre-fetched ranked list in decreasing order of $P(ret \mid q', D)$ – which we name MasterList – of documents pertaining to the same topic as q, we proceed as illustrated in Algorithm 1 (Fig. 1) and Fig. 2.

```
MasterList = docs ranked in decreasing
order of P(ret | Q,D);
MasterList= MasterList minus docs that
appeared in the ranked list of only one
possible query q' in our sample space;
    DocsToShuffle={}

    //record docs that must be shuffled
    For each doc_i in baseline
        If  doc_i is in MasterList
            Add doc_i to DocsToShuffle;

    //proceed to the actual shuffling
    nextIndex=0;
    For i in baseline.size()
        If  doc(i) is in DocsToShuffle
            doc(i)= DocsToShuffle(nextIndex);
            nextIndex++;
```

Fig. 1. Algorithm 1

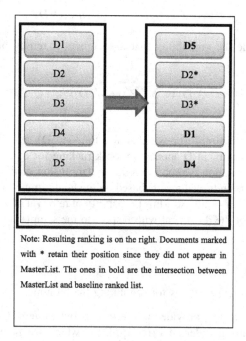

Note: Resulting ranking is on the right. Documents marked with * retain their position since they did not appear in MasterList. The ones in bold are the intersection between MasterList and baseline ranked list.

Fig. 2. Illustration of Method 1

```
MasterList = docs ranked in decreasing
order of P(ret | Q,D);

MasterList= MasterList minus docs that
appeared in the ranked list of only one
possible query q' in our sample space;

IntersectDocs={}

/*record   docs   that   are   both   in
MasterList and in baseline ranking */
For each doc_i in baselineList
    If  doc_i is in MasterList
        Add doc_i to IntersectDocs;

/* Start by first ranking the docs from
IntersectDocs in  the  order  the  appear
in  (decreasing  order  of  P(ret | q',
D) */
nextIndex=0;
For i in IntersectDocs.size()
    doc(i)= IntersectDocs (i);

For    j=  1    to        rankCutoff   +
IntersectDocs.size()
    If  IntersectDocs  does  not  contain
baselineList(j)
        Doc(j+      IntersectDocs.size())=
(baselineList(j)

        doc(i)= IntersectDocs (i);
```

Fig. 3. Algorithm 2

The idea is to keep, in our final ranked list, all documents that do not appear in MasterList, at the same position where they appeared in the baseline ranked list (RL). The only documents to be shuffled are the ones that:

- appeared in both the MasterList and the baseline ranked list;
- appeared in the ranked list of more than one possible query q' from the sample space Q.

The shuffling will be done such that the documents with higher P(ret | Q, D) – as recorded in the MasterList – will be ranked higher than the documents with lower P(ret | Q, D) in the final ranked list. But again, the documents that are in the baseline RL but not in the MasterList, remain at their original rank.

Method 2. Given a query q provided by the user to our system and a MasterList that contains a pre-fetched list of documents ranked in decreasing order of P(ret | q', D), we proceed as illustrated in Algorithm 2 (Fig. 3) and Fig. 4.

The essential idea in this method is to ensure that the documents with higher P(ret | Q, D) – as recorded in the MasterList – will be ranked higher than the

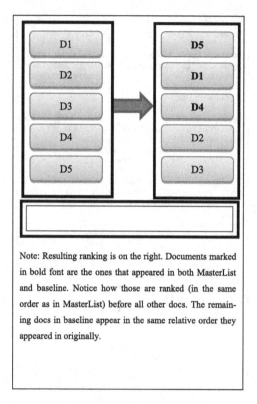

Fig. 4. Illustration of Method 2

documents with lower P(ret | $\mathcal{Q}$, D) in the final ranked list. And, in this method – unlike in Method 1 – all the documents that are at the intersection of MasterList and the baseline RL will have precedence over all other documents in the baseline RL. That is, every document that appeared in the ranked list of more than one possible query q′ from the sample space $\mathcal{Q}$, and that also appeared in the baseline RL, will be ranked before all other documents.

4   Experiments and Results

4.1   Data

Each of the TREC Web track 2013 and 2014 datasets contains 50 queries [10, 11]. Those queries were created after perusing candidate topics from query logs from commercial search engines. Some topics were faceted (with several possible subtopics), others were non-faceted with single intents, and a few others were ambiguous (queries with several intents). The task of the participants is to provide a diversified ranking of no more than 10000 documents per query for the 50 queries. Unlike in previous versions of the track that ran from 2009 to 2011 where the emphasis was on diversity ranking in addition to ad-hoc ranking, in these two versions, the emphasis was on risk-minimization and ad-hoc rankings. However the queries and relevance judgments were still suitable for diversity retrieval evaluations, and participants were also provided with diversity-based results once the competition was over. In this paper, we use diversity measures to show how effective our method is, and we use risk-sensitive measures adopted by TREC Web track organizers to show robustness.

For this experiment, in order to show that our method can be applied to many retrieval models, we use many baseline ranked lists (RL). Specifically, we use RLs submitted by TREC Web 2013 and 2014 participants, each RL is created by a different search system. For the 2013 TREC Web track, there were a total of 60 runs, while for 2014, there were a total of 30 ad-hoc runs and 12 risk-sensitive runs. All runs were supposed to be risk-sensitive, although some participants did not specifically optimize for risk-sensitivity [10, 11].

Baselines for computing risk-sensitive measures were also made public by the track organizers, and include Indri, Terrier, Indri-with-spam-filtering. The idea behind using several baselines is to see how truly robust a system is with respect to various baselines, and mitigate the bias that gets introduced when using only one baseline (in such a case, there would be bias towards systems that are built on top of a ranker similar to the baseline)

4.2   Evaluation Measures

We use four evaluation measures (two for diversity evaluation and two for traditional non-diversity evaluation). For non-diversity measure, we opted for using the two measures adopted by TREC Web track organizers: ERR and nDCG. nDCG rewards documents with high relevance grades and discounts the gains of documents that are ranked at lower positions [14]. ERR is defined as the expected reciprocal length of time it takes the user to find a relevant document [8], and it takes into account the position of the document as well as the relevance of the documents shown above it. For diversity measures, we opted for using α-nDCG and ERR-IA. α-nDCG is an extension of nDCG that rewards novelty and diversity by penalizing redundancy and rewarding systems for including new subtopics [9]. Similarly, ERR-IA is an extension of ERR to compute the expectation of ERR over the different intents [7].

As for measuring the robustness of the systems, we adopted the risk-sensitive measures proposed by TREC Web track organizers as well [11]. For each run, we create two new runs using the re-ranking methods Method 1 and Method 2 respectively. For each query of each new run, we compute the absolute difference (Δ) between the effectiveness of the new run and that of the baseline provided by the track organizers – as mentioned in Sect. 4.1, that can be either Indri or Terrier or Indri-with-spam-filtering. When the difference is positive the new run has a win over the baseline. When it is negative, it has a loss over the baseline, otherwise it is a tie.

Let $\Delta(q) = R_A(q) - R_{BASE}(q)$ be the absolute win or loss for query q with system retrieval effectiveness $R_A(q)$ relative to the baseline's effectiveness $R_{BASE}(q)$ for the same query. We define the risk-sensitive utility measure $U_{RISK}(q)$ of a system over a set of queries Q as:

$$U_{RISKaq}(Q) = \frac{1}{N}\left[\sum_{q \in Q_+} \Delta(q) - (\alpha + 1)\sum_{q \in Q_-} \Delta(q)\right]$$

where Q_+ is the set of queries for which $\Delta(q) > 0$ and Q_- is the set of queries for which the $\Delta(q) < 0$.

It is important to note that we did not need to apply the guideline given by [13] to mitigate bias by using the mean within-topic system effectiveness as a baseline. In fact, we focus on showing that, for most rankings obtained using specific retrieval models, we can apply our method to improve the robustness of the system as well as the overall effectiveness. And we do so by comparing the risk-sensitive measure between the original ranking and the baseline to the risk-sensitive measure between the new re-ranking and the baseline (Table 1).

Table 1. TREC 2013 results for **Method 2**. Bold font denotes positive difference between Method 2 and the submitted-run, in terms of ERR-IA@20.+denotes statistical significance

runID	baselineERR-IA@20	ERR-IA@20	runID	baselineERR-IA@20	ERR-IA@20
clustmrfaf	0.5540	**0.5701**	udemQlml1FbR	0.4620	**0.5769+**
clustmrfbf	0.4888	**0.5266**	udemQlml1R	0.4597	**0.5706+**
cwiwtl3cpe	0.4082	**0.5575+**	UDInfolabWEB1	0.4856	**0.6013+**
cwiwtl3cps	0.4726	**0.5692+**	UDInfolabWEB1R	0.4856	**0.6013+**
cwiwtl3kld	0.3274	**0.5525+**	UDInfolabWEB2	0.5738	**0.6013**
dlde	0.0453	**0.2411+**	UDInfolabWEB2R	0.5738	**0.6013**
ICTNET13ADR1	0.4743	**0.5777+**	UJS13LCRAd1	0.4265	**0.5266+**
ICTNET13ADR2	0.4925	**0.5712+**	UJS13LCRAd2	0.4580	**0.5266**
ICTNET13ADR3	0.4415	**0.5784+**	UJS13Risk1	0.4435	**0.5266+**
ICTNET13RSR1	0.5185	**0.5799**	UJS13Risk2	0.4606	**0.5266**
ICTNET13RSR2	0.4847	**0.5915+**	uogTrADnLrb	0.5123	**0.5645**
ICTNET13RSR3	0.5420	**0.5777**	uogTrAIwLmb	0.5391	**0.5516**
mmrbf	0.4980	**0.5266**	uogTrAS1Lb	0.5041	**0.5399**
msr_alpha0	0.3409	**0.5449+**	uogTrAS2Lb	0.5064	**0.5378**
msr_alpha0_95_4	0.3576	**0.5624+**	uogTrBDnLaxw	0.5297	0.5276
msr_alpha1	0.3565	**0.5713+**	uogTrBDnLmxw	0.5252	**0.5276**
msr_alpha10	0.3510	**0.5607+**	ut22base	0.5066	**0.6211+**
msr_alpha5	0.3515	**0.5692+**	ut22spam	0.4368	**0.6241+**
RMITSC	0.3758	**0.4488+**	ut22xact	0.5005	**0.6211+**
RMITSC75	0.3762	**0.4466+**	UWCWEB13RISK01	0.2872	**0.5767+**
RMITSCTh	0.3757	**0.5531+**	UWCWEB13RISK02	0.3152	**0.5767+**
udelCombUD	0.4913	**0.5701+**	webishybrid	0.3516	**0.5720+**
udelManExp	0.5086	**0.5682**	webismixed	0.4092	**0.5798+**
udelPseudo1	0.4556	**0.5701+**	webisnaive	0.3658	**0.5803+**
udelPseudo1LM	0.3758	**0.5701+**	webiswikibased	0.3812	**0.5808+**
udelPseudo2	0.5163	**0.5701**	webiswtbaseline	0.3733	**0.5809+**
udemFbWikiR	0.4746	**0.5703+**	wistud.runA	0.4379	**0.5830+**
udemQlml1	0.4597	**0.5706+**	wistud.runB	0.4523	**0.5642+**
udemQlml1Fb	0.4014	**0.5705+**	wistud.runC	0.3870	**0.3871**
udemQlml1FbWiki	0.4746	**0.5703+**	wistud.runD	0.5026	**0.5731**

4.3   Results

Effectiveness. Effectiveness results, given by diversity measures α-nDCG@20 and ERR-IA@20, show that our approach is very promising. The results for applying Methdod 2 on the 2013 runs show that, in most cases, there are large improvements. In fact, out of the 60 runs, only one of them saw a slight decrease in ERR-IA@20 using Method 2. The general trend, indeed, is that Method 2 performs well on both datasets – 59 runs improved out of 60 for the 2013 dataset, 28 runs improved out of the 30 runs for 2014 dataset (ad-hoc runs category) and 9 runs improved out of the 12 runs for 2014 dataset (risk-sensitive runs category). It is worth noting that the runs from 2014 dataset that Method 2 failed to improve are all from the same participating group that performed the best. This could have to do with the set of documents retrieved for their runs. Results for α-nDCG@20 have similar trends as results for ERR-IA, as shown in Table 2.

Table 2. Summary of effectiveness. This shows, for each measure for each dataset, the number of runs for which a given method performs better (on average) than another method – and vice versa. Alg1 stands for method 1. Base is short for baseline

	α-nDCG@20 all 2013	ERR-IA@20 all 2013	α-nDCG@20 adhoc 2014	ERR-IA@20 adhoc 2014	α-nDCG@20 risk-runs 2014	ERR-IA@20 risk-runs 2014
# (alg1 > base)	57	55	26	27	8	7
# (alg1 < base)	3	5	4	3	4	5
# (alg2 > base)	58	59	27	28	9	9
# (alg2 < base)	2	1	3	2	3	3

Table 3. Summary of Risk-sensitive results. This shows, for each measure for each dataset and with respect to a specific baseline, the number of runs for which a given method is more robust than another method – and vice versa. $\alpha = 5$

	U-ERR indri all 2013	U-ERR indri-filt all 2013	U-ERR terrier all 2013	U-ERR indri adhoc 2014	U-ERR indri-filt adhoc 2014	U-ERR terrier adhoc 2014	U-ERR indri risk-runs '14	U-ERR indri-filt risk-runs '14	U-ERR terrier risk-runs '14
#(alg1 > base)	57	56	56	19	13	12	7	3	2
#(alg1 < base)	3	4	4	11	17	18	5	9	10
#(alg2 > base)	51	55	48	15	22	15	8	7	4
#(alg2 < base)	9	5	12	5	8	15	4	5	8
#(alg2 > alg1)	46	52	39	15	26	27	7	7	7
#(alg2 < alg1)	14	8	21	5	4	3	5	5	5

Table 4. Risk sensitive measures using indri as baseline for each run on 2013 datasets. Alg1. UERR@20 is the risk-sensitive measure, using Method 1. Alg1-base.UERR@20 is the difference between Method 1 and the submitted-run, in terms of $U_{RISK}(ERR@20)$. $\alpha = 5$

runID	base. UERR@20	alg1. UERR@20	alg1-base. UERR@20	alg2. UERR@20	alg2-base. UERR@20	alg2-alg1. UERR@20
clustmrfaf	−0.0247	−0.0014	+	0.003	+	+
clustmrfbf	−0.1729	−0.1450	+	−0.180	−	−
cwiwt13cpe	−0.1102	−0.0947	+	−0.004	+	+
cwiwt13cps	−0.0655	−0.0491	+	0.003	+	+
cwiwt13kld	−0.1586	−0.0747	+	−0.008	+	+
dlde	−0.5465	−0.5444	+	−0.410	+	+
ICTNET13ADR1	−0.1166	−0.0232	+	−0.017	+	+
ICTNET13ADR2	−0.1221	−0.0184	+	−0.017	+	+
ICTNET13ADR3	−0.1790	−0.0369	+	−0.018	+	+
ICTNET13RSR1	−0.1291	−0.0431	+	−0.007	+	+
ICTNET13RSR2	−0.1116	−0.0404	+	−0.002	+	+
ICTNET13RSR3	−0.0783	0.0066	+	−0.017	+	−
mmrbf	−0.1727	−0.1441	+	−0.180	−	−
msr_alpha0	−0.1967	−0.0261	+	−0.025	+	+
msr_alpha0_95_4	−0.1781	−0.0249	+	−0.023	+	+
msr_alpha1	−0.1750	−0.0254	+	−0.023	+	+
msr_alpha10	−0.1812	−0.0261	+	−0.025	+	+
msr_alpha5	−0.1880	−0.0335	+	−0.032	+	+
RMITSC	−0.0202	−0.0118	+	−0.094	−	−
RMITSC75	−0.0213	−0.0140	+	−0.103	−	−
RMITSCTh	−0.0200	−0.0006	+	−0.018	+	−
udelCombUD	−0.1267	−0.0032	+	0.003	+	+
udelManExp	−0.1002	−0.0087	+	−0.006	+	+
udelPseudo1	−0.1933	−0.0040	+	0.003	+	+
udelPseudo1LM	−0.2632	−0.0050	+	0.003	+	+
udelPseudo2	−0.1238	−0.0380	+	0.003	+	+
udemFbWikiR	−0.0610	−0.0430	+	0.003	+	+
udemQlm11	−0.0570	−0.0107	+	0.004	+	+
udemQlm11Fb	−0.0990	−0.0375	+	0.004	+	+
udemQlm11FbWiki	−0.0610	−0.0430	+	0.003	+	+
udemQlml1FbR	−0.1037	−0.0223	+	−0.019	+	+
udemQlml1R	−0.0570	−0.0107	+	0.004	+	+
UDInfolabWEB1	−0.2053	−0.0722	+	−0.039	+	+
UDInfolabWEB1R	−0.2053	−0.0722	+	−0.039	+	+
UDInfolabWEB2	−0.0897	−0.0425	+	−0.039	+	+
UDInfolabWEB2R	−0.0897	−0.0425	+	−0.039	+	+
UJS13LCRAd1	−0.2446	−0.1838	+	−0.180	+	+
UJS13LCRAd2	−0.2116	−0.1603	+	−0.180	+	−
UJS13Risk1	−0.2340	−0.1638	+	−0.180	+	−
UJS13Risk2	−0.2206	−0.1528	+	−0.180	+	−
uogTrADnLrb	−0.0425	−0.0416	+	−0.030	+	+

(Continued)

Table 4. (*Continued*)

runID	base. UERR@20	alg1. UERR@20	alg1-base. UERR@20	alg2. UERR@20	alg2-base. UERR@20	alg2-alg1. UERR@20
uogTrAIwLmb	−0.0662	−0.0823	−	−0.017	+	+
uogTrAS1Lb	−0.0514	−0.0451	+	−0.042	+	+
uogTrAS2Lb	−0.0673	−0.0738	−	−0.044	+	+
uogTrBDnLaxw	−0.1671	−0.1460	+	−0.180	−	−
uogTrBDnLmxw	−0.1641	−0.1245	+	−0.180	−	−
ut22base	−0.0651	−0.0534	+	−0.080	−	−
ut22spam	−0.1842	−0.1517	+	−0.061	+	+
ut22xact	−0.0453	−0.0362	+	−0.076	−	−
UWCWEB13RISK01	−0.2978	−0.0301	+	−0.021	+	+
UWCWEB13RISK02	−0.2362	−0.0449	+	−0.021	+	+
webishybrid	−0.1240	−0.1055	+	−0.008	+	+
webismixed	−0.1069	−0.0917	+	−0.001	+	+
webisnaive	−0.1356	−0.1095	+	−0.001	+	+
webisrandom	−0.1182	−0.0867	+	0.003	+	+
webiswikibased	−0.1255	−0.1052	+	−0.001	+	+
webiswtbaseline	−0.1333	−0.1048	+	−0.001	+	+
wistud.runA	−0.1299	−0.0410	+	−0.038	+	+
wistud.runB	−0.1656	−0.0380	+	−0.036	+	+
wistud.runC	−0.2830	−0.2900	−	−0.286	−	+
wistud.runD	−0.0442	0.0232	+	−0.001	+	−

Table 2 also shows results produced when applying Method 1. They are very close to the ones obtained using Method 2, albeit slightly lower – using ERR-IA@20, 55 results improved vs 58 for Method 2 on 2013 runs, 27 vs 28 on 2014 ad-hoc runs, and 7 vs 9 on 2014 risk-sensitive runs. However, the actual numbers for Method 1 effectiveness (not shown here due to space constraints) are much lower than the numbers for Method 2.

Risk Analysis. Summary of risk-sensitive measures shown in Table 3 are evidence that our methods, overall, improve robustness as well. The effectiveness measure (R) used in the delta formula ($\Delta_q = R_A(q) - R_{BASE}(q)$) is ERR@20. We use $\alpha = 5$ for U_{RISK}.

However, robustness does not go up whenever effectiveness goes up. In fact, although an overwhelming number of runs witness an improvement of their robustness, there are far less improvement in risk-sensitive measure than there were for average effectiveness measures for the 2014 runs. For the 2013 dataset, the number of improved runs based on risk-sensitive measures is very close to the number improved using average effectiveness measures: with respect to Terrier, Method 1 improved robustness of 56 out of 60 runs and Method 2 improved 48 out of 60. But for the TREC Web 2014 dataset, the number is much lower, especially for the runs submitted to the risk-sensitive track (Table 4).

There are also stark differences in risk-sensitive measures depending on whether Terrier, Indri or Indri-with-spam-filtering is being used. The least significant improvements – as well as the most decrease in risk-sensitive measures – again are observed on the runs from the 2014 dataset. For instance, using Method 1, 10 out of the 12 runs submitted for the risk-sensitive track see a decrease in risk-sensitivity utility measure with respect to Terrier, and 18 out-of-the 30 for the runs submitted to the 2014 ad-hoc track with respect to Terrier. This is not very surprising since Indri – rather than Terrier – was used to obtain our pre-fetched list of documents sorted by popularity.

Also, even though Method 1 re-ranking is more robust than the original ranking more often than Method 2, there are more cases where Method 2 is more robust than Method 1.

5 Conclusion and Future Work

In this paper, we proposed two re-ranking approaches based on exploiting document popularity across a topic, and show that these methods can help improve average overall effectiveness as well as robustness. Using the runs submitted to TREC Web track 2013 and 2014 as baselines, we show that, after our re-ranking, overall effectiveness gets improved in an overwhelming number of cases, and robustness gets improved in a large number of cases but fewer than for overall effectiveness. Our future efforts will focus on establishing a principled framework for better exploiting the popularity of documents as well as other features to improve robustness of systems.

References

1. Bah, A., Carterette, B.: Aggregating results from multiple related queries to improve web search over sessions. In: Jaafar, A., Mohamad Ali, N., Mohd Noah, S.A., Smeaton, A.F., Bruza, P., Bakar, Z.A., Jamil, N., Sembok, T.M.T. (eds.) AIRS 2014. LNCS, vol. 8870, pp. 172–183. Springer, Heidelberg (2014)
2. Bhattacharjee, R., Goel, A.: Algorithms and incentives for robust ranking. In: SODA (2007)
3. Burges, C., Shaked, T., Renshaw, E., Lazier, A., Deeds, M., Hamilton, N., Hullender, G.: Learning to rank using gradient descent. In: ICML (2005)
4. Büttcher, S., Clarke, C. L., Lushman, B.: Term proximity scoring for ad-hoc retrieval on very large text collections. In: SIGIR (2006)
5. Carbonell, J., Goldstein, J.: The use of MMR, diversity-based reranking for reordering documents and producing summaries. In: SIGIR (1998)
6. Carterette, B., Chandar, P.: Probabilistic models of ranking novel documents for faceted topic retrieval. In: CIKM (2009)
7. Chapelle, O., Ji, S., Liao, C., Velipasaoglu, E., Lai, L., Wu, S.L.: Intent-based diversification of web search results: metrics and algorithms. IR **14**(6), 572–592 (2011)
8. Chapelle, O., Metlzer, D., Zhang, Y., Grinspan, P.: Expected reciprocal rank for graded relevance. In: CIKM (2009)
9. Clarke, C.L., Kolla, M., Cormack, G.V., Vechtomova, O., Ashkan, A., Büttcher, S., MacKinnon, I.: Novelty and diversity in information retrieval evaluation. In: SIGIR (2008)

10. Collins-Thompson, K., Bennett, P., Diaz, F., Clarke, C.L., Voorhees, E.M.: TREC 2013 web track overview. In: TREC (2013)
11. Collins-Thompson, K., Bennett, P., Diaz, F., Clarke, C.L., Voorhees, E.M.: TREC 2014 web track overview. In: TREC (2014)
12. Collins-Thompson, K.: Reducing the risk of query expansion via robust constrained optimization. In: CIKM (2009)
13. Macdonald, C., Ounis, I., Dinçer, B.: Tackling biased baselines in the risk-sensitive evaluation of retrieval systems. In: de Rijke, M., Kenter, T., de Vries, A.P., Zhai, C., de Jong, F., Radinsky, K., Hofmann, K. (eds.) ECIR 2014. LNCS, vol. 8416, pp. 26–38. Springer, Heidelberg (2014)
14. Järvelin, K., Kekäläinen, J.: Cumulated gain-based evaluation of IR techniques. TOIS 20(4), 422–446 (2002)
15. Kang, C., Wang, X., Chen, J., Liao, C., Chang, Y., Tseng, B., Zheng, Z.: Learning to re-rank web search results with multiple pairwise features. In: WSDM 2011 (2011)
16. Liu, T.Y.: Learning to rank for information retrieval. FnTIR 3(3), 225–331 (2009)
17. Lv, Y., Zhai, C., Chen, W.: A boosting approach to improving pseudo-relevance feedback. In: SIGIR (2011)
18. Metzler, D., Croft, W.B.: A markov random field model for term dependencies. In: SIGIR (2005)
19. Ponte, J.M., Croft, W.B.: A language modeling approach to information retrieval. In: SIGIR (1998)
20. Radlinski, F., Dumais, S.: Improving personalized web search using result diversification. In: SIGIR (2006)
21. Robertson, S.E., Walker, S., Jones, S., Hancock-Beaulieu, M. M., Gatford, M.: Okapi at TREC-3. In: TREC (1994)
22. Santos, R.L., Macdonald, C., Ounis, I.: Exploiting query reformulations for web search result diversification. In: WWW (2010)
23. Tao, T., Zhai, C.: An exploration of proximity measures in information retrieval. In: SIGIR (2007)
24. Wang, J., Zhu, J.: Portfolio theory of information retrieval. In: SIGIR (2009)
25. Wang, L., Bennett, P.N., Collins-Thompson, K.: Robust ranking models via risk-sensitive optimization. In: SIGIR (2012)
26. Zhai, C.X., Cohen, W.W., Lafferty, J.: Beyond independent relevance: methods and evaluation metrics for subtopic retrieval. In: SIGIR (2003)
27. Zhu, J., Wang, J., Cox, I.J., Taylor, M.J.: Risky business: modeling and exploiting uncertainty in information retrieval. In: SIGIR (2009)

Heading-Aware Snippet Generation
for Web Search

Tomohiro Manabe[✉] and Keishi Tajima

Graduate School of Informatics, Kyoto University, Sakyo, Kyoto 606-8501, Japan
manabe@dl.kuis.kyoto-u.ac.jp, tajima@i.kyoto-u.ac.jp

Abstract. We propose heading-aware methods of generating search result snippets of web pages. A heading is a brief description of the topic of its associated sentences. Some existing methods give priority to sentences containing many words that also appear in headings when selecting sentences to be included in snippets with limited length. However, according to our observation, words in heading are very often omitted from their associated sentences because readers can understand the topic of the sentences by reading their heading. To score sentences considering such omission, our methods count keyword occurrences in their headings as well as in the sentences themselves. Our evaluation result indicated that our methods were effective only for queries with clear intents or containing four or more keywords. To discuss the statistical significance of the result, another evaluation with more queries is needed.

Keywords: Snippet generation · Query-biased summarization · Web search result snippets · Heading structure

1 Introduction

Most web pages contain hierarchical heading structure [11]. The structure is composed of nested logical blocks and each block is associated with a heading that briefly describes the topic of the block. Because of this feature of headings, to fully understand sentences in web pages, readers should first read the contextual headings of the sentences. The contextual headings (or merely headings) of a sentence are the headings associated with either the block containing the sentence or its hierarchical ancestor blocks. Therefore, the contextual heading words (or merely heading words), i.e. the words in the contextual headings, are important for understanding their associated sentences.

For this reason, there have been several studies on heading-aware snippet generation [13,17]. These methods assign higher scores to headings themselves [17] or sentences containing their heading words [13]. However, contextual heading words are very often omitted from their associated sentences because human readers can recognize the topic of the sentences by reading the headings first. For example, in the example page in Fig. 1, by the sentence "It has risks as well

---

T. Manabe—Research Fellow of Japan Society for the Promotion of Science.

G. Zuccon et al. (Eds.): AIRS 2015, LNCS 9460, pp. 188–200, 2015.
DOI: 10.1007/978-3-319-28940-3_15

> # Outline of exercise
> Aerobic exercise
> **Swimming**
> It is one of the aerobic exercises. It has risks as well as big benefits.
> **Running**
> Jogging is exercise at a gentle pace, and sprint is exercise at top speed.
> The main benefit is to increase physical fitness.
> Anaerobic exercise
> **Strength training**
> Its benefit is to induce muscular contraction.

Fig. 1. Example web page with hierarchical heading structure.

as big benefits", the author is writing about swimming without the word swimming. For a query "swimming risks", the existing methods cannot assign higher relevance scores for such sentences.

To solve the problem, we develop a new method of heading-aware snippet generation that takes the omission of heading words into account. Our method assign higher scores to sentences that either include query keywords within themselves or have contextual headings including query keywords. Our new approach does not conflict with the existing approach that uses heading-word occurrences in sentences. Therefore, we also consider another method which combines the two types of evidences, namely heading-word occurrences in sentences themselves and query keyword occurrences in the contextual headings of sentences.

2   Related Work

Generally, snippet generation methods uses some types of important words and document fragments. Almost all methods count the occurrences of query keywords. Additionally, some methods use pseudo relevance feedback to expand queries and obtain more keywords [8,19]. Frequently occurring words in a page may also be important for the page [13,17,19]. The first paragraph of a page [17] or the first sentence of a paragraph [13] may also be important. As listed above, most summarization methods do not focus on heading words and headings.

As explained in Sect. 1, two heading-aware summarization methods exist. The method by Tombros and Sanderson regards headings as important sentences and assigns higher scores to headings than to other sentences [17]. However, as discussed in Sect. 1, headings are also important for scoring other sentences. The method by Pembe and Güngör counts heading-word occurrences in sentences to score the sentences [13]. However, as also discussed in Sect. 1, their method does not take the omission of heading words into account.

Some snippet generation methods focus on the locations of the occurrences of query keywords. Some methods count the occurrences in document titles, which are a type of headings [17,19]. The method by Zhang et al. distinguishes

```
┌─────────────────────────────────────────────────────────────────────┐
│ Outline of exercise(0)                                               │
│ ┌─────────────────────────────────────────────────────────────────┐ │
│ │ Aerobic exercise(1)                                             │ │
│ │ ┌─────────────────────────────────────────────────────────────┐ │ │
│ │ │ Swimming(2)                                                 │ │ │
│ │ │ It is one of the aerobic exercises. It has risks as well as big benefits. │ │ │
│ │ └─────────────────────────────────────────────────────────────┘ │ │
│ │ ┌─────────────────────────────────────────────────────────────┐ │ │
│ │ │ Running(2)                                                  │ │ │
│ │ │ Jogging is exercise at a gentle pace, and sprint is exercise at top speed. │ │ │
│ │ │ The main benefit is to increase physical fitness.          │ │ │
│ │ └─────────────────────────────────────────────────────────────┘ │ │
│ └─────────────────────────────────────────────────────────────────┘ │
│ ┌─────────────────────────────────────────────────────────────────┐ │
│ │ Anaerobic exercise(1)                                          │ │
│ │ ┌─────────────────────────────────────────────────────────────┐ │ │
│ │ │ Strength training(2)                                        │ │ │
│ │ │ Its benefit is to induce muscular contraction.             │ │ │
│ │ └─────────────────────────────────────────────────────────────┘ │ │
│ └─────────────────────────────────────────────────────────────────┘ │
└─────────────────────────────────────────────────────────────────────┘
```

Fig. 2. Hierarchical heading structure of page in Fig. 1. Each rectangle encloses block, each text with subscript is heading and each subscript number represents depth of block in the hierarchy.

attribute names, which are also a type of headings [20]. These methods, however, do not count query keyword occurrences in general headings.

Outside the field of web search, many snippet generation methods for XML documents are based on XML element retrieval [8,19], and many XML element retrieval methods take the hierarchical ancestors of elements into account [2,3]. However, unlike our methods, most XML element retrieval methods do not distinguish headings from other components of elements. The BM25E function for element scoring distinguish headings from other components [10]. However, the application of the function to snippet generation has not been discussed.

3 Heading Structure Extraction

Hierarchical heading structure of web pages is not obvious. In this section, we introduce an outline of HEPS, our previously proposed method for extracting the implicit hierarchical heading structure from HTML web pages [11]. Throughout this paper, we assume that the hierarchical heading structure of web pages are already extracted by this method. See our previous paper for the detailed design decisions and evaluation results of HEPS itself [11].

First we define the hierarchical heading structure and its components.

Heading: In our definition, a *heading* is a highly summarized description of the topic of a part of a web page.

Block: As explained above, a heading is associated with a *block*, a clearly specified region in a web page. We consider neither a block that consists of its heading only nor a block without its heading. A whole web page is also a block because it is clearly specified and we can regard its title (or URL) as its heading.

Hierarchical Heading Structure: A block may contain another block entirely, but two blocks never partially overlap. All blocks in a page form a hierarchical heading structure whose root is the block representing the entire page. In Fig. 2, we show the hierarchical heading structure in the example page in Fig. 1. Each block (including the page) is enclosed by a rectangle, and its heading is associated with a subscript representing its depth in the hierarchy.

The HEPS method involves pre-processing and three main steps. In the first main step, it classifies DOM nodes into sets of nodes sharing the same visual style (e.g., font size and font weight). Second, it sorts the sets in descending order of visual significance of their elements. Third, it determines the actual heading set of the highest significance and divides the page into blocks. The third step is recursively repeated to divide a page into nested blocks.

4   Snippet Generation Methods

In this section, we explain four snippet generation methods for web search.

4.1   Basic Snippet Generation Method

Generally, the quality of document summaries relies on three factors [1]. The *readability* of a summary is how easy it is for humans to read [5,7], its *representativeness* is how well it represents the contents of the original document [9], and its *judgeability* is to what extent it helps users to judge the relevance of the original document to the users' informational needs [9]. Among the three factors, judgeability is the most important for search result snippets.

Basically, search result snippets are generated from web pages by search systems in three steps as described below [13,17,19]. First, the system splits the page into text fragments. To generate readable snippets, many systems split it into semantically coherent fragments such as sentences. Second, the system scores the fragments based on the numbers of the occurrences of important words in the fragments. The occurrences of the query keywords directly indicate the relevance of the original page to the users' intent behind the query. Therefore, almost all systems take keyword occurrences into account for higher judgeability. On the other hand, other important words (see Sect. 2) in a document represent the contents of the original document better than other words. Therefore, many systems take important-word occurrences into account for higher representativeness. Third, the system selects the top-ranked sentences into the summary. In this step, the system selects the sentences in descending order of their scores until the length of the summary reaches the limit. Our *baseline* method also consists of these three steps. The method is described as below.

Input: A web page with its DOM tree structure. Note that we consider only documents in English throughout this paper. This is merely because we use some language-dependent libraries for sentence segmentation and stemming. Note that our heading-aware sentence scoring methods are language-independent.

Outline of exercise

... It has risks as well as big benefits. ... Running ... The main benefit is to increase physical fitness. ... Its benefit is to induce muscular contraction.

Fig. 3. Example baseline snippet for example query "benefits running".

Sentence Segmentation: First, the method extracts the text contents of the page, then segments the contents into sentences. As the text contents of a page, we extract the text contents of all text and IMG (image) nodes under the BODY (content body) node of the page, and concatenate them in the document order. As the text contents of IMG nodes, we extract their alternate text. From IMG nodes without alternate text, we extract the URLs of the images. We split the text contents of the page into sentences by the Stanford CoreNLP toolkit [12].

Sentence Scoring: We score the sentences based on the number of keyword occurrences in them by a variant of the BM25 function [15]. The function calculates the score of a sentence s for keyword query q by the following formula:

$$\text{score}(q, s) = \sum_{\kappa \in q} \frac{\text{weight}(\kappa, s)}{k_1 + \text{weight}(\kappa, s)} \log \frac{N - \text{sf}(\kappa) + 0.5}{\text{sf}(\kappa) + 0.5} \tag{1}$$

where κ is a keyword in q, k_1 is a parameter to modify the scaling of occurrence frequency, N is the number of all sentences, and $\text{sf}(\kappa)$ is the number of sentences containing κ in the page. The $\text{weight}(\kappa, s)$ is defined as

$$\text{weight}(\kappa, s) = \frac{\text{occurs}(\kappa, s)}{\left((1 - b) + b \cdot \frac{\text{length}(s)}{\text{avgLength}} \right)} \tag{2}$$

where $\text{occurs}(\kappa, s)$ is the number of occurrences of κ in s, b is the parameter to modify the strength of length normalization, $\text{length}(s)$ is the length of s in number of words, and avgLength is the average length of sentences in the page. We count $\text{occurs}(\kappa, s)$ after the basic pre-processing, i.e. stemming by the Porter stemming algorithm [14] and removal of 33 default stop words of Apache Lucene.

Sentence Selection: To select the sentences, we simply scan the sentences in descending order of their scores, and if there still remains the space to include the sentence into the snippets, we include it. We can also adopt advanced methods for the selection, such as Maximal Marginal Relevance [4,20], however, we adopt this simple method because this step is not the main topic of this paper.

Output: The generated snippet and the title of the input page. If there is no page title specified, we output the page URL. Figure 3 is an example output.

4.2 Occurrences of Heading Words in Sentences

Heading words are important to represent their associated blocks because the words are selected by the authors to describe the topics of the blocks briefly.

Outline of exercise

> **Aerobic exercise** > **Swimming**
It is one of the aerobic exercises. ...
> **Aerobic exercise** > **Running**
Jogging is exercise at a gentle pace, and sprint is exercise at top speed. The main
benefit is to increase physical fitness.

Fig. 4. Example heading-aware snippet for example query "benefits running".

As discussed in Sect. 1, we consider the contextual headings and heading words
of sentences. For example, the heading words of the sentence "It is one of the
aerobic exercises" in Fig. 2 are outline, of, exercise, aerobic, and swimming.

One promising way to generate representative snippets is to extract sentences containing many occurrences of their contextual heading words. Pembe
and Güngör [13] proposed such a method. We also use this idea for our *existing*
method, which is based on summation of the BM25 scores for two types of words,
namely query keywords and heading words. However, a weighted summation of
BM25 scores produces a worse ranking in case that they count occurrences of the
same words [16]. Therefore, we split the words into three types, namely narrow
query keywords (NK-words), narrow heading-words (NH-words), and heading
keywords (HK-words). The NK-words are query keywords which are not heading words, and the NH-words are heading words which are not query keywords.
The HK-words are the words which are heading words and also query keywords.
We modify the baseline method explained before as described below.

Sentence Segmentation: Because headings and blocks are semantically coherent fragments and no sentence should overlap the boundaries of them, we segment the text contents of pages into text fragments by all their boundaries, and
then segment the fragments into sentences by the Stanford CoreNLP toolkit [12].
Because we show headings in a different way from other components of snippets
(as discussed later), we separately extract headings and other sentences.

Sentence Scoring: The new $\text{score}(q, s)$ and $\text{weight}(w, s)$ are calculated by:

$$\text{score}(q, s) = \sum_{w \in q \cup h(s)} \frac{\text{weight}(w, s)}{k_1 + \text{weight}(w, s)} \log \frac{N - \text{sf}(w) + 0.5}{\text{sf}(w) + 0.5}, \quad (3)$$

$$\text{weight}(w, s) = \frac{\text{occurs}(w, s) \cdot boost^{\text{typeof}(w)}}{\left((1 - b) + b \cdot \frac{\text{length}(s)}{\text{avgLength}} \right)} \quad (4)$$

where $h(s)$ is heading words of s, w is a word in q or $h(s)$, and typeof
$(w) \in \{\text{NH-words}, \text{NK-words}, \text{HK-words}\}$ is the type of w. The parameter
$boost^{\text{typeof}(w)}$ represents the importance of the occurrences of the words whose
type is typeof(w).

Output: The generated text snippets and their headings including the title or
URL of the input page. In case that heading structure of documents are given,

we can improve readability of snippets by showing sentences and their headings separately [13]. We also adopt this idea. Figure 4 shows an example output.

The other steps of this method are same as those of the baseline method. We call this method the *existing* method.

4.3 Keyword Occurrences in Headings

Our observation is that the heading words are very often omitted from sentences. Despite such omission, heading words are important to clarify the topic of their associated sentences. Therefore, to select sentences that well represent the original document considering such omission, we must count keyword occurrences in the contextual headings of the sentences as well as in the sentences themselves.

Sentence Scoring: Based on this idea, we regard that each sentence comprises two fields, namely the contents of the sentence itself and its contextual headings, and adopt a variant of BM25F, a scoring function for documents comprising multiple fields [16]. The function calculates the score of a sentence S comprising two fields for keyword query q by the following formulas:

$$\text{score}(q, S) = \sum_{\kappa \in q} \frac{\text{weight}(\kappa, S)}{k_1 + \text{weight}(\kappa, S)} \log \frac{N - \text{sf}(\kappa) + 0.5}{\text{sf}(\kappa) + 0.5}, \tag{5}$$

$$\text{weight}(\kappa, S) = \sum_{f \in S} \frac{\text{occurs}(\kappa, f, S) \cdot \text{boost}_f}{\left((1 - b) + b \cdot \frac{\text{length}(f,S)}{\text{avgLength}(f)}\right)} \tag{6}$$

where f is a field in S, $\text{occurs}(\kappa, f, S)$ is the number of occurrences of κ in f of S, boost_f is the weight of keyword occurrences in f, $\text{length}(f, S)$ is the length of f in S, and $\text{avgLength}(f)$ is the average length of f. The other steps of this method are same as those of the existing method. We call this method *our* method.

4.4 Combination of Two Advanced Methods

Above two modifications can be applied independently. Therefore, we can consider the fourth method which adopts both of them.

Sentence Scoring: We calculate the combined score and weight by:

$$\text{score}(q, S) = \sum_{w \in q \cup h(S)} \frac{\text{weight}(w, S)}{k_1 + \text{weight}(w, S)} \log \frac{N - \text{sf}(w) + 0.5}{\text{sf}(w) + 0.5}, \tag{7}$$

$$\text{weight}(w, S) = \sum_{f \in S} \frac{\text{occurs}(w, f, S) \cdot \text{boost}_f^{typeof(w)}}{\left((1 - b) + b \cdot \frac{\text{length}(f,S)}{\text{avgLength}(f)}\right)} \tag{8}$$

where $\text{boost}_f^{typeof(w)}$ is the weight of occurrences of w in f. The other steps are same as those of our method. We call this method the *combination* method.

Table 1. Boost for occurrence of words of each type in each field.

Parameter name	Value	Parameter name	Value
$boost^{\text{HK-words}}_{\text{headings}}$	3.0	$boost^{\text{HK-words}}_{\text{sentence}}$ ($boost^{\text{HK-words}}$)	4.0
$boost^{\text{NH-words}}_{\text{headings}}$	0	$boost^{\text{NH-words}}_{\text{sentence}}$ ($boost^{\text{NH-words}}$)	1.0
$boost^{\text{NK-words}}_{\text{headings}}$ ($boost_{\text{headings}}$)	3.0	$boost^{\text{NK-words}}_{\text{sentence}}$ ($boost^{\text{NK-words}}$, $boost_{\text{sentence}}$)	3.0

4.5 Parameters and Fine Tuning

These scoring functions require three types of parameters: The saturation factor k_1 controls scaling of weighted term frequency, b controls the strength of length normalization, and *boost* controls the weights of term occurrences of each type of words in each field. Because the scaling and normalization are not the main topic of this paper, we use the default values 2.0 for k_1 and 0.75 for b [16].

The setting of *boost* is important for effective heading-aware snippet generation. According to the observation by Pembe and Güngör [13], occurrences of query keywords are three times more important than those of heading words. Therefore, we use 3.0 for all $boost^{\text{NK-words}}_{\text{sentence}}$ in Sect. 4.4, $boost_{\text{sentence}}$ in Sect. 4.3, and $boost^{\text{NK-words}}$ in Sect. 4.2 while we use 1.0 for all $boost^{\text{NH-words}}_{\text{sentence}}$ in Sect. 4.4 and $boost^{\text{NH-words}}$ in Sect. 4.2. Because there is no existing observation about the balance of weights of the keyword occurrences in sentences and in their contextual headings, we simply use 3.0 (same as $boost^{\text{NK-words}}_{\text{sentence}}$) for $boost^{\text{NK-words}}_{\text{headings}}$ in Sect. 4.4 and $boost_{\text{headings}}$ in Sect. 4.3. Because heading words always occur in headings, we use 0 for $boost^{\text{NH-words}}_{\text{headings}}$. As the weight of HK-words, we use the summations of the weight of NH-words and the weight of NK-words. All the *boost* values are listed in Table 1 for reference.

5 Evaluation

In this section, we evaluate each snippet generation method.

5.1 Evaluation Methodology

As discussed in Sect. 4.1, judgeability is the most important property of effective search result snippets. Therefore, to measure the effectiveness of snippet generation methods, we measure the judgeability of their output snippets. To measure the judgeability, in the INEX snippet retrieval track [18], the results of relevance judgments under two different conditions are compared. One judgment is performed based on the entire documents while the other is only based on their snippets. If they agree, the snippets provided high judgeability and the snippet generation method was effective. We use this measure and also their length limit of snippets, which is 180 letters for a page.

However, the target of INEX is XML documents while our target is web pages. Therefore, we used a data set for text retrieval conference (TREC) 2014 web track ad-hoc task [6].

Table 2. Comparison of average evaluation scores of four methods.

Method	Recall	NR	GM
Baseline	**.475**	**.828**	**.512**
Exist.	.373	.780	.386
Ours	.438	.777	.456
Combi.	.396	.776	.401

Table 3. Average evaluation scores of four methods for each type of queries.

(A) For 24 *faceted* queries.

Method	Recall	NR	GM
Baseline	**.524**	**.806**	**.539**
Exist.	.416	.737	.378
Ours	.509	.723	.470
Combi.	.290	.737	.257

(B) For 24 *single* queries.

Method	Recall	NR	GM
Baseline	.431	**.837**	.488
Exist.	.336	.804	.392
Ours	.375	.816	.443
Combi.	**.491**	.795	**.530**

5.2 Data Set and Evaluation Measures

Queries and Intents: Fifty keyword queries and their intent descriptions.

Document Collection: ClueWeb12 B13, a web snapshot crawled in 2012. We extracted top-20 pages for each query (total 1,000 pages) from the official baseline search result for the TREC task. The result is generated by the default scoring by Indri search engine and filtered by Waterloo spam filter.

Page-based Relevance Judgment Data: The TREC official graded relevance of the entire pages to the intents. We simply regarded that documents whose grades are more than 0 as relevant to the intent, and the others are irrelevant.

Snippet-based Relevance Judgment Data: We carried out a user experiment with four participants. They are all non-native English readers familiar with web search. In each period of the experiment, each participant is required to read the intent description behind a query first. Next, he is required to scan top-20 search result items containing the snippets generated by a method and to judge whether each original page is relevant to the intent. We broke out the search results to participants by Graeco-Latin square, therefore each snippet was not judged more than once, and each participant did not judge a page more than once and used all methods almost evenly. As described above, we adopted binary relevance. It is because the user of a real web search engine must decide to read or not for each original page based on its snippets and there is no intermediate choice.

Evaluation Measures: We use three evaluation measures from the INEX track: Recall, negative recall (NR), and the geometric mean (GM) of them. Recall is the ratio of pages correctly judged as relevant on their snippets to pages relevant as a whole. It is calculated by $|Correctly\ judged\ pages\ relevant\ as\ a\ whole|/|Pages\ relevant\ as\ a\ whole|$. On the other hand, NR is the ratio of pages correctly judged as irrelevant on their snippets to pages irrelevant as a whole. It is calculated by $|Correctly\ judged\ pages\ irrelevant\ as\ a\ whole|/|Pages\ irrelevant\ as\ a\ whole|$. GM is the primary evaluation measure of the INEX track and our evaluation. It is calculated by $\sqrt{Recall \cdot NR}$. To integrate the evaluation scores for multiple queries, we calculated the arithmetic mean of them.

Table 4. Average GM scores of four methods and query length excluding stopwords.

$|Keywords|$	$|Queries|$	Baseline	Exist.	Ours	Combi.
2	25	**.585**	.406	.543	.393
3	10	**.503**	.387	.378	.388
4 or more	15	.394	.350	.362	**.425**

5.3 Evaluation Results and Discussion

Comparison of Snippet Generation Methods: First, we compared the average evaluation scores of four snippet generation methods. Table 2 lists the results. The baseline method achieved the top scores by all the evaluation measures. Our heading-aware method achieved the second GM score. The existing heading-aware method achieved the worst GM score and its difference from the baseline method was statistically significant ($p < 0.05$) according to Student's paired t-test where each pair is composed of the evaluation scores of the baseline and heading-aware methods for a query. Hereafter in this paper, we discuss statistical significance based on the same test procedure. There was no statistically significant difference from the baseline to the other methods. As shown in this result, the heading-aware methods were not effective for general queries. The difference of the GM scores was mainly caused by the difference of the recall scores. In fact, the best method improved the recall score by 27.3 % from the worst while the NR score by only 6.70 %. In other words, the effectiveness of the methods mainly depends on how many relevant pages its output snippets can indicate to the users. This tendency was seen through all evaluations.

Effect of Query Type: For detailed evaluation, TREC splits queries into several types. *Faceted* queries are underspecified, while there are clear and focused intents behind *single* queries [6]. The data set contains 24 queries of each type. It also contains only two *ambiguous* queries, however we ignored them. The scores for the faceted queries are listed in Table 3 (A) and the scores for the single queries are listed in (B). As shown in these tables, the baseline method achieved the best scores for faceted queries while the combination method achieved the best recall and GM scores for single queries. Only the GM score difference between the baseline and combination methods for faceted queries was statistically significant. This fact suggests that heading-aware snippet generation methods may be effective for clearly specified intents. To indicate the relevance of a page to a clearly specified intent, small number of sentences and their rich contextual information, i.e. their headings, may be important. In the other cases, it may be important to show a larger number of sentences in the page. For further discussion, another evaluation with more queries is needed.

Effect of Query Length: When a user inputs multiple keywords, the user is probably requesting pages in which all the keywords occur in relation to each other. On the other hand, as discussed in Sect. 4.3, contextual heading words

Table 5. Median amount of required time in second to check snippets of 20 pages for one query.

Table 6. Comparison of average evaluation scores of four participants.

(A) By each method.

Method	Time in sec.
Baseline	411.5
Exist.	**308.5**
Ours	315.5
Combi.	349.0

(B) By each participant.

Participant	Time in sec.
A	297.5
B	429.0
C	347.5
D	293.0

Participant	Recall	NR	GM
A	.367	.787	.376
B	.482	.783	.498
C	.459	.815	.451
D	.375	.776	.430

have semantic relationship to their associated sentences. Therefore, the heading-aware methods must be more useful for queries containing more keywords. In other words, there are usually less sentences containing more different keywords directly. However, considering the contextual headings of the sentences, heading-aware methods can detect more of relevant sentences. Based on this idea, we classified the queries by their numbers of keywords excluding stopwords. Table 4 lists the numbers of queries in each class and the GM scores of each method for each class. For the queries with two keywords, the baseline method achieved the best GM score and its differences from the existing and combination methods were statistically significant. However, only the combination method retained its score for the longer queries while the other three methods lost their scores. The correlation coefficient of the GM score and the number of pairs of different query keywords for each query was .247 for the combination method while –0.105 for the baseline. Especially, for four or more keywords, the combination method achieved the best score. It supports the above discussion about longer queries. For further discussion, another evaluation with more queries is needed.

Query Type and Query Length: Query type depends on query length because more query keywords specify the intents of the query more clearly. In fact, the average length of single queries was 3.54 words while that of faceted queries was 2.25 words. Note that this dependence might affect our evaluation results.

Required Time Analysis: We also measured the median required time for checking 20 pages for a query. Table 5 (A) lists the results. Intuitively, the assessors took much more time for our evaluation tasks than practical search tasks. It may be because they are non-native English reader, and/or because they read snippets more carefully for more accurate judgment than usual. Generally, heading-aware snippets significantly reduced the required time. It must be because the users can read structured text more easily than plain text.

Effect of Assessors: We also compared the required time and evaluation scores for each assessor. Table 5 (B) lists the median time in second required for checking 20 pages and Table 6 lists the average evaluation scores for each assessor. As shown in Table 5 (B), the required times are quite different for each assessor.

The difference also affects the average GM scores of them, that is, the most and second-most careful assessors, B and C, achieved the best and second-best GM scores respectively. Note that the effect of the assessors for the other comparative evaluations is limited because each assessor uses each methods almost evenly.

6 Conclusion

We introduced a novel idea for heading-aware snippet generation and compared one baseline and three heading-aware snippet generation methods. The idea is that sentences whose contextual headings contain query keywords provide judgeability as well as sentences containing query keywords directly. Our evaluation result indicated that the heading-aware methods were not effective for general queries. Only for queries representing its intents clearly or containing four or more keywords, the heading-aware combination method achieved the best score. This fact suggests that heading-aware snippet generation is useful for such queries. However, to discuss the statistical significance of the result, an additional evaluation with more queries is needed.

Acknowledgment. This work was supported by JSPS KAKENHI Grant Number 13J06384, 26540163.

References

1. Ageev, M., Lagun, D., Agichtein, E.: Towards task-based snippet evaluation: preliminary results and challenges. In: MUBE (SIGIR Workshop), pp. 1–2 (2013)
2. Amer-Yahia, S., Lalmas, M.: XML search: languages, INEX and Scoring. SIGMOD Rec. **35**(4), 16–23 (2006)
3. Arvola, P., Kekäläinen, J., Junkkari, M.: Contextualization models for XML retrieval. Inf. Process. Manage. **47**(5), 762–776 (2011)
4. Carbonell, J., Goldstein, J.: The use of MMR, diversity-based reranking for reordering documents and producing summaries. In: SIGIR, pp. 335–336 (1998)
5. Clarke, C.L.A., Agichtein, E., Dumais, S., White, R.W.: The influence of caption features on clickthrough patterns in web search. In: SIGIR, pp. 135–142 (2007)
6. Collins-Thompson, K., Macdonald, C., Bennett, P.N., Diaz, F., Voorhees, E.M.: TREC 2014 web track overview. In: TREC (2014)
7. Kanungo, T., Orr, D.: Predicting the readability of short web summaries. In: WSDM, pp. 202–211 (2009)
8. Leal Bando, L., Scholer, F., Thom, J.: RMIT at INEX 2011 snippet retrieval track. In: Geva, S., Kamps, J., Schenkel, R. (eds.) INEX 2011. LNCS, vol. 7424, pp. 300–305. Springer, Heidelberg (2012)
9. Liang, S.F., Devlin, S., Tait, J.I.: Evaluating web search result summaries. In: Lalmas, M., MacFarlane, A., Rüger, S.M., Tombros, A., Tsikrika, T., Yavlinsky, A. (eds.) ECIR 2006. LNCS, vol. 3936, pp. 96–106. Springer, Heidelberg (2006)
10. Lu, W., Robertson, S., MacFarlane, A.: Field-weighted XML retrieval based on BM25. In: Fuhr, N., Lalmas, M., Malik, S., Kazai, G. (eds.) INEX 2005. LNCS, vol. 3977, pp. 161–171. Springer, Heidelberg (2006)

11. Manabe, T., Tajima, K.: Extracting logical hierarchical structure of HTML documents based on headings. VLDB **8**(12), 1606–1617 (2015)
12. Manning, C.D., Surdeanu, M., Bauer, J., Finkel, J., Bethard, S.J., McClosky, D.: The Stanford CoreNLP natural language processing toolkit. In: ACL, pp. 55–60 (2014)
13. Pembe, F.C., Güngör, T.: Structure-preserving and query-biased document summarisation for web searching. Online Info. Rev. **33**(4), 696–719 (2009)
14. Porter, M.F.: An algorithm for suffix stripping. In: Readings in information retrieval, pp. 313–316. Morgan Kaufmann Publishers (1997)
15. Robertson, S., Walker, S., Jones, S., Hancock-Beaulieu, M., Gatford, M.: Okapi at TREC-3. In: TREC, pp. 109–126 (1996)
16. Robertson, S., Zaragoza, H., Taylor, M.: Simple BM25 extension to multiple weighted fields. In: CIKM, pp. 42–49 (2004)
17. Tombros, A., Sanderson, M.: Advantages of query biased summaries in information retrieval. In: SIGIR, pp. 2–10 (1998)
18. Trappett, M., Geva, S., Trotman, A., Scholer, F., Sanderson, M.: Overview of the INEX 2013 snippet retrieval track. In: CLEF (2013)
19. Wang, S., Hong, Y., Yang, J.: PKU at INEX 2011 XML snippet track. In: Geva, S., Kamps, J., Schenkel, R. (eds.) INEX 2011. LNCS, vol. 7424, pp. 331–336. Springer, Heidelberg (2012)
20. Zhang, L., Zhang, Y., Chen, Y.: Summarizing highly structured documents for effective search interaction. In: SIGIR, pp. 145–154 (2012)

Text Processing, Understanding and Categorization

Smoothing Temporal Difference for Text Categorization

Fumiyo Fukumoto[✉] and Yoshimi Suzuki

Graduate Faculty of Interdisciplinary Research,
University of Yamanashi, Kofu, Japan
{fukumoto,ysuzuki}@yamanashi.ac.jp

Abstract. This paper addresses text categorization problem that training data may be derived from a different time period than test data. We present a method for text categorization that minimizes the impact of temporal effects by using term smoothing and transfer learning techniques. We first used a technique called Temporal-based Term Smoothing (TTS) to replace those time sensitive features with representative terms, then applied boosting based transfer learning algorithm called TrAdaBoost for categorization. The results using a 21-year Japanese Mainichi Newspaper corpus showed that integrating term smoothing and transfer learning improves overall performance, especially it is effective when the creation time period of the test data differs greatly from the training data.

Keywords: Temporal adaptation · Term smoothing · Text categorization · Transfer learning

1 Introduction

A basic assumption in text categorization is that the creation time period of training data is the same as test data. When the assumption does not hold, traditional classification methods might not work well. However, it is often the case that the term distribution in the training data is different from that of the test data when the training data may drive from a different time period from the test data. For instance, the term "Alcindo" frequently appeared in the documents tagged "Sports" category in 1994. This is reasonable because Alcindo is a Brazillian soccer player and he was one of the most loved players in 1994. However, the term did not occur more frequently in the Sports category since he retired in 1997. The observation shows that the informative term such as "Alcindo" appeared in the training data with Sports category, is not informative in the test data when training data may derive from a different time period from the test data. If we can have a large number of training documents with the same time period of test data, we can classify test documents with high accuracy. However, manual annotation of tagged new data is very expensive and time-consuming. The methodology for accurate classification of the new test data by

© Springer International Publishing Switzerland 2015
G. Zuccon et al. (Eds.): AIRS 2015, LNCS 9460, pp. 203–214, 2015.
DOI: 10.1007/978-3-319-28940-3_16

making the maximum use of tagged old data is needed to improve categorization performance.

In this paper, we present a method for text categorization that minimizes the impact of temporal effects. We focused on organization and person names which frequently appear in a specific category. These often appeared in specific time period, *e.g.*, "Alcindo" occurred in 1994, and "Ronaldo" first appeared in 2004 in the documents. We identified these terms in the documents and replaced these to a representative term in order to regard that these terms are equally salient for a specific category, *i.e.* sports across full temporal range of training documents. We call this procedure, Temporal-based Term Smoothing (TTS). Each document is represented by using a vector of terms including representative terms, and classifiers are trained. We applied boosting based transfer learning, called TrAdaboost [3] in order to minimize the impact of temporal effects, *i.e.* it decreases the weights of training instances that are very different from the test data.

The rest of the paper is organized as follows. The next section describes an overview of existing related work. Section 3 presents our approach, especially describes how to adjust temporal difference between training and test documents. Finally, we report some experiments with a discussion of evaluation.

2   Related Work

The analysis of temporal aspects is widely studied since corpora from the WWW became a popular source for text processing tasks. One attempt is detection of concept or topic drift [6,13,16]. The earliest known approach is the work of [14]. They presented a method to handle concept changes with SVMs. They used $\xi\alpha$-estimates to select the window size so that the estimated generalization error on new examples is minimized. The results which were tested on the TREC show that the algorithm achieves a low error rate and selects appropriate window sizes. He *et al.* proposed a method to find bursts, periods of elevated occurrence of events as a dynamic phenomenon instead of focusing on arrival rates [11]. They used Moving Average Convergence/Divergence (MACD) histogram which was used in technical stock market analysis [21] to detect bursts. They tested the method using MeSH terms and reported that the model works well for tracking topic bursts. Most of these focused just on identifying the increase of a new context, and not relating these contexts to their chronological time. Wang *et al.* developed the continuous time dynamic topic model (cDTM) [25]. The cDTM is an extension of the discrete dynamic topic model (dDTM). The dDTM is a powerful model. However, the choice of discretization affects the memory requirements and computational complexity of posterior inference. cDTM replaces the discrete state space model with its continuous generalization, Brownian motion.

Another attempt is domain adaptation. The goal of this attempt is to develop learning algorithms that can be easily ported from one domain to another, *e.g.*, from newswire to biomedical documents [4]. Domain adaptation is particularly interesting in Natural Language Processing (NLP) because it is often the case that we have a collection of labeled data in one domain but truly desire a model that works well for another domain. Several authors e.g., Xiao et al. [26] for part-of-speech tagging, Daume [4] for named-entity, and Glorot et al. [10] for sentiment classification have attempted to improve results by using domain adaptation techniques. One approach to domain adaptation is to use transfer learning. The transfer learning is a learning technique that retains and applies the knowledge learned in one or more domains to efficiently develop an effective hypothesis for a new domain. The earliest discussion is done by ML community in a NIPS-95 workshop[1], and more recently, transfer learning techniques have been successfully applied in many applications. Blitzer et al. proposed a method for sentiment classification using structual correspondence learning that makes use of the unlabeled data from the target domain to extract some relevant features that may reduce the difference between the domains [1]. Several authors have attempted to learn classifiers across domains using transfer learning in the text classification task [2,22,24]. Dai et al. proposed a co-clustering based classification algorithm to propagate the label information across different domains [2]. The method first, the in-domain data provide the class structure, which defines the classification task, by propagating label information. Then, co-clustering is extended for out-of-domain data to obtain out-of-domain document and word clusters. The extension is that class labels in the in-domain data can constrain the word clusters, which is shared among the two domains. This allows each out-of-domain cluster to be mapped to a corresponding class label based on their correlation with the document categories in the in-domain data. They showed that the algorithm improves the classification performance over the traditional learning algorithms including Transductive Support Vector Machines. All of these approaches mentioned above aimed at utilizing a small amount of newly labeled data to leverage the old labeled/unlabeled data to construct a high-quality classification model for the new data. However, the temporal effects are not explicitly incorporated into their models.

To our knowledge, there have been only a few previous work on temporal-based text categorization. Mourao et al. investigated the impact of temporal evolution of document collections based on three factors: (i) the class distribution, (ii) the term distribution, and (iii) the class similarity. They reported that these factors have great influence in the performance of the classifiers throughout the ACM-DL and Medline document collections that span across more than 20 years [20]. Salles et al. presented an approach to classify documents in scenarios where the method uses information about both the past and the future, and this information may change over time [23]. They addressed the drawbacks of which instances to select by approximating the Temporal Weighting Function (TWF) using a mixture of two Gaussians. They applied TWF to every training document. However, it is often the case that terms with informative for a specific time period and informative across the full temporal range of training documents are both included in the training data that affects overall performance of text categorization as these terms are equally weighted in their approach.

---

[1] http://socrates.acadiau.ca/courses/comp/dsilver/NIPS95_LTL/transfer.workshop. 1995.html.

In contrast with the aforementioned works, here we propose a method for text categorization that minimizes the impact of temporal effects by using term smoothing and transfer learning techniques. Our experimental results show the effectiveness of the method, especially when the creation time period of the test data differs greatly from the training data.

3 Temporal-based Term Smoothing

For reasons of both efficiency and accuracy, feature selection techniques such as χ^2 statistics, mutual information, and information gains are often used in order to select informative terms since the early 1990s when applying machine learning methods to text categorization. We note that the selected terms by using these feature selection techniques often include organization and person names. When the creation time period of training data is the same as test data, these terms are key features to classify test documents correctly. However, this situation is often hampered by the assumption that the test data may be derived from a different time period than training data, although these terms are still informative terms for a specific category. We then identified words with semantically related with each other, and replaced these to a representative word in order to regard that these words are equally informative across training and test sets.

We firstly used named entities recognition programme, and extracted person name and organization name, and make a named entity list for each category. Next, we collected semantically related words. To do this, we used word2vec tool released by Google in 2013. The word2vec first constructs a term from the training text data and then learns vector representation of words. It is provided two main model architectures, continuous bag-of-words and skip-gram. We used skip-gram model as it gives better word representations when the data is small [19]. The skip-gram model's objective funcion L is to maximize the likelihood of the prediction of contextual words given the center word. Given a sequence of training words $w_1, w_2, \cdots, w_T$, the objective of the model is to maximized L:

$$L = \frac{1}{T} \sum_{t=1}^{T} \sum_{-k \leq j \leq k, j \neq 0} \log p(w_{t+j} \mid w_t)$$

where k is a hyperparameter defining the window of the training words. Every word w is associated with two learnable parameter vectors, input vector I_w and output vector O_w of the w. The probability of predicting the word w_i given the word w_j is defined as:

$$p(w_i \mid w_j) = \frac{\exp(I_{w_i}{}^\top O_{w_j})}{\sum_{l=1}^{V} \exp(I_l{}^\top O_{w_j})}$$

where V refers to the number of words in the vocabulary. For larger vocabulary size, it is not efficient for computation, as it is proportional to the number of words in the V. Word2vec uses the hierarchical softmax objective function to solve the problem. The learned vector representations can be used to find the closest words for a user-specified word. For each category, we collected a small number of documents and created a training data. We applied word2vec to each training data. As a result, we obtained the number of n models where n is the number of category. Then, for each term in the named entity list Cl_i, ($1 \leq i \leq n$), we collected a certain number of related terms according to the similarity value, and created a set. Finally, for each set, we regarded the first order term as a representative term. If each term appeared in the documents is listed in a set, we replaced the term in the documents to its representative term.

4   Document Categorization

So far, we made use of the maximum amount of tagged data in temporal-based term smoothing. The final step is document categorization. We trained the model and classified documents by using TrAdaBoost [3]. TrAdaBoost extends AdaBoost [8] which aims to boost the accuracy of a weak learner by adjusting the weights of training instances and learn a classifier accordingly. TrAdaBoost uses two types of training data. One is so-called *same-distribution* training data that has the same distribution as the test data. In general, the quantity of these data is often limited. In contrast, another data called *diff-distribution* training data whose distribution may differ from the test data is abundant. The TrAdaBoost aims at utilizing the diff-distribution training data to make up the deficit of a small amount of the same-distribution to construct a high-quality classification model for the test data. TrAdaBoost is the same behavior as boosting for same-distribution training data. The difference is that for diff-distribution training instances, when they are wrongly predicted, we assume that these instances do not contribute to the accurate test data classification, and the weights of these instances decrease in order to weaken their impacts. Dai *et al.* applied TrAdaBoost to three text data, 20 Newsgroups, SRZZ, and Reuters-21578 which have hierarchical structures. They split the data to generate diff-distribution and same-distribution sets which contain data in different subcategories. We used TrAdaBoost as a learning technique to classify documents. We note that we used two types of labeled training data: One is the same creation time period with the test data. Another is different creation time period from the test data. We call the former *same-period* training, and the latter *diff-period* training data. In the TrAdaBoost, we replaced same-distribution data to same-period data, and diff-distribution data to diff-period data. TrAdaBoost is illustrated in Fig. 1.

Tr_d shows the diff-period training data that $Tr_d = \{(x_i^d, c(x_i^d))\}$, where $x_i^d \in X_d$ ($i = 1, \cdots, n$), and X_d refers to the diff-period instance space. Similarly, Tr_s represents the same-period training data that $Tr_s = \{(x_i^s, c(x_i^s))\}$, where $x_i^s \in X_s$ ($i = 1, \cdots, m$), and X_s refers to the same-period instance space. n and m are the number of documents in T_d and T_s, respectively. $c(x)$ returns a label for the input instance x. The combined training set $T = \{(x_i, c(x_i))\}$ is given by:

$$x_i = \begin{cases} x_i^d & i = 1, \cdots, n \\ x_i^s & i = n+1, \cdots, n+m \end{cases}$$

Input {
    The diff-period training data Tr_d, the same-period training data Tr_s,
    and the maximum number of iterations N.
}
Output {

$$h_f(x) = \begin{cases} 1, \prod_{t=N/2}^{N} \beta_t^{-h_t(x)} \leq \prod_{t=N/2}^{N} \beta_t^{-\frac{1}{2}} \\ 0, \text{otherwise} \end{cases}$$

}
Initialization {
    $\mathbf{w}^1 = 1/n$.
}

TrAdaBoost {
For $t = 1, \cdots, N$
  1. Set $\mathbf{P^t} = \mathbf{w^t} / (\sum_{i=1}^{n+m} w_i^t)$.
  2. Train a weak learner on the combined training set Tr_d and Tr_s with
     the distribution $\mathbf{P^t}$, and create weak hypothesis h_t: $X \to Y$
  3. Calculate the error of h_t on Tr_s:
     $\epsilon_t = \sum_{i=n+1}^{n+m} \frac{w_i^t \cdot |h_t(x_i) - c(x_i)|}{\sum_{i=n+1}^{n+m} w_i^t}$.
  4. Set $\beta_t = \epsilon_t / (1 - \epsilon_t)$ and $\beta = 1/(1 + \sqrt{2\ln n/N})$.
  5. Update the new weight vector:

$$w_i^{t+1} = \begin{cases} w_i^t \beta^{|h_t(x_i) - c(x_i)|}, & 1 \leq i \leq n \\ w_i^t \beta_t^{-|h_t(x_i) - c(x_i)|} & n+1 \leq i \leq n+m \end{cases}$$

}

Fig. 1. Flow of the algorithm

We used the Support Vector Machines (SVM) as a learner. We represented each training and test document as a vector, each dimension of a vector is a term/representative term appeared in the document, and each element of the dimension is a term frequency. We applied the algorithm shown in Fig. 1. After several iterations, a learner model is created, and a test document is classified using a learner.

5   Experiments

We evaluated our method by using the Mainichi Japanese newspaper documents.

5.1   Experimental Setup

We choose the Mainichi Japanese newspaper corpus from 1991 to 2012. The corpus consists of 2,883,623 documents organized into 16 categories. We selected 8 categories, "International", "Economy", "Home", "Culture", "Reading", "Arts",

Table 1. The # of classes and the averaged # of terms in each classes

Cat	Class	Avg	Cal	Class	Avg
International	2,556	21	Economy	2,516	21
Home	2,560	27	Culture	2,152	32
Reading	2,074	34	Arts	2,649	31
Sports	2,722	21	Local news	3,636	20

"Sports", and "Local news", each of which has sufficient number of documents. The total number of documents assigned to these categories are 787,518. All documents were tagged by using Yet Another Japanese Dependency Structure Analyzer, CaboCha [15] including named entity recognition. We selected noun words including person names and organization names.

For each category within each year, we divided documents into two folds. The first fold is used for temporal-based term smoothing. We further divided second fold into three: 2 % of documents are used as the same-period training data, 50 % of documents are the diff-period training data, and the remains are used to test our classification method. When the creation time period of the training data is the same as the test data, we used only the same-period training data. Table 1 shows the number of classes (representative terms) and the average number of terms in each class.

We used LIBLINEAR [9] as a basic learner in the experiments. We compared our method, TbTS &TAB (TrAdaBoost) with five baselines: (1) SVM without TbTS (SVM/wo), (2) SVM with TbTS (SVM/w), (3) biased-SVM [18] by SVM-Light [12] without TbTS (bSVM/wo), (4) biased-SVM with TbTS (bSVM/w), and (5) TrAdaBoost without TbTS (TAB/wo). Biased-SVM (b-SVM) is known as the state-of-the-art SVMs method, and often used for comparison [5]. Similar to SVM, for biased-SVM, we merged the first two folds, $i.e.$ 2 % of documents with the same-period and 50 % of documents with diff-period, and used them as a training data. We classified test documents directly, $i.e.$ we used closed data. We empirically selected values of two parameters, "c" (trade-off between training error and margin) and "j", $i.e.$ cost (cost-factor, by which training errors on positive instances) that optimized result obtained by classification of test documents. Similar to [18], "c" is searched in steps of 0.02 from 0.01 to 0.61. "j" is searched in steps of 5 from 1 to 200. As a result, we set c and j to 0.01 and 30, respectively. To make comparisons fair, all five methods including our method are based on linear kernel. Throughout the experiments, the number of iterations is set to 20. We used error rate as an evaluation measure [3].

5.2   Results

Categorization results for 8 categories are shown in Table 2.

Each value in Table 2 shows macro-averaged error rate across 22 years. "Macro Avg" in Table 2 refers to macro-averaged error rate across categories. The results

Table 2. Categorization results

Cat	SVM/wo	SVM/w	bSVM/wo	bSVM/w	TAB/wo	TAB/w
International	0.247	0.023	0.271	0.059	0.234	0.044
Economy	0.266	0.073	0.295	0.118	0.183	0.056
Home	0.344	0.268	0.365	0.138	0.249	0.109
Culture	0.379	0.376	0.145	0.049	0.157	0.139
Reading	0.245	0.081	0.070	0.018	0.102	0.092
Arts	0.429	0.398	0.351	0.156	0.310	0.202
Sports	0.139	0.018	0.162	0.053	0.128	0.039
Local news	0.213	0.016	0.477	0.145	0.189	0.031
Macro Avg.	0.283	0.157	0.267	0.092	0.194	0.089

Table 3. Selected terms by TbTS

Cat	Rep terms	Terms
International	Obama	Bush, Clinton, Biden, Gore, Gibbs
Economy	BMW	Volkswagen, Audi, Porsche, Opel, Mercedes-benz
Home	Liqueur	Spirits, Cocktail, Sake, Soda, Sherry
Culture	Bazaar	Free Market, Charity, Sale, Used clothes, Campaign
Reading	Fitzgerald	Scott, Capote, Kafka, Hemingway, Carver
Arts	Shostakovich	Prokofiev, Mahler, Debussy, Schubert, Argerich
Sports	Ivanišević	Sabatini, Henman, Frazier, Agassi, Björkman
Local news	Hybrid	Prius, Immobilizer, Gasoline engine, EcoCAR, HV

obtained by biased-SVM indicate the maximized F-score obtained by varying the parameters, "c" and "j". As can be seen clearly from Table 2, the overall performance obtained by TAB were better than the results obtained by other methods including biased-SVM except for "Culture" and "Reading", although biased-SVM were the results by using closed data. The results obtained by SVM with and without TbTS was the worst result among other methods. These observations show that once the training data drive from a different time period from the test data, the distributions of terms between training and test documents are not identical.

The overall performance with TbTS were better to those without TbTS in all methods. This shows that temporal-based term smoothing contributes classification performance. Table 3 shows some examples obtained by TbTS. Each representative term is randomly selected, and each term in a class is within the topmost five terms according to the representative term. As we can see from Table 3 that semantically similar words such as car names in Economy category, and novelists in Reading category are identified. Moreover, each term is salient for a specific year. For example, Obama, Bush, and Clinton is a USA president

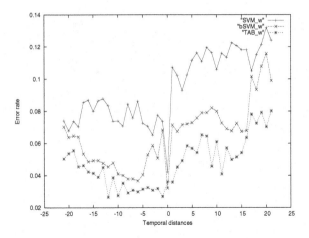

Fig. 2. Error rate with TbTS

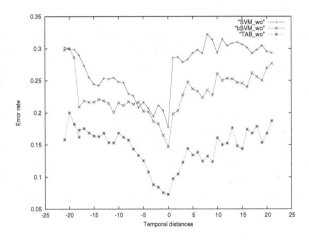

Fig. 3. Error rate without TbTS

from 2009 to 2013, 2001 to 2009, and 1993 to 1997, respectively. These terms are equally salient for a specific category, *i.e.* international across full temporal range of documents.

Figures 2 and 3 illustrate F-score with/without temporal-based term smoothing against the temporal difference between training and test data. Both training and test data are the documents from 1991 to 2012. For instance, "5" of the x-axis in Figs. 2 and 3 indicate that the test documents are created 5 years later than the training documents. We can see from Figs. 2 and 3 that the results with TbTS were better to those without TbTS in all of the methods. Moreover, the result obtained by "TAB/w" in Fig. 2 was the best in all of the temporal distances. There are no significant differences among three methods when the test and training data are the same time period in Fig. 2. The performance of these

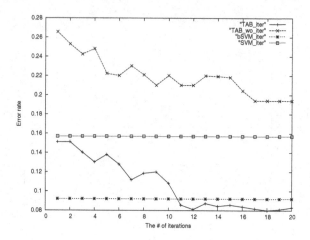

Fig. 4. Error rate against the # of iterations

methods including "SVM" drops when the period of test data is far from the training data in both of the Figs. 2 and 3. However, the performance of "TAB" was still better to those obtained by other methods. This demonstrates that the algorithm which applies temporal-based term smoothing and TrAdaboost learning is effective for categorization.

Figure 4 shows the averaged error rate with and without TbTS against the number of iterations. We can see from Fig. 4 that the curve obtained by the method with TbTS was better to biased-SVM after 11 iterations. Although the curves obtained by both with and without TbTS are not quite smooth, they converge around 20 iterations. This indicates that term-based smoothing method itself does not significantly contribute to the fast convergence, while it is effective to improve overall performance of categorization.

6   Conclusion

We proposed an approach for text categorization that training data may derive from a different time period from the test data. The basic idea is to minimize the impact of temporal effects in both term representation and learning techniques. The results by using Japanese Mainichi newspaper corpus show that combination of temporal-based term smoothing and learning method works well for categorization, especially when the creation time of the test data differs greatly from the training data.

There are a number of interesting directions for future work. We showed that word2vec is effective for term smoothing. However, it requires tagged corpora across full temporal range of training documents. Such corpora are often annotated by hand, and manual annotation of corpora is extremely expensive and time-consuming. In the future, we will try to extend our framework to address this issue. We used TrAdaboost as a learning technique which needs at least two improvements. Firstly, TrAdaboost is not explicitly incorporated temporal effects into the model as it is a discontinuous model. More precisely, once diff-period training instances are wrongly predicted, we assume that these instances do not contribute to the accurate test data classification, and the weights of these instances decrease in order to weaken their impacts. However, the weight is equally decreased regardless of the temporal difference between training and test data. Therefore, it is necessary to develop the continuous temporal model for further improvement. Secondly, as Dai et al. mentioned that the rate of convergence $(O(\sqrt{\mathrm{In}n/N}))$ is slow. Here, n is the number of training data, and N is the number of iterations. This is a rich space for further improvement. We used Japanese newspaper documents in the experiments. For quantitative evaluation, we need to apply our method to other data such as ACM-DL and a large heterogeneous collection of web content in addition to the experiment to examine the performance against the ratio between same-period and diff-period training data.

References

1. Blitzer, J., McDonald, R., Pereira, F.: Domain adaptation with structural correspondence learning. In: Proceedings of the Conference on Empirical Methods in Natural Language Processing, pp. 120–128 (2006)
2. Dai, W., Xue, G.R., Yang, Q., Yu, Y.: Co-clustering based classification for out-of-domain documents. In: Proceedings of the 13th ACM SIGKDD International Conference on Knowledge Discovery and Data Mining, pp. 210–219 (2007)
3. Dai, W., Yang, Q., Xue, G.R., Yu, Y.: Boosting for transfer learning. In: Proceedings of the 24th International Conference on Machine Learning, pp. 193–200 (2007)
4. Daume III, H.: Frustratingly easy domain adaptation. In: Proceedings of the 45th Annual Meeting of the Association of computational Linguistics, pp. 256–263 (2007)
5. Elkan, C., Noto, K.: Learning classifiers from only positive and unlabeled data. In: Proceedings of the KDD 2008, pp. 213–220 (2008)
6. Folino, G., Pizzuti, C., Spezzano, G.: An adaptive distributed ensemble approach to mine concept-drifting data streams. In: Proceedings of the 19th IEEE International Conference on Tools with Artificial Intelligence, pp. 183–188 (2007)
7. Forman, G.: An extensive empirical study of feature selection metrics for text classification. Mach. Learn. Res. 3, 1289–1305 (2003)
8. Freund, Y., Schapire, R.E.: A decision-theoretic generalization of on-line learning and an application to boosting. J. Comput. Syst. Sci. 55(1), 119–139 (1997)
9. Fan, F.E., Chang, K.W., Hsieh, C.J., Wang, X.R., Lin, C.J.: LIBLINEAR: a library for large linear classification. Mach. Learn. 9, 1871–1874 (2008)
10. Glorot, X., Bordes, A., Bengio, Y.: Domain adaptation for large-scale sentiment classification: a deep learning approach. In: Proceedings of the 28th International Conference on Machine Learning, pp. 97–110 (2011)
11. He, D., Parker, D.S.: Topic dynamics: an alternative model of bursts in streams of topics. In: Proceedings of the 16th ACM SIGKDD Conference on Knowledge discovery and Data Mining, pp. 443–452 (2010)

214   F. Fukumoto and Y. Suzuki

12. Joachims, T.: SVM Light Support Vector Machine. Department of Computer Science Cornell University (1998)
13. Kleinberg, M.: Bursty and hierarchical structure in streams. In: Proceedings of the Eighth ACM SIGKDD International Conference on Knowledge Discovery and Data Mining, pp. 91–101 (2002)
14. Klinkenberg, R., Joachims, T.: Detecting concept drift with support vector machines. In: Proceedings of the 17th International Conference on Machine Learning, pp. 487–494 (2000)
15. Kudo, T., Matsumoto, Y.: Fast methods for kernel-based text analysis. In: Proceedings of the 41st Annual Meeting of the Association for Computational Linguistics, pp. 24–31 (2003)
16. Lazarescu, M.M., Venkatesh, S., Bui, H.H.: Using multiple windows to track concept drift. Intel. Data Anal. 8(1), 25–59 (2004)
17. Li, Y., Yang, M., Zhang, Z.: Scientific articles recommendation. In: Proceedings of the ACM International Conference on Information and Knowledge Management CIKM 2013, pp. 1147–1156 (2013)
18. Liu, B., Dai, Y., Li, X., Lee, W.S., Yu, P.S.: Building text classifiers using positive and unlabeled examples. In: Proceedings of the ICDM 2003, pp. 179–188 (2003)
19. Mikolov, T., Chen, K., Corrado, G., Dean, J.: Efficient estimation of word representations in vector space. In: Proceedings of the International Conference on Learning Representations Workshop (2013)
20. Mourao, F., Rocha, L., Araujo, R., Couto, T., Goncalves, M., Meira, Jr., W.: Understanding temporal aspects in document classification. In: Proceedings of the 1st ACM International Conference on Web Search and Data Mining, pp. 159–169 (2008)
21. Murphy, J.: Technical Analysis of the Financial Markets, Prentice Hall, New Jersey (1999)
22. Raina, R., Ng, A.Y., Koller, D.: Constructing informative priors using transfer learning. In: Proceedings of the 23rd International Conference on Machine Learning, pp. 713–720 (2006)
23. Salles, T., Rocha, L., Pappa, G.L.: Temporally-aware algorithms for document classification. In: Proceedings of the 33rd Annual International ACM SIGIR Conference on Research and Development in Information Retrieval, 307–314 (2010)
24. Sparinnapakorn, K., Kubat, M.: Combining subclassifiers in text categorization: a dst-based solution and a case study. In: Proceedings of the 13th ACM SIGKDD International Conference on Knowledge Discovery and Data Mining, pp. 210–219 (2007)
25. Wang, C., Blei, D., Heckerman, D.: Continuous time dynamic topic models. In: Proceedings of the 24th Conference on Uncertainty in Artificial Intelligence, pp. 579–586 (2008)
26. Siao, M., Guo, Y.: Domain adaptation for sequence labeling tasks with a probabilistic language adaptation model. In: Proceedings of the 30th International Conference on Machine Learning, pp. 293–301 (2013)

Charset Encoding Detection of HTML Documents

A Practical Experience

Shabanali Faghani[✉], Ali Hadian, and Behrouz Minaei-Bidgoli

Department of Computer Engineering, Iran University of Science
and Technology, Tehran, Iran
{shabanali.faghani,ali.hadian}@gmail.com,
b_minaei@iust.ac.ir

Abstract. Charset encoding detection is a primary task in various web-based systems, such as web browsers, email clients, and search engines. In this paper, we present a new hybrid technique for charset encoding detection for HTML documents. Our approach consists of two phases: "Markup Elimination" and "Ensemble Classification". The Markup Elimination phase is based on the hypothesis that charset encoding detection is more accurate when the markups are removed from the main content. Therefore, HTML markups and other structural data such as scripts and styles are separated from the rendered texts of the HTML documents using a decoding-encoding trick which preserves the integrity of the byte sequence. In the Ensemble Classification phase, we leverage two well-known charset encoding detection tools, namely Mozilla CharDet and IBM ICU, and combine their outputs based on their estimated domain of expertise. Results show that the proposed technique significantly improves the accuracy of charset encoding detection over both Mozilla CharDet and IBM ICU.

Keywords: Charset encoding · HTML markups · Multilingual environments

1 Introduction

Since the beginning of the information age, many coding schemes have been designed to support different languages. However, early character encodings were designed to support a specific language, or a family of similar languages, hence almost all of them were bilingual, supporting Latin characters plus the specific local script. Those schemes were incompatible with each other, and were dependent on the operating system and the software being used. With the advent of globalization and development of the Internet, the variety of existing coding schemes had become a barrier in front of the information exchange. For a web-based application, e.g. a web browser, it is hard to support and integrate documents from different types of encodings. To address this issue, everyone should use a standard multilingual coding scheme, so that any application can easily decode it. Unicode is the de-facto universal charset encoding which provides a universal coding scheme, particularly in the web. However, despite many efforts to use Unicode as universal charset encoding, not every web page is converted to Unicode so far, and other local coding schemes are still being widely used.

© Springer International Publishing Switzerland 2015
G. Zuccon et al. (Eds.): AIRS 2015, LNCS 9460, pp. 215–226, 2015.
DOI: 10.1007/978-3-319-28940-3_17

From the language point of view, charset encodings can be categorized into three types: *universal*, *multilingual*, and local. Universal encodings, such as UTF-8 and UTF-16, are extensively used for the majority of contents in many languages. Likewise, multilingual charset encodings, such as ISO-8859 (for European languages) and ISO-2022 (for East Asian languages) families, are used for a class of languages whose individual orthographies are sufficiently similar. The most well-known multilingual charset encoding is ISO-8859 which supports languages with Latin scripts, such as European languages [1]. Local charset encodings are specific to a single language. For instance, Shift_JIS, EUC-KR, GB18030 and Windows-1256 are local encodings used for Japanese, Korean, Chinese and Arabic languages respectively.

In order to process a HTML web page, the charset encoding of the page should be detected at first. Charset encoding detection is the single point of failure in many web-based systems; if the detected charset is wrong, the results of any further processing on the page turns to be unreliable. In some web pages, the character encoding is explicitly specified in the Meta tag. Moreover, some HTTP servers provide clients with the information about the charset encoding of the requested web pages in the HTTP headers. As shown further in Sect. 5, approximately half of the famous web sites do not explicitly declare the encodings. In these situations, automatic identification of the charset encoding of the web pages is inevitable.

The remainder of this paper is organized as follows: A brief review of related works is presented in Sect. 2. We give an overview of the charset encodings in Sect. 3. Section 4 describes the new approach presented in this paper. Section 5 states test bed specification and results of the evaluations. Finally, Sect. 6 provides some concluding remarks.

2 Related Work

Only few works have been done on charset encoding detection for HTML documents, of which the most are based on machine leaning techniques [2]. For example, Russell et al. propose a Language and Encoding Identification (LEI) system, named SILC. They use Naïve Bayes classifier along with a combined trigram/unigram model [1]. Kim et al. investigated different feature sets, such as byte-level and character-level N-grams, for language and encoding detection. Also, they tested different learning algorithms, such as Support Vector Machine (SVM) and Naïve Bayes (NB). Their results show that a NB classifier with character-level features yields the best accuracy for charset detection and SVM with character-level features is the best configuration for language detection [2]. Similarly, Kikui suggests a similar approach and states that using word-unigrams for single-byte charsets and character-unigram for multi-byte charsets gives the best accuracy [5].

In addition to academic research, there are two open source implementations for charset encoding detection, namely Mozilla CharDet [3] and IBM ICU [4], which are widely used in practice. However, when applied to HTML web pages, their accuracy does not meet the requirements of the target applications.

Russell et al. also discuss about the relation between encoding and language detection. They state that in many cases, the language can be determined if the charset encoding is known. If the detected encoding is EUC-KR, for example, we are confident that the language is Korean. Therefore, in some specific cases, the encoding and language are two sides of the same coin. However, since language detection falls outside the scope of this paper, we only focus on charset encoding detection. Interested readers are referred to [1, 2] for more details. Also of note, we do not use machine learning techniques in this paper. Instead, we suggest a pre-processing phase, i.e. markup elimination, to improve the accuracy of charset detection tools. Using the markup-eliminated version of HTML documents, we suggest that the two charset detectors can be combined in a way that significantly improves the accuracy.

3   Charset Encodings

A character set, or so-called charset, is defined as a set of characters to be used in a special coding scheme, while encoding is defined as a mapping from an abstract character repertoire to a serialized sequence of bytes. In some cases a single charset is used in various encodings. For instance, the Unicode charset is used for UTF-8, UTF-16, etc. However, in many cases charsets and encodings have one-to-one mapping. Therefore, while the two terms are used interchangeably, they are different in nature. Nonetheless, charset encoding detection and charset detection refer to the same task, which is to heuristically guess the character encoding of a series of bytes that represent text [6].

In practice, charset encoding is more complex than a simple mapping from a character repertoire to the corresponding codes. IBM, for example, uses a three level encoding model named CDRA (Character Data Representation Architecture) to organize and catalog its own vendor-specific array of character encodings [7]. Similarly, Unicode uses a four level coding architecture [8]. In this section, we describe Unicode character encoding model based on a technical report by Unicode organization [8]. Further details on the literature and terminology about character encoding issues and a deep study about glyph, character and encoding can be found in [9].

A generic text T is composed of a sequence of characters $\{c_1, c_2, ..., c_n\}$ which can be encoded as a serialized sequence of bytes $\{b_1, b_2, ..., b_m\}$. The transition between characters and bytes in Unicode is mediated by four levels of representation, namely ACR, CCS, CEF, and CES.

3.1   ACR

Abstract Character Repertoire is the set of characters to be encoded, i.e. some alphabet or symbol set. It determines which potential characters can be represented in a special charset. The word 'abstract' implies that the characters typically have varying graphical representations.

3.2   CCS

Coded Character Set is a mapping from an abstract character repertoire to a set of nonnegative integers. A coded character set may also be known as a character encoding, a coded character repertoire, or a code page. The name 'code page' is used for East Asian character inventories. A code page is customarily presented in a tabular 'row-cell' form, where each <row, cell> index pair corresponds to a distinct integer.

3.3   CEF

Character Encoding Form is a mapping from a set of nonnegative integers that are elements of a CCS to a set of sequences of particular code units. A code unit is an integer occupying a specified binary width in computer architecture, which can be an 8-bit byte or a 16-bit and even a 32-bit integer. The sequences of code units do not necessarily have the same length. The code units of UTF-8, for example, vary from one to four 8-bit units.

3.4   CES

Character Encoding Scheme is a reversible transformation from a set of sequences of code units (i.e. from one or more CEFs) to a serialized sequence of bytes. A CES can be simple, compound or compressing. A simple CES uses a mapping of each code unit of a CEF into a unique serialized byte sequence in order, while a compound CES uses two or more simple CESs plus a mechanism to shift between them. A compressing CES maps a code unit sequence to a byte sequence while minimizing the length of the byte sequence.

It is important not to confuse a CEF and a CES; the CEF maps code points to code units, while the CES transforms sequences of code units to byte sequences. Also it should be noted that the CES must take into account the byte-order serialization of all code units used in the CEF that are wider than a byte. Table 1 presents four HTML elements saying 'Hello' in different languages along with the hexadecimal representation of three charset encodings for each element. The texts exhibited by h1 elements and their corresponding code units are shown in boldface. Also, the code units of non-ASCII characters are underlined. For each h1 element in the Table 1, we use charset encodings that support characters of the corresponding language. For example, GB18030 supports Chinese characters, but ISO-8859-1 does not.

4   The Proposed Method

The intuition behind our proposed method is that structural data in HTML documents, i.e. HTML markups and scripts, can drastically decrease the precision of the charset detector. This is due to the fact that most charset detection algorithms use statistical analysis to detect the charset of the given byte stream. At the same time, a HTML document contains HTML markups along with the contents of the document, each of which have different statistical properties. Markups are composed of English characters and specific symbols, such as wickets, slash, etc. The HTML markups use ASCII-supported symbols and characters and can be stored using ASCII character encoding. Besides, ASCII characters are stored with the same byte patterns in every character encoding scheme for the sake of compatibility. As a result, if there are lots of HTML markups in the text, it becomes hard to detect the encoding of the rest of the content.

Table 1. Four h1 elements along with their hexadecimal representation in three charset encodings

HTML Elements	Charset Encoding Details	
	Charset Encodings	Sequences of bytes (hexadecimal representation)
`<h1 lang="en-US">Hello world</h1>`	UTF-8	3c 68 31 20 6c 61 6e 67 3d 22 65 6e 2d 55 53 22 3e **48 65 6c 6c 6f 20 77 6f 72 64** 3c 2f 68 31 3e
	ISO-8859-1	3c 68 31 20 6c 61 6e 67 3d 22 65 6e 2d 55 53 22 3e **48 65 6c 6c 6f 20 77 6f 72 6c 64** 3c 2f 68 31 3e
	Windows-1252	3c 68 31 20 6c 61 6e 67 3d 22 65 6e 2d 55 53 22 3e **48 65 6c 6c 6f 20 77 6f 72 6c 64** 3c 2f 68 31 3e
`<h1 lang="fa-IR"> سلام علیکم</h1>`	UTF-8	3c 68 31 20 6c 61 6e 67 3d 22 66 61 2d 49 52 22 3e **d8 b3 d9 84 d8 a7 d9 85 20 d8 b9 d9 84 d9 8a d9 83 d9 85** 3c 2f 68 31 3e
	Windows-1256	3c 68 31 20 6c 61 6e 67 3d 22 66 61 2d 49 52 22 3e **d3 e1 c7 e3 20 da e1 ed df e3** 3c 2f 68 31 3e
	MacArabic	3c 68 31 20 6c 61 6e 67 3d 22 66 61 2d 49 52 22 3e **d3 e4 c7 e5 20 d9 e4 ea e3 e5** 3c 2f 68 31 3e
`<h1 lang="de-DE">Grüß Gott</h1>`	UTF-8	3c 68 31 20 6c 61 6e 67 3d 22 64 65 2d 44 45 22 3e 47 72 **c3 bc c3 9f** 20 47 6f 74 74 3c 2f 68 31 3e
	Windows-1252	3c 68 31 20 6c 61 6e 67 3d 22 64 65 2d 44 45 22 3e 47 72 **fc df** 20 47 6f 74 74 3c 2f 68 31 3e
	ISO-8859-9	3c 68 31 20 6c 61 6e 67 3d 22 64 65 2d 44 45 22 3e 47 72 **fc df** 20 47 6f 74 74 3c 2f 68 31 3e
`<h1 lang="zh-CN">你好</h1>`	UTF-8	3c 68 31 20 6c 61 6e 67 3d 22 7a 68 2d 43 4e 22 3e **e4 bd a0 e5 a5 bd** 3c 2f 68 31 3e
	GB18030	3c 68 31 20 6c 61 6e 67 3d 22 7a 68 2d 43 4e 22 3e **c4 e3 ba c3** 3c 2f 68 31 3e
	Big5	3c 68 31 20 6c 61 6e 67 3d 22 7a 68 2d 43 4e 22 3e **a7 41 a6 6e** 3c 2f 68 31 3e

For example, in Table 1, each HTML snippet is encoded in different encodings, but the standard ASCII characters of the snippets are encoded with the same code points and code units, regardless of the encoding being used. The first ten bytes in all charset encodings, i.e. **3c 68 31 20 6c 61 6e 67 3d 22** which corresponds to the substring **<h1 lang = "**, are equally encoded in every character encoding. This is why the structure of a web page do not become thoroughly corrupted when the browser detects a wrong charset encoding. Note that there are some exceptions for some charset encodings, such as UTF-16 and UTF-32 in which though ASCII characters have the same code points; ASCII characters have different code units due to the different CEFs of these encodings.

From the statistical point of view, the statistical properties of characters in an HTML documents is a linear combination of the encoding of the contents and the encoding of the markups. Since the markups are usually encoded in US-ASCII, the two distributions are essentially different when the content has an encoding different from US-ASCII. The markups disturb the statistical charset detectors because they bias all HTML document towards the distribution of US-ASCII. For example, if a HTML documents have a small non-English UTF-8 content with huge lines of HTML codes, a charset detector will classify the documents as ASCII, since it is statistically closest to ASCII. In short, when a large portion of document is markups, it is hard to detect the encoding of the document. Moreover, with the increase of Internet bandwidths and the processing speed of computers, the portion of structural data like scripts, styles, menus, navigational links, etc. in HTML web pages is increasing. Therefore, the large amount of structural data can potentially affect the precision of charset detection tools.

Our core idea is to remove the markups from the HTML document, so that only the content is considered for character detection. Markup elimination is not trivial, because in order to remove the markups, one should know in advance the charset of the page. In this regard, we propose two methods for eliminating the markups without knowing the encoding of the page.

4.1   Direct Markup Elimination

As mentioned earlier, almost all charset encoding schemes use the same code points and the same code units for the traditional US-ASCII characters. Merely knowing the ASCII code of the characters and keywords being used in HTML markups, we can identify and remove the byte sequences corresponding to the markups. However, this is practically a challenge, because there are many malformed HTML documents in the web, e.g. pages with broken tags, odd structures, etc. For example, there are pages with more than one body elements. Even though this method essentially works, it is a highly error-prone method because the program may face many unseen conditions in HTML documents.

4.2   ISO-8859-1 Decoding-Encoding Markup Elimination

Among the existing charset encoding schemes, ISO-8859-1 has an interesting feature. Having a byte sequence of a document with an arbitrary encoding, if we decode the byte sequence using ISO-8859-1 and then re-encode the content with ISO-8859-1, we exactly get the same byte sequence. In other words, ISO-8859-1 preserves code units of any charset encoding in the decoding-encoding process, while others do not necessarily do so. We may leverage this feature in order to remove the markups from the document without caring about the encoding of the page. It can be done in three steps. In step 1, the document is decoded using ISO-8859-1. Then, in step 2 the HTML markups are removed using a HTML parser. The out-of-the-box benefit of using a custom HTML parser is that it can repair malformed HTML documents, so we feel free about exceptions that potentially could be caused by these documents in HTML markup elimination process. Having the markup-eliminated document, we re-encode it using ISO-8859-1 in step 3. Then, a regular charset encoding detector can be used to identify

Table 2. ISO-8859-1 decoding-encoding markup elimination

HTML Elements	Charset Encoding Details & Steps of Markup Elimination Using ISO-8859-1				
	Charset Encoding details		Step1	Step2	Step3
	Charset Encoding	Sequence of bytes (hex. Rep.)	Decoded text using ISO-8859-1	Extracted content using a html parser	Encoded content using ISO-8859-1
\<h1>Hello world\</h1>	UTF-8	3c 68 31 3e **48 65 6c 6c 6f 20 77 6f 72 6c 64** 3c 2f 68 31 3e	\<h1>Hello world\</h1>	Hello world	**48 65 6c 6c 6f 20 77 6f 72 6c 64**
	ISO-8859-1	3c 68 31 3e **48 65 6c 6c 6f 20 77 6f 72 6c 64** 3c 2f 68 31 3e	\<h1>Hello world\</h1>	Hello world	**48 65 6c 6c 6f 20 77 6f 72 6c 64**
\<h1> سلام علیکم\</h1>	Windows-1256	3c 68 31 3e **d3 e1 c7 e3** 20 **da e1 ed df e3** 3c 2f 68 31 3e	\<h1>ÓáÇã Úáíßã\</h1>	ÓáÇã Úáíßã	**d3 e1 c7 e3 20 da e1 ed df e3**
	MacArabic	3c 68 31 3e **d3 e4 c7 e5** 20 **d9 e4 ea e3 e5** 3c 2f 68 31 3e	\<h1>ÓäÇå Ùäêåå\</h1>	ÓäÇå Ùäêåå	**d3 e4 c7 e5 20 d9 e4 ea e3 e5**
\<h1>Grüß Gott\</h1>	UTF-8	3c 68 31 3e 47 72 **c3 bc c3 9f** 20 47 6f 74 74 3c 2f 68 31 3e	\<h1>GrÃ¼ÃŸ Gott\</h1>	GrÃ¼ÃŸ Gott	47 72 **c3 bc c3 9f** 20 47 6f 74 74
	Windows-1252	3c 68 31 3e 47 72 **fc df** 20 47 6f 74 74 3c 2f 68 31 3e	\<h1>Grüß Gott\</h1>	Grüß Gott	47 72 **fc df** 20 47 6f 74 74
\<h1>你好\</h1>	GB18030	3c 68 31 3e **c4 e3 ba c3** 3c 2f 68 31 3e	\<h1>Äã°Ã\</h1>	Äã°Ã	**c4 e3 ba c3**
	Big5	3c 68 31 3e **a7 41 a6 6e** 3c 2f 68 31 3e	\<h1>§A¦n\</h1>	§A¦n	**a7 41 a6 6e**

the charset of the resulting contents. Table 2 illustrates markup elimination from four HTML snippet using ISO-8859-1 decoding-encoding method.

Note that, by using ISO-8859-1 to decode a byte sequence we might get gibberish or mojibake text in many cases; however we still have a valid HTML string in which we could look for a potential charset encoding in its Meta tags. This can be considered as an option just to use in practical environment and we did not use Meta tag information in the evaluations presented in Sect. 5.

Approximately 80 % of the websites use UTF-8 as their charset encoding [10], alt-hough this ratio varies greatly across regions, from the low ratio in East Asian countries to a very high ratio, say near the 100 %, in countries like Iran. Anyway, due to the high usage rate of UTF-8, we should be as accurate as possible about UTF-8 pages. In this regard, our primitive tests show that Mozilla CharDet works very accurate for UTF-8 documents, but IBM ICU shows a better performance when dealing with other formats. Based on primitive tests, to maximize the precision we combine these tools in a particular way as shown in Fig. 1.

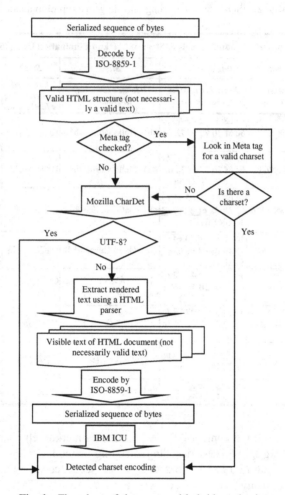

Fig. 1. Flowchart of the proposed hybrid mechanism

5   Evaluations

We evaluated our hybrid mechanism via two test scenarios: encoding-wise and language-wise. In the former test scenario we used a corpus of HTML documents with various encodings and in the latter test scenario we used a collection of URLs pointing to pages with different languages. Also of note, we tried to select these test pages and URLs from a real-word sample, and not from specific domains like Wikipedia. Typically, special websites and domains have uniform structure and special features which can potentially affect the credibility of evaluations. The source code and test data we used for our experiments is freely available[1].

---

[1] https://github.com/shabanali-faghani/IUST-HTMLCharDet .

5.1 Encoding-Wise Comparison

In this test scenario we test Mozilla CharDet, IBM ICU and the hybrid mechanism against a corpus of HTML Documents. Unfortunately, there is no evaluation dataset for charset encoding detection. It is because, as we mentioned in Sect. 2, there are few works in this area and each of the existing works used its own specific benchmark for evaluation. Hence, we collected a corpus of nearly 2700 HTML pages with various charset encoding types.

To create the corpus, we wrote a multi-threaded crawler and used a simple version of Tunneling [12] to crawl the Web. Tunneling enables us to gather more diverse web pages, consequently the more diverse web pages to be in the corpus, the more reliable results will be produced. For the sake of diversity, we used various seeds from different TLDs (Top Level Domains). By the way, the main criteria for adding a page to the corpus is the information provided in the charset field of the HTTP header. The evaluation results on our corpus is presented in Table 3.

Table 3. Histogram of true vs. detected charsets using IBM ICU, Mozilla CharDet and Hybrid

Encoding-Wise Comparison	True Encodings														
	UTF-8			Windows-1251			GBK			Windows-1256			Shift_JIS		
Detected Encodings	ICU	Ch.D.	Hyb.	ICU	Ch.D.	Hyb.	ICU	Ch.D.	Hyb.	ICU	Ch.D.	Hyb.	ICU	Ch.D.	Hyb.
UTF-8	549	657	657					3	3	5	14	14			
Windows-1251						313									
GB18030							414	145	407		1			636	
Windows-1256										26		616			
Shift_JIS					10						227		639	3	636
ISO-8859-1	108			311			5			573		6			
Windows-1252				3		1				31	28	8			2
Big5								73	4						
GB2312								197							
UTF-16LE											4			1	
UTF-16BE								1			28				
Other Encodings									5	10		1	1		2
No Match				304							343				
Accuracy (%)	84	100	100	0	0	100	99	35	97	4	0	96	100	1	100

Results show that the mean average precision of the hybrid mechanism in detecting charset encoding of pages that are encoded by Windows-1251, Windows1256 and UTF-8 is improved. The average overall accuracy of IBM ICU, Mozilla CharDet, and our hybrid mechanism are 61 %, 30 %, and 99 %, respectively.

Note that for a fair judgment on the accuracy of a charset detection tool in a multi-encoding environment, like our case in Table 3, one should consider all charset encodings together. For example, while the accuracy of Mozilla CharDet for UTF-8 is 100 %, it performs poorly for other charset encodings. On the other hand, if IBM ICU is used instead, it works very accurate for East Asian languages, which use charsets like GBK, Shift_JIS and also EUC-KR, but it is very inaccurate for Cyrillic-based and Arabic languages which extensively use Windows-1251 and Windows-1256 charsets respectively. All this, while the accuracy of the hybrid mechanism in detecting Windows-1251, Windows1256 and UTF-8 is near 100 %, its accuracy in detecting GBK and Shift_JIS do not meaningfully drop from the accuracy of IMB ICU on raw HTML pages.

5.2 Language-Wise Comparison

In the language-wise test scenario, we compare our hybrid mechanism with the two other charset detector tools from language point of view. We collected a list of URLs that are pointing to various web pages with different languages. The URLs are selected from the top one million websites visited from all over the world, as reported by Alexa (www.alexa.com). In order to collect HTML documents in a specific language, we investigated webpages with the internet domain name of that language. For example, Japanese web pages are collected from .jp domains. The results of evaluation for eight different languages are shown in details in Table 4.

Table 4. Accuracy of three charset encoding detectors in monolingual and multilingual environments

Language-Wise Comparison	Details of Fetched Web Pages & Accuracy of Charset Detector Tools					
Language(s)	TLD(s)	# primitive URLs	# sites with valid charset in HTTP header	IBM ICU	Mozilla CharDet	Hybrid
English	.us, .uk	21565	13223	9895 = %75	12373 = %94	12586 = %95
Russian	.ru	39453	28257	21716 = %77	21840 = %77	27900 = %99
Japanese	.jp	20339	6833	6461 = %95	6051 = %86	6741 = %99
Arabic	.iq, 11 more	1904	1093	888 = %81	1015 = %93	1064 = %97
Chinese	.cn	10086	3776	3596 = %95	3449 = %91	3640 = %96
German	.de	35318	23225	20115 = %87	20137 = %87	21371 = %92
Persian	.ir	7396	4018	3896 = %97	3971 = %99	4003 = %100
Indian, English	.in	12236	4867	3203 = %66	4640 = %95	4677 = %96
Total (multilingual)	all above	148297	85292	69770 = %82	73476 = %86	81982 = %96

In our test set, the English pages are selected from .uk and .us domains. Similarly, Arabic pages are selected from 12 different country-specific TLDs. Note that, the .in domain is quite different from the others, because the pages in this domain are not purely in Indian, but also include contents in English and some pages have both Indian and English contents. As illustrated in Table 4, for all languages there is a significant difference between the number of primitive URLs and the number of sites with valid charset in HTTP header. This difference is mainly due to that the HTTP servers do not provide clients with the information about charset encodings for approximately half of all requested web pages, and sometimes a result of networking problems.

As illustrated in Table 4, the accuracy of our hybrid mechanism is significantly higher than the two other tools. Besides, unlike IBM ICU and Mozilla CharDet, the accuracy of this mechanism is not volatile in different languages and regions. Also, it should be noted that the high accuracy of Mozilla CharDet over the IBM ICU in this test is due to the high usage of UTF-8 in various languages and regions. In absence or low usage of UTF-8, Mozilla CharDet is not likely to outperform IBM ICU.

In addition to UTF-8 there are three charset encodings GB2312, GBK and GB18030 which are extensively used in China. GBK is backward compatible with GB2312, because it was defined as an extension of GB2312 in 1993. Later on, GB18030 became compatible with GBK in 2000. Hence, there is forward compatibility between both GB2312 and GBK with GB18030. It seems that neither IBM ICU nor Mozilla CharDet has GBK in its charset list, but this charset is frequently appearing in HTTP header of the Chinese web pages. Therefore, to have a fair comparison we considered the actual direction of compatibility among these charsets as accurate detection in this evaluation.

Except for Chinese language, equality between the charset in HTTP header and the detected charset by a tool was the accuracy measure in this test. However as we know there is a pretty cool difference between **equality** and **equity**. Equity for two charsets on a web page can be defined as the validity of both of them for decoding that page, i.e. if we use either charset for decoding that web page the generated html documents are quite equal. Unfortunately the validity check for a charset on a web page could not be done automatically, because often, a charset validator does not complain even if a wrong encoding is detected or selected. Hence, the validator cannot decide whether the decoded text makes sense or not [11]. The reason behind this is that there is a close similarity between many of charset encoding types. Windows-1252, for example, is an extension on ISO-8859-1. On the other hand there is quite some overlap between ISO-8859-1 (Western Europe) and ISO-8859-2 (Eastern Europe) and other charset encodings in this series. Since typically the difference of similar charsets falls on unused code points and obsolete characters, as about the Windows-1252 and ISO-8859-1, in many cases either charset is valid to decoding a text that is encoded by one of them. Altogether, in addition to the charsets in HTTP headers and considering the right direction of compatibility between charsets, only visual inspection can make sure that whether a detected charset is valid or not. However, visual inspection is impossible for large collections like our case in Table 4.

6 Conclusions and Future Work

In this paper, we propose a hybrid mechanism for charset encoding detection. The proposed method heavily relies on the fact that removing structural data from HTML documents can significantly increase the precision of the charset detector tools. In addition to high accuracy, our method is able to cope with normal HTML web pages, as well as the noisy and malformed ones, in multi-encoding and multilingual environments. The proposed method is compared against two test scenario with real-world HTML. Results show a significant improvement over the two most famous charset encoding detection tools, namely IBM ICU and Mozilla CharDet.

Charset detection is not a foolproof process, because it is essentially based on statistical data and what actually happens is guessing not detecting. Both IBM ICU and Mozilla CharDet provide the probability of accuracy along with the guessed charset. Using these probabilities in a special way seems to be useful to yet increasing precision of the hybrid approach. Also, since the more critical mission web applications brings with it the need to more reliability and precision, developing special-purpose charset validator may assure these applications. In this connection, inspecting special characters in decoded text, like the replacement character (U+FFFD: ◆) in Unicode, could help the validator.

Acknowledgements. The authors would like to thank Hamed Kordestanchi, who proposed the language-wise evaluation, and Mojtaba Akbarzade for his guidance during this work.

References

1. Russell, G., Lapalme, G., Plamondon, P.: Automatic identification of language and encoding. In: Rapport Scientifique, Laboratoire de Recherche Appliquée en Linguistique Informatique (RALI), Université de Montréal, Canada (2003)
2. Kim, S., Park, J.: Automatic Detection of Character Encoding and Language, Technical Report, Machine Learning, Stanford University (2007)
3. Tang, F.Y.F.: Mozilla Charset Detectors, Mozilla (2008). http://www-archive.mozilla.org/projects/intl/chardet.html
4. ICU - International Components for Unicode, IBM (2014). http://site.icu-project.org
5. Kikui, G.I.: Identifying, the coding system and language, of on-line documents on the internet. In: Proceedings of the 16th Conference on Computational linguistics, vol. 2, pp. 652–657 (1996)
6. Charset detection, Wikipedia (2014). http://en.wikipedia.org/wiki/Charset_detection
7. Character Data Representation Architecture, IBM (2013). http://www.ibm.com/software/globalization/cdra/index.jsp
8. Whistler, K., Davis, M., Freytag, A.: Unicode Technical Report# 17, Character Encoding Model, The Unicode Consortium (2008). http://www.unicode.org/reports/tr17
9. Dürst, M.J., Yergeau, F., Ishida, R., Wolf, M., Texin, T.: Character Model for the World Wide Web 1.0: Fundamentals, W3C (2005). http://www.w3.org/TR/charmod
10. Historical trends in the usage of character encodings for websites, W3Techs (2015). http://w3techs.com/technologies/history_overview/character_encoding
11. Dürst, M.: Checking the character encoding using the validator, W3C (2003). http://www.w3.org/International/questions/qa-validator-charset-check.en
12. Bergmark, D., Lagoze, C., Sbityakov, A.: Focused crawls, tunneling, and digital libraries. In: Agosti, M., Thanos, C. (eds.) ECDL 2002. LNCS, vol. 2458, pp. 91–106. Springer, Heidelberg (2002)

Structure Matters: Adoption of Structured Classification Approach in the Context of Cognitive Presence Classification

Zak Waters[1]($\boxtimes$), Vitomir Kovanović[2], Kirsty Kitto[1], and Dragan Gašević[2]

[1] Queensland University of Technology, Brisbane, Australia
{z.waters,kirsty.kitto}@qut.edu.au
[2] The University of Edinburgh, Edinburgh, UK
v.kovanovic@ed.ac.uk, dgasevic@acm.org

Abstract. Within online learning communities, receiving timely and meaningful insights into the quality of learning activities is an important part of an effective educational experience. Commonly adopted methods–such as the Community of Inquiry framework–rely on manual coding of online discussion transcripts, which is a costly and time consuming process. There are several efforts underway to enable the automated classification of online discussion messages using supervised machine learning, which would enable the real-time analysis of interactions occurring within online learning communities. This paper investigates the importance of incorporating features that utilise the structure of online discussions for the classification of "cognitive presence"–the central dimension of the Community of Inquiry framework focusing on the quality of students' critical thinking within online learning communities. We implemented a Conditional Random Field classification solution, which incorporates structural features that may be useful in increasing classification performance over other implementations. Our approach leads to an improvement in classification accuracy of 5.8 % over current existing techniques when tested on the same dataset, with a precision and recall of 0.630 and 0.504 respectively.

Keywords: Text classification · Conditional random fields · Online learning · Online discussions

1 Introduction

The classification of social interactions occurring among individuals who participate in an online community is an important research problem. Not all participant contributions have the same value, with some being more thoughtful than others. This problem is particularly important in an educational domain, where online discussions are often being used to support both fully online and blended models of learning [7]. A substantial body of research aims to foster higher-order thinking among students in online learning communities. One prominent framework for approaching this problem is the Community of Inquiry (CoI) model [8]

© Springer International Publishing Switzerland 2015
G. Zuccon et al. (Eds.): AIRS 2015, LNCS 9460, pp. 227–238, 2015.
DOI: 10.1007/978-3-319-28940-3_18

which describes the important dimensions of learning in online communities, and provides a quantitative coding scheme for their assessment. This coding scheme provides a method for categorising various interactions between participants within a particular online community, which is traditionally conducted by two human "coders" who manually label discussion messages for post hoc analysis.

Despite wide adoption by online education researchers, coding online discussion transcripts is a manual and labor-intensive task, often requiring several coders to dedicate significant amounts of time to code each of the discussion messages. This approach (i) does not enable for a real-time feedback on the quality of learning interactions, and (ii) limits the wider adoption of the CoI framework by educational practitioners. This problem makes the task an ideal candidate for automation, and a number of approaches aimed at automating the process of coding transcripts using machine learning techniques are in development [2,17,22]. While these approaches have produced promising results, their text classification models currently make class predictions on a per-message basis, using only features derived from a single post, without consideration of the context of a post or of the preceding classification sequence. Given that human coders take discussion context into account during the classification process, and that the underlying construct of cognitive presence develops over time [7,9], it seems likely that structural classification features can be used to model context in a similar fashion, and that these might improve classification accuracy.

This paper presents the preliminary results of an alternate approach to the automated analysis of online discussions within online learning communities using Conditional Random Fields (CRFs) [26], which is a novel extension of previous work that aims to automate the text-classification of online discussions using the CoI framework. Our results show that the use of structural features in combination with a CRF model produce a higher classification accuracy than currently available methods. In Sect. 2, the CoI model is briefly introduced, and examines current approaches of analysing community participants' "cognitive presence". Related applications of CRFs to online discussions are also reviewed. Section 3 outlines our approach, which aims to improve on existing approaches by combining structural features with a Linear-Chain CRF model. The results of this experiment are presented in Sect. 4, where they are compared against current approaches and human accuracies. Structural features and their potential use across a number of contexts and discussion media are discussed in Sect. 5, along with the limitations of the current study, which form the basis of the future work directions. Finally, the research and key contributions are summarised in Sect. 6.

2   Background Work

2.1   The Community of Inquiry (CoI) Framework

Overview. The Community of Inquiry (CoI) framework [7,8] proposes three important dimensions (presences) of inquiry-based online learning:

1. **Teaching presence** defines the role of instructors before and for the duration of a course, consisting of (i) direct instruction, (ii) course facilitation, and (iii) course organization and design.
2. **Social presence** provides insights into the social climate between course participants. It consists of (i) affective communication, (ii) group cohesion, and (iii) interactivity of communication.
3. **Cognitive presence** is a central component of the framework and defines phases in the development of cognitive and deep thinking skills in online learning community [8].

The CoI framework defines multi-dimensional content analysis schemes [4] for the coding of student discussion messages, which is the main unit of analysis used to assess the level of the three presences. This framework has gained considerable attention in the educational research community, with a large number of replication studies and empirical validations (cf. [9,10]). Overall, the CoI framework and its coding schemes show sufficient levels of robustness (see Sect. 3.1 for an example) resulting in widespread adoption of the framework in the online education research community [10].

Of particular interest is the level of cognitive presence exhibited by the community members, due to its indication of their critical thinking. It is defined as the *"extent to which the participants in any particular configuration of a community of inquiry are able to construct meaning through sustained communication."* [8, p. 11], and is operationalized through a practical inquiry model which defines the four phases of the inquiry process that occurs during learning [8]:

1. **Triggering**: In the first phase, students are faced with some problem or dilemma which triggers a learning cycle. This typically results in messages asking questions and expressing a sense of puzzlement.
2. **Exploration**: This phase is primarily characterized by the exploration–both individually and in group–of different ideas and solutions to the problem at hand. Brainstorming, questioning, leaping into conclusions, and information exchange are the primary activities in the exploration phase.
3. **Integration**: After exploring different ideas, students synthesize the relevant ideas which ultimately leads to construction of meaning [8]. From the perspective of an instructor, this is the most difficult phase to detect as integration of ideas is often not clearly visible in discussion transcripts.
4. **Resolution**: In the final phase, students apply the newly constructed knowledge to the original problem, typically in the form of hypothesis testing or the building of a consensus.

Challenges of CoI Framework Adoption. One of the biggest practical challenges in adoption of the CoI framework – and other transcript analysis methods– is that it requires experienced coders and substantial labor-intensive work to code (i.e. categorise) discussion messages for the levels of three presences [4,17]. As such, it is argued that this and similar approaches have had very little practical impact upon current educational practices [4]. To enable for a more proactive use of the Community of Inquiry framework by the course instructors, there is a need for an automated content analysis of online discussions that would provide instructors with a real-time feedback about student learning activities [15].

2.2 Automated Classification of Student Discussion Messages

Despite the labor intensive nature of manually coding online discussion messages, human coders that categorise online discussion messages into the phases of cognitive presence typically achieve very high intersubjective agreements. Moreover, the high levels of agreement among coders suggests that humans can identify the latent phases of cognitive presence from text-based discussions with relative ease. On the other hand, using machine learning to classify student messages in a similar manner is a challenging task. Where humans construct meaning from text using various inferences and abstractions that manifest as complex higher-order cognitive processes, machine learning approaches require meticulously constructed feature spaces, which are representative of the problem task. Kovanović et al. [17] presented an approach to classifying cognitive presence from online discussions, using a Support Vector Machine (SVM) classification model, which achieved classification accuracy of 58.84 %. While the results of this work are promising, the overall performance of this approach is substantially less accurate than what can be achieved by human coders, which provides further evidence of the overall complexity of this task. In this approach, Kovanović et al. [17] made use of lexical features derived from the content of each individual discussion message that are prominent within the literature. These features consisted of various N-grams, POS tags, name entity counts and dependency tuples, as well as intuitive features such as whether a post or reply is the first in a discussion thread. In contrast, human coders may typically utilise contextual information when making their coding decisions, such as the structure the discussion or the sequence in which discussion messages appear. Because of this, it is worth investigating how structural features about a discussion in addition to considering discussion messages in sequence may further improve classification performance.

Beyond the CoI framework, many studies have acknowledged that accounting for the relationships between individual messages and the latent structure of discussions may improve classification performance for transcript analysis [5,23,25]. Specifically, Ravi and Kim [23] suggests that using features derived from a previous message can be a positive indicator for classification of the next post along in a discussion. Other related work in threaded-discussion classification that seeks to incorporate the structural features of discussions is becoming increasingly common [6,14,28]. The most common type of structural features utilised include a post's position relative to others in a discussion, whether a post is the first or the last in a thread, how similar a post is as compared to its neighbours, and how many replies a post accrued. For this study, we attempt to account for the latent structure between posts in a discussion by incorporating these features into a Conditional Random Field approach.

2.3 Conditional Random Fields for Automated Detection of Cognitive Presence

We have implemented a Conditional Random Field (CRF) classification model [26] to annotate posts within a discussion with the phases of cognitive presence. Unlike traditional text classification methods, Conditional Random Fields consider the label sequence of a data set. Because of this, Conditional Random Fields have found numerous applications in natural language processing (NLP) tasks, such as part-of-speech (POS) tagging [18], document segmentation and summarisation [24], as well as gene prediction from biological sequence data [3].

Recent related research has extended CRFs to online forum discussions, where posts and interactions between participants are sequential in nature. Wang et al. [28] applied CRFs to discussion forums to learn the reply structure of forum interactions. This was achieved by using rich features that capture both short and long range dependencies within posts of an online discussion such as the lexical content similarity between two neighbouring posts. Similarly, FitzGerald et al. [6] combined the lexical features of posts with a Linear-Chain CRF to detect high quality comments in blog discussions, such as the word and sentence count of the post. Moreover, FitzGerald et al. [6] postulates that there exists sequential dependencies between posts in a forum, which emphasises the usefulness of structural features derived from the entire discussion, as well as lexical features from a single post. To date, CRF classification has not been applied to the problem of automating the detection of Cognitive Presence in online discussion transcripts. Here, we show that making this step improves the accuracy of classification when compared with the current best practices.

3 Methods

3.1 Dataset

The data used in this study comes from six offerings of a fully-online masters-level research-oriented course in software engineering at a Canadian public university. This is the same dataset as was used in the study by Kovanović et al. [17] which makes for more accurate and direct comparison between the two different classification approaches. In total, the data consists of 1,747 messages produced by 81 students. Each message was coded by two experienced coders who achieved an excellent level of coding agreement of 0.97 Cohen's Kappa, which is a measure commonly used to measure inter-rater reliability between coders using a quantitative categorisation scheme. Table 1 shows the distribution of messages in different phases of cognitive presence. The details of course structure and organization are explained in detail in Kovanović et al. [16], Gašević et al. [12].

3.2 Classifier Implementation

For this study, we implemented a Linear-Chain Conditional Random Field (LCCRF) model to predict the phases of cognitive presence occurring in online

Table 1. Cognitive presence coding

ID	Phase	Messages	(%)
0	Other (no signs of cognitive presence)	140	8.01 %
1	Triggering event	308	17.63 %
2	Exploration	684	39.17 %
3	Integration	508	29.08 %
4	Resolution	107	6.12 %
	All phases	1747	100 %

discussions. This LCCRF was implemented in Java using the Mallet library [21], which is a widely used open source toolkit for machine learning. This library was extended as needed to suit our experimental requirements.

3.3 Data Preprocessing

In this dataset, online discussions form a tree-like hierarchical structure (i.e., each discussion message can receive replies which can also receive replies). This presents a problem; in order to train and test our LCCRF implementation, the structure of the data must be linear, as opposed to the current tree structure. In order to obtain appropriate sequences of data, sub-threads were extracted such that every sequence of posts from the root node to every leaf node in a tree was obtained. To obtain reliable results, these sub-threads must be remerged after classification to produce one classification per message in a discussion; this remerging process in described in Sect. 4.1. While other CRF models will accept hierarchical structures (e.g., such as Tree-Structured and Hierarchical CRFs), we chose a linear-chain model over other approaches due to the size constraints imposed by the dataset, which had only 84 coded discussion threads in total to use for training and testing a tree-structured model. Breaking these up into linear chains produced more message sequences that could be used to train our linear model.

In addition to the extraction of linear sequences, the discussion threads in the data set were split into two sets; one for training and testing the CRF model, the other for validation from which our results are derived. These threads were split 70/20/10 % for training, testing and validation, respectively.

3.4 Classification Features

Many of the features used for the purpose of this study were extracted using the various functionalities of the Stanford CoreNLP Java library [20], and are derived from the related work in our literature review. Each post in the discussion is described by a feature vector that attempts to encapsulate both lexical and structural features. In addition to word unigrams, lexical features were derived from the text content of a post itself, and structural features were used to indicate where a post resides in the context of the entire discussion thread. These features are presented below:

1. **Entity Count** is the number of entities within a post as found by the Stanford CoreNLP Named Entity Recognition (NER) tool. The rationale behind using this feature is that discussion participants posting exploration comments are more likely to introduce a number of entities through their exploration of ideas.
2. **First Post** and **Last Post** are boolean features that are set to true when a post is the first and last in a discussion respectively. This feature represents the implicit structure of the discussion, where it is intuitive to believe that most Triggering phases occur at the start of a discussion.
3. **Comment Depth** is the number assigned to a post based on its chronological order within a discussion thread.
4. **Post Similarity** of the previous and next post in a discussion is calculated by obtaining the cosine similarity of two TF-IDF weighted vectors. The post similarity features assist in incorporating the local structure of the discussions, where it is expected that some phases of cognitive presence differ significantly from one another, and some only slightly.
5. **Word and Sentence counts** capture the number of words and sentences within a particular post. It is expected that when a discussion is reaching the integration and resolution phases, there is a lot more content due to the synthesis and integration of ideas.
6. **Number of Replies** to a post, which provides the classifier with the intuition that the earlier phases of cognitive presence (Triggering and Exploration) will have more replies than the later phases. Additionally, this feature also helps model the implicit structure within a discussion, giving the classifier an indication of how large the discussion is. The rationale behind this feature is that the triggering and exploration phases would generally have more replies than the integration and resolution phases.

These features form a feature vector for each message in a discussion thread. Because our classifier is sequential, these feature vectors are combined to form a feature vector sequence used in Mallet for training and testing our CRF classification model.

4   Results

The aim of this study was to investigate whether classifying posts in sequence, with the addition of structural features improves upon the current approach to identifying cognitive presence in online learning discussions. In order to evaluate the effectiveness of our approach we use Cohen's Kappa, which is a metric often used for judging the reliability of a categorisation scheme. Cohen's Kappa is advantageous as it allows for a genuine comparison between the performance of human coders and our approach. A comparison between this experiment and the approach with the current highest accuracy is described in Table 2.

Before remerging the discussion threads, the CRF model achieved an accuracy of 67.2 %, and 0.515 and 0.620 precision and recall respectively and a F-measure of 0.562. Because sub-threads were extracted for this experiment (detailed in Sect. 3.3), messages found earlier in the discussion threads have been classified multiple times. As a result of this, these accuracies are optimistically high due to multiple correct classifications diluting the overall classification accuracies. This problem was fixed by re-merging the discussion threads back into their original hierarchical form in order using a majority vote mechanism.

Table 2. Comparison of results

Approach	Cohen's Kappa	Accuracy
Kovanović et al. [17]	0.410	58.4 %
LCCRF	0.482	64.2 %
Human	0.97	NA

4.1 Re-Merging Discussion Threads

As mentioned earlier in Sect. 3.3, every message sequence from a root post to every leaf node in a discussion was extracted to produce an appropriate linear sequence to train the LCCRF. This means that the earlier posts in a discussion may have been classified multiple times. Furthermore, the predicted phase need not necessarily be the same for these multiple classifications; a post that was classified as Triggering in one sequence might be classified as Exploration in the next sequence that it appears in. In order to obtain one classification result for each message in a threaded discussion, the sub-threads were remerged using a majority vote mechanism. This method of remerging posts results in a final accuracy of 64.2 % for the validation set. A large majority of posts that were classified multiple times belonged to the Triggering label, but many of these multiple classifications were correctly identified. Thus, the resulting small drop in performance is representative of the general classification accuracy obtained by the LCCRF. It seems that this implementation performs well at this type of classification task, with an overall precision and recall of 0.630 and 0.504 respectively and a F-measure of 0.559. Moreover, our implementation achieves a Cohen's Kappa value of 0.482, which gives us a comparison with the human coding according to this widely used metric for judging the overall reliability of a coding or categorisation scheme. Table 2 demonstrates that while an improvement has been obtained, more work needs to be completed before we can be sure that an automated approach is performing at a level similar to human coders in this task.

5 Discussion

Our LCCRF approach shows promise for the automated classification of cognitive presence in discussions occurring within an online learning community.

Moreover, the results of this work show a modest improvement over the work conducted by Kovanović et al. [17], who presented an accuracy of 58.4 % as seen in Table 2. The key differences in these two approaches is clear: our approach considers discussion messages in sequence, modelled via the CRF, utilising features that attempt to convey the context of the discussion. In contrast the work presented by Kovanović et al. [17] considers each message separately, relying on primarily lexical features and a SVM.

These results suggest that a CRF utilising structural features is well suited to this text classification task. Using this approach, the classifier may more appropriately model the dependencies between messages in online discussions. The structurally oriented feature-set allows for a contrast between posts that would otherwise contain very similar lexical features. By combining these features, the probabilistic CRF implementation appears to better model the dependencies between posts, leading to increased predictive performance. This improvement provides preliminary evidence of how modelling the structure of discussions, and considering discussion posts in sequence may be an important factor in further improving the automated detection of cognitive presence. Further studies using our approach will seek to confirm this theory by exploring alternate features and CRF implementations.

5.1 Limitations and Future Work

One key limitation of this work is contextual, our results may be biased towards the single course from which the dataset was derived. Moreover, there are a number of different platforms in which online learning discussions can take place. For example, a learning community using Social Media may be more informal in nature than one conducted in an institutes formal discussion forum. Using a model trained on one community may not produce reliable results for another community. Future research needs to consider data sets from courses in other subject areas and delivery modes (i.e., blended learning). One potential advantage of a structural approach is that it may perform more consistently across different datasets. A classification based upon structural features is more likely to prove robust under changed conditions than specific lexical characteristics, and so there is the possibility that the CRF approach will achieve better performance at text annotation across multiple discussion groups and fora. Further research and new datasets will be required to investigate whether this claim holds merit.

Other approaches to move towards automating the coding process will be investigated as future work. Because this approach uses a linear-chain model, some dependencies between messages in an online discussion may be missed. However, this linear model allows for the implementation of coding practice rules used by various CoI coding schemes, such as "coding up"– i.e., when a message has traces of two phases of cognitive presence, it is coded with the higher phase [16]. Despite this, approaches that might better model dependencies across hierarchical structures, such as a tree CRF may further improve on our current accuracy. As seen in Table 1, the distribution of phases (class labels) in our dataset is largely uneven. This disparity between the individual phases of cognitive presence is seen in the predictive performance of our classifier, where the lowest represented phases are typically classified correctly less often than that of their higher represented counterparts. Unfortunately, collaboration within online learning communities commonly takes this form, where learners typically do not progress to the resolution phase of cognitive presence [11,12]. Future attempts at automation may benefit from a method of accounting for this uneven distribution of class labels.

In order to replace the current approach to analysing online learning communities with manual hand-coding transcripts, we aim to achieve Cohen's Kappa value of close to 0.80, which indicates an almost perfect agreement among coders according to the Landis and Koch [19] interpretation of Cohen's Kappa. Our CRF approach achieved a Kappa value of 0.482, which indicates a moderate agreement, but will require further improvement before machine learning techniques can replace hand coders. Future work will aim to further improve our classifier's performance. Specifically, we plan to further improve our model by: (i) evaluating our model on another, larger dataset with a more even distribution of phases; (ii) seeking additional features that may improve upon our current accuracies, such as Coh-Metrix [13] and features derived from the Linguistic Inquiry and Word Count (LIWC) framework [27] that are commonly used to characterise cognitive processing associated with comprehending and producing text and discourse, and; (iii) better modelling the dependencies between threaded discussions using a Tree-Structured CRF model approach.

6   Conclusion

In this work, we presented a new approach to automating the detection of the four phases of cognitive presence arising in online discussions. By reconceptualising online discussions as a sequence prediction problem, we predicted a sequence of labels (i.e. the phases of cognitive presence) for a sequence of messages. This allowed us to use a linear chain Conditional Random Field model for classification, which incorporates structural features of online discussions rather than just the lexical features that have previously been applied to solving this problem. This approach to automating the detection of cognitive presence has shown promise, with moderate improvements over alternative approaches with an accuracy of 64.2 % and a Cohen's Kappa value of 0.482. However, classification accuracies are not yet high enough to replace the current approach of manually coding transcripts. Further improving this model is a priority for future work where we aim to further evaluate the model on alternative datasets, investigate additional features, and attempt to better model the dependencies between posts using a tree-structured CRF model.

References

1. Arbaugh, J., Cleveland-Innes, M., Diaz, S.R., Garrison, D.R., Ice, P., Richardson, J.C., Swan, K.P.: Developing a community of inquiry instrument: testing a measure of the community of inquiry framework using a multi-institutional sample. Internet High. Educ. **11**(3–4), 133–136 (2008)
2. Corich, S., Hunt, K., Hunt, L.: Computerised content analysis for measuring critical thinking within discussion forums. J. e Learn. Knowl. Soc. **2**(1), 47–60 (2012)
3. Culotta, A., Kulp, D., McCallum, A.: Gene prediction with conditional random fields (2005)
4. Donnelly, R., Gardner, J.: Content analysis of computer conferencing transcripts. Interact. Learn. Environ. **19**(4), 303–315 (2011)
5. Feng, D., Shaw, E., Kim, J., Hovy, E.: An intelligent discussion-bot for answering student queries in threaded discussions. In: Proceedings of the 11th International Conference on Intelligent User Interfaces, pp. 171–177. ACM (2006)
6. FitzGerald, N., Carenini, G., Murray, G., Joty, S.: Exploiting conversational features to detect high-quality blog comments. In: Butz, C., Lingras, P. (eds.) Canadian AI 2011. LNCS, vol. 6657, pp. 122–127. Springer, Heidelberg (2011)
7. Garrison, D.R.: Thinking Collaboratively: Learning in a Community of Inquiry. Routledge, New York (2015)
8. Garrison, D.R., Anderson, T., Archer, W.: Critical thinking, cognitive presence, and computer conferencing in distance education. Am. J. Distance Educ. **15**(1), 7–23 (2001)
9. Garrison, D.R., Anderson, T., Archer, W.: The first decade of the community of inquiry framework: a retrospective. Internet High. Educ. **13**(1–2), 5–9 (2010a)
10. Garrison, D.R., Arbaugh, J.: Researching the community of inquiry framework: review, issues, and future directions. Internet High. Educ. **10**(3), 157–172 (2007)
11. Garrison, R., Cleveland-Innes, M., Fung, T.S.: Exploring causal relationships among teaching, cognitive and social presence: student perceptions of the community of inquiry framework. Internet High. Educ. **13**(1–2), 31–36 (2010b)
12. Gašević, D., Adesope, O., Joksimović, S., Kovanović, V.: Externally-facilitated regulation scaffolding and role assignment to develop cognitive presence in asynchronous online discussions. Internet High. Educ. **24**, 53–65 (2015)
13. Graesser, A.C., McNamara, D.S., Kulikowich, J.M.: Coh-metrix providing multilevel analyses of text characteristics. Educ. Res. **40**(5), 223–234 (2011)
14. Jin, W.: Blog comments classification using tree structured conditional random fields. Ph.D. Thesis, University of British Columbia (Vancouver (2012)
15. Kovanovic, V., Gasevic, D., Hatala, M.: Learning analytics for communities of inquiry. J. Learn. Analytics **1**(3), 195–198 (2014)
16. Kovanović, V., Gašević, D., Joksimović, S., Hatala, M., Adesope, O.: Analytics of communities of inquiry: effects of learning technology use on cognitive presence in asynchronous online discussions. Internet High. Educ. **27**, 74–89 (2015)
17. Kovanović, V., Joksimović, S., Gašević, D., Hatala, M.: Automated contentanalysis of online discussion transcripts. In: Proceedings of the Workshops at the LAK 2014 Conferenceco-Located with 4th International Conference on Learning Analytics and Knowledge (LAK 2014). Indianapolis, March 2014. http://ceur-ws.org/Vol-1137/
18. Lafferty, J., McCallum, A., Pereira, F.C.N.: Conditional random fields: probabilistic models for segmenting and labeling sequence data (2001)

19. Landis, J.R., Koch, G.G.: The measurement of observer agreement for categorical data. Biometrics **33**(1), 159–174 (1977)
20. Manning, C.D., Surdeanu, M., Bauer, J., Finkel, J., Bethard, S.J., McClosky, D.: The stanford CoreNLP natural language processing toolkit. In: Proceedings of 52nd Annual Meeting of the Association for Computational Linguistics: System Demonstrations, pp. 55–60 (2014). http://www.aclweb.org/anthology/P/P14/P14-5010
21. McCallum, A.K.: Mallet: A machine learning for language toolkit. Technical report (2002). http://mallet.cs.umass.edu
22. McKlin, T., Harmon, S., Evans, W., Jones, M.: Cognitive presence in web-based learning: a content analysis of students' online discussions. In: IT Forum, vol. 60 (2002)
23. Ravi, S., Kim, J.: Profiling student interactions in threaded discussions with speech act classifiers. Front. Artif. Intell. Appl. **158**, 357 (2007)
24. Shen, D., Sun, J.T., Li, H., Yang, Q., Chen, Z.: Document summarization using conditional random fields. In: IJCAI International Joint Conference on Artificial Intelligence, pp. 2862–2867 (2007)
25. Soller, A., Lesgold, A.: A computational approach to analyzing online knowledge sharing interaction. In: Proceedings of Artificial Intelligence in Education, pp. 253–260 (2003)
26. Sutton, C., McCallum, A.: An introduction to conditional random fields. Found. Trends Mach. Learn. **4**(4), 267–373 (2011)
27. Tausczik, Y.R., Pennebaker, J.W.: The psychological meaning of words: LIWC and computerized text analysis methods. J. Lang. Soc. Psychol. **29**(1), 24–54 (2010)
28. Wang, H., Wang, C., Zhai, C.X., Han, J.: Learning online discussion structures by conditional random fields. In: SIGIR 2011 - Proceedings of the 34th International ACM SIGIR Conference on Research and Development in Information Retrieval, Beijing, pp. 435–444 (2011)

Topics and Models

A Sequential Latent Topic-Based Readability Model for Domain-Specific Information Retrieval

Wenya Zhang[1], Dawei Song[1,2], Peng Zhang[1（⊠）], Xiaozhao Zhao[1], and Yuexian Hou[1]

[1] Tianjin Key Laboratory of Cognitive Computing and Application, Tianjin University, Tianjin, China
{wenyazhang,dwsong,pzhang,zxz,yxhou}@tju.edu.cn
[2] The Computing Department, The Open University, Buckinghamshire, UK

Abstract. In domain-specific information retrieval (IR), an emerging problem is how to provide different users with documents that are both relevant and readable, especially for the lay users. In this paper, we propose a novel document readability model to enhance the domain-specific IR. Our model incorporates the coverage and sequential dependency of latent topics in a document. Accordingly, two topical readability indicators, namely *Topic Scope* and *Topic Trace* are developed. These indicators, combined with the classical *Surface*-level indicator, can be used to rerank the initial list of documents returned by a conventional search engine. In order to extract the structured latent topics without supervision, the hierarchical Latent Dirichlet Allocation (hLDA) is used. We have evaluated our model from the user-oriented and system-oriented perspectives, in the medical domain. The user-oriented evaluation shows a good correlation between the readability scores given by our model and human judgments. Furthermore, our model also gains significant improvement in the system-oriented evaluation in comparison with one of the state-of-the-art readability methods.

Keywords: Domain-specific retrieval · Readability · Documents reranking

1 Introduction

Conventional search engines aim to return relevant documents based on "similarity" (between a document and a query) and "popularity" (with respect to the hyperlink structure). A recently emerging relevance criteria is document readability. With the diversification of web resources and users, it is increasingly difficult for a search engine to provide different users, especially lay users, with documents in a specific domain that are not only relevant but also readable [10]. The readability plays an important role in assessing documents' relevance [21,22], quality [1] and utility [19]. However, traditional similarity and popularity measures do not necessarily reflect the readability of the returned documents [10].

© Springer International Publishing Switzerland 2015
G. Zuccon et al. (Eds.): AIRS 2015, LNCS 9460, pp. 241–252, 2015.
DOI: 10.1007/978-3-319-28940-3_19

In a typical human reading process, the readability of documents can be interpreted at different levels [16,17]. It is argued in [13] that humans' minds appear to go far beyond the data available, which means there will be a complicated process of abstraction in humans' understanding. Thus, we propose to measure documents readability from two levels. The first is the surface level readability that relates to the surface content. It can be assessed by a series of classical readability features. Beyond the surface content, a higher level, namely the topic level readability, reflects whether it is easy for a user to comprehend the hidden topics in documents. Thus, we propose a topic-based readability method, which can be used to enhance domain-specific IR by considering both the surface and topic level readability of documents.

2   Related Work

There have been various general-purpose readability measures in the literature, such as the Flesh-Kincaid Grade Level and SMOG Index [3,11]. Based on the surface-level features of a document, e.g., word length, sentence length, etc., these classical measures usually generate a numeric score that maps onto an educational grade level. To further improve the accuracy of readability computation, various statistical, semantic and syntactic features of documents have been used [9,15]. However, they are designed for traditional general-purpose texts, thus insufficient to deal with domain-specific documents. Most of the existing measures do not consider the documents' readability at a topic level, which is indeed important for domain-specific documents which often contain a large amount of domain related topics and concepts.

A concept-based approach has been proposed by Yan et al. [16,17], which takes into account the coverage (Scope) and relatedness (Cohesion) of domain topics (concepts) within a document, with reference to a domain taxonomy. In the taxonomy, the topics are at different abstraction levels and their relationships are organized into a hierarchical tree structure [12]. Topic taxonomy encodes high-quality domain knowledge and can be used to improve a user's understanding of the content of the text [2]. The general hypothesis is that the more abstract a topic is, the more general and easier to understand the topic tends to be. A limitation is that explicit domain taxonomy may not be always available. Recently, the hierarchical Latent Dirichlet Allocation (hLDA) has been widely used to discover the latent topics from large scale data [2,20]. Thus in this paper we propose to automatically build latent topic structures to represent the domain. Moreover, Yan's model does not take into account the sequential dependency between adjacent topics which is important in understanding documents content easily and logically. Different from Yan's work, in [6,7,14] a readability measure based on the term embedding and sequential discourse cohesion is proposed. However, it does not refer to a domain taxonomy. Nonetheless, their thought about sequential discourse cohesion gives us an inspiration for incorporating sequential dependency information within a latent topic based approach.

In this paper, we propose a novel readability enhanced domain-specific information retrieval model. Specifically, two latent topical indicators, i.e., *Topic*

Scope and *Topic Trace*, are proposed. They capture the sequential dependency of topics at different granularities, through mapping a document onto an automatically constructed topic taxonomy. *Topic Scope*, as originally proposed by Yan et al. [16], reflects the overall coverage of domain topics in a document. *Topic Trace* tracks how the sequence of topics occurring in a document traverses on the topic taxonomy. Additionally, we use the ratio of complex words as an indicator of the document's surface level readability. The individual indicators and their combinations can be used to measure, from different perspectives, the readability of a document. Based on the documents' readability scores, we can then rerank the initial list of results generated by a convention search engine.

3 Sequential Latent Topic Based Readability Computation

3.1 Topic Taxonomy Extraction and Topic Identification

A topic taxonomy can be extracted from a collection of documents. As a tree structure, it consists of topics (nodes) that are at different abstraction levels and connected by the subsumption relationships (edges). In this paper, we use a nonparametric generative procedure, namely the hierarchical Latent Dirichlet Allocation (hLDA), to generate a tree structure of topics by means of a nested Chinese Restaurant Process (nCRP) and Bayesian nonparametric inference. Each topic can be represented as a probability distribution over words in the vocabulary. In the extracted topic taxonomy, the deeper a topic is, the more specific it tends to be. Thus, the root has the broadest meaning, while the leaves are the most specific ones. Figure 1 shows a fragment of topic taxonomy extracted from the CLEF eHealth 2013 medical collection [5].

A domain-specific document can then be mapped onto the topic taxonomy through a topic identification process. In this paper, we identify topics contained in a document based on the occurrence of top 10 probability words from the underlying distributions of topics. Therefore, a document can be represented as a sequence of identified topics, i.e., $d = (t_1, t_2, ..., t_n)$,as illustrated in Fig. 3.

3.2 Topical Readability Indicators

After the topic identification, we propose two topical readability indicators. *Topic Trace*, tracks the identified topics sequentially on the taxonomy. Another indicator, *Topic Scope* reflects the coverage of the identified topics in a document.

Topic Trace (TT). This indicator is based on the hypothesis that the topical line to compose a document is like the planning of travels among a number of scenery spots. A good traveling plan can help tourists visit as many scenery spots as possible with as little expense as possible and as small bumpy leap as possible. Similarly, a well-organized (thus more readable) document should introduce the related topics sequentially with little *Topical Expense* and small

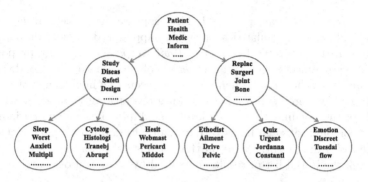

Fig. 1. A fragment of automatically constructed topic taxonomy from CLEF eHealth 2013 medical collection (construction details will be shown in Sect. 4). In this fragment, the root topic consists of top 4 high probability words of "Patient, Health, Medic Inform" (stemmed by the Porter stemmer) which are general concepts (topics) in medical domain. Its children nodes "Study, Diseas, Safeti" and "Replac, Surgeri, Joint" have relatively specific meaning, while its grandchildren nodes are more specific.

Topical Leap, which can reflect the coherence among sequential topics defined as *Topic Trace* here. Thus, the *Topic Trace* for a document d_i can be calculated as in Eq. (1),

$$Trace(d_i) = Expense(d_i)^{-1} * e^{-\lambda * Leap(d_i)} \tag{1}$$

where $Expense(d_i)$ and $Leap(d_i)$ refer to the *Topical Expense* and *Topic Leap*, respectively, and λ is a parameter to control the influence of the *Leap* on the trace score ($\lambda = 0.001$, the optimal values by experiments). A high trace score means the high readability of the document.

Topical Leap means the bumpiness when the identified topics sequentially traverse on the topical taxonomy, as defined in Eq. (2). H_{t_j} denotes the depth of topic t_j in the taxonomy.

$$Leap(d_i) = \sum_{t_j, t_{j+1}} \left| H_{t_j} - H_{t_{j+1}} \right| \tag{2}$$

Topical Expense reflects the difficulty to parse the identified topics sequentially. Hypothesizing that the topical expense of a document is inversely related to the overall coherence among the topics within the document, we measure it as follows:

$$Expense(d_i) = \left(\frac{\sum_{t_j} ConCoh(t_j)}{|MC| - 1} \right)^{-1} \tag{3}$$

where MC is the size of the set of identified topics and $ConCoh(t_j)$ computes the contextual coherence, simplified as cc_{t_j}, of t_j in term of its average topical similarity with its surrounding topics (i.e., context).

Specifically, to compute $ConCoh(t_j)$, we use a sliding window [4] with fixed size M (an odd number, $M = 5$ is the optimal value by experiments in this paper) which takes the center topic as the current topic, while the other surrounding topics within the window as contextual topics, as illustrated in Fig. 2. The contextual coherence of the current topic can be computed as in Eq. (4):

$$ConSim(C_j) = \frac{\sum_{m=-\frac{M-1}{2}}^{\frac{M-1}{2}} e^{-|m|} * Sim(C_j, C_{j+m})}{|M|} \qquad (4)$$

$Sim(t_j, t_{j+1})$ calculated as in Eq. (5), is the topical similarity between the current topic t_j and a context term t_{j+1} within the sliding window. $e^{|m|}$ means sequential dependency between ti and ti+m gets stronger when they are closer in the sliding window. m the relative distance between the two topics in the window. Thus, we can get a global topic trace vector, i.e., $tv(d) = (cc_{t_1}, cc_{t_2}, ..., cc_{t_n})$, for each document.

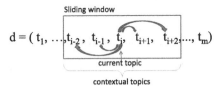

Sliding window

$$d = (\ t_1, ..., t_{i-2},\ t_{i-1},\ t_i,\ t_{i+1},\ t_{i+2}, ..., t_m)$$

current topic

contextual topics

Fig. 2. Sliding window for contextual coherence with $M = 5$.

One way for calculating similarity between two topics has been shown in Eq. (5) [8]. L means the shortest path, and H is the depth of the most specific subsumer. The constants α and β are set to 0.2 and 0.6, respectively (the optimal values by experiments).

$$Sim(t_i, t_{i+m}) = e^{-\alpha L} \frac{e^{\beta H} - e^{-\beta H}}{e^{\beta H} + e^{-\beta H}} \qquad (5)$$

By now we have defined all the components in Eq. (1) for calculation of document trace, i.e., $Trace(d_i)$. The score of $Trace(d_i)$ falls into the range of (0,1). Figure 3 gives an example, where the $Trace(d_i) = 0.42 * e^{-0.005} = 0.417$. For d_j, the $Trace(d_j) = 0.17 * e^{-0.005} = 0.169$. It turns out that d_i is more readable. Furthermore, from the structure perspective, d_i would seem to be more concise and logical than d_j.

Topic Scope (TS). Based on a general hypothesis that the overall lower taxonomy depths of identified topics in the taxonomy would indicate a better document readability, we also employ the average tree depth of the identified topics to calculate the topic scope. Compared with Yan's work [16], we measure the document scope on topic level rather than conceptual level. As shown in Eq. (6), n_t is the number of identified topics, while $depth(t_i)$ represents the depth of the identified topic t_i on the topic taxonomy. Falling in (0,1), the higher the scope score is, the more readable the document tends to be.

$$Scope(d_i) = e^{-\left(\frac{\sum_{i=1}^{n} depth(t_i)}{n_t}\right)} \qquad (6)$$

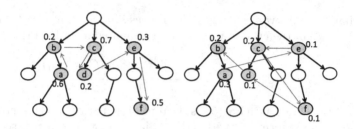

Fig. 3. Topic sequence for $d_i = (t_a, t_b, t_c, t_d, t_e, t_f)$ (left) and $d_j = (t_a, t_e, t_c, t_f, t_d, t_b)$ (right), and their corresponding global topic trace are $tv(d_i) = (0.6, 0.2, 0.7, 0.2, 0.3, 0.5)$ and $tv(d_j) = (0.3, 0.1, 0.2, 0.1, 0.1, 0.2)$, respectively. Thus $Expense(d_i) = [(0.6 + 0.2 + 0.7 + 0.2 + 0.3 + 0.5)/6]^{-1} = 0.42^{-1}$ and $Expense(d_j) = 0.17^{-1}$; $Leap(d_i) = 0.001 * 5 = 0.005$ and $Leap(d_j) = 0.005$.

3.3 Document Reranking Based on Readability

We combine the two levels of readability to calculate the overall readability score of d_i as follows:

$$ReadScore(d_i) = \frac{x * Scope(d_i) + y * Trace(d_i)}{1 + z * Surface(d_i)} \tag{7}$$

where $Surface(d_i)$ measures the surface level readability of the document. Specifically, x, y and z are explored to control the weight of three readability indicators, respectively. Both limited to $(0,1)$, $x + y = 1$, and z is 0 or 1. Thus, $ReadScore$ can be normalized into $(0,1)$. The larger the $ReadScore$ is, the more readable the document will be.

As shown in Eq. (8), we employ the ratio of complex words that are not in the Dale-Chall word list [3] to calculate the surface level readability, where $ComplexWords$ is the number of complex words and $TotalWords$ is the number of total words in the document.

$$Surface(d_i) = \frac{ComplexWords}{TotalWords} \tag{8}$$

After we get the readability score, in the same way as in [16], we use Eq. (9) to compute the total score for reranking, where $RelScore(d_i)$ is the relevance score returned by a conventional search engine. m controls the weight of relevance score in documents reranking, while n controls the weight of readability score.

$$Score(d_i) = RelScore(d_i, Q)^m \cdot e^{-(ReadScore(d_i))^n} \tag{9}$$

4   Experiments and Results

In order to evaluate our proposed model, both user-oriented and system-oriented evaluations have been carried out. The former aims to find out how well our model's prediction is correlated with human judgment on document readability, while the latter aims to evaluate how effectively our model can improve document ranking in medical information retrieval.

Aiming to provide valuable and relevant documents to lay users, CLEF eHealth 2013 dataset [5] contains 50 test queries and one million English documents covering a broad set of medical topics. The initial search results were returned by the TF-IDF model in Lemur. All documents have been stemmed by Porter stemmer and filtered by SMART 571 stop word list. As a comparison, MeSH (Medical Subject Headings), an existing medical taxonomy, had been used to calculate Yan's model (*Scope*, the most effective indicator). Since it is expensive to construct taxonomy by hLDA on all documents, we employed the top 20 returned documents for all queries as the same method used in [18]. Specifically, we limited the vocabulary to be the 29795 words that appeared in more than 5 documents and a number of meaningless symbols were removed, such as "[", "_", "&", "$" etc. As a result, 634 topics have been nested in a topic taxonomy with a depth of 8, of which a fragment has been shown in Fig. 1.

Topic: Crohn's disease

A: Keeping your Crohn's disease under control can feel like a full time job. That's because it is. Avoiding potential triggers can help prevent a flare up. You may be doing things that are detrimental to your health. Do you know what they are? (1) Smoking cigarettes has a significant effect on Crohn's disease. The entire body is affected by smoking, including the digestive tract. (2) If all you give your body all day is coffee and diet cola, you can't expect your digestive tract to treat you well.

B: Talk to just about anyone with Crohn's disease and it's likely that they have had surgery. Approximately 75 percent of Crohn's disease patients who have disease in the small bowel will have surgery in the first 10 years after diagnosis. Resections to remove diseased tissue are common, and may be repeated as the disease recurs in different sections of the intestine. Crohn's can cause narrowing of the bowel, also called a stricture, which may require stricture plasty. Complications from Crohn's such as abscesses or fissures can also require surgery.

Fig. 4. Sample pair of medical passages with different topic scopes for Query1-Crohn's disease in the first task of user-oriented evaluation.

User-oriented Evaluation. In this evaluation, users were instructed to answer a series of questions related to the readability of the passages selected from CLEF eHealth 2013. We only selected 6 simple queries (Query1-Crohn's disease; Query2-Scar; Query3-Lightheaded; Query4-Liver transplantation; Query5-C.diff; Query6-Cardiac arrest) to avoid exhausting users. For each query, two user tasks, corresponding to topic scope and topic trace respectively, were performed independently with different sets of users to avoid the learning effect. In the first task, one pair of medical passages with different topic scopes (preselected from the top returned documents for the query in initial search results and labeled as passage "A" and "B", each of which are limited to 80-90 words, as shown in Fig. 4) are presented to a set of users. Through actual reading, the users were asked to answer the following questions: (1) Filtering question;

Table 1. Calculation matrix for similar rates of users' judgements.

	PassageA	PassageB
Topic Scope (TS, by users)	n_{as}	n_{bs}
Topic Trace (TT, by users)	n_{at}	n_{bt}
Average Readability Score (by users)	$Read(A)$	$Read(B)$

---

1. Which of the following statements are correct according to the passages?

☐ We can infer that controlling Crohn's disease needs much attention.
☐ Resections to remove diseased tissue can be done once for all.
☐ Complications from abscesses or fissures also need surgery.

2. Which passage is more specific, in relation to the topic?

○ A

○ B

3. Please assign a readability score for A:

○ "5": Very easy to read, all content can be understood effectively.

○ "4": Easy to read, useful information can be found and most of the content can be understood effectively except for some complex words.

○ "3": Average level, useful information can be found, but some parts can not be understood.

○ "2": Hard to read, only the main structure and the rough idea of the passage can be understood, hard vocabulary.

○ "1": Very hard to read, mostly not understandable.

Fig. 5. Detailed questions for the first task in user-oriented evaluation.

(2) Scope related question; (3) Readability score for A; (4) Readability score for B. Detailed information has been shown in Fig. 5.

$$SimilarRate(TS) = \begin{cases} n_{as}/(n_{as} + n_{bs}), & \text{if } Read(A) > Read(B) \\ n_{bs}/(n_{as} + n_{bs}), & \text{if } Read(A) < Read(B) \end{cases} \tag{10}$$

In the second task, another pair of passages, also manually selected from the top returned documents, with different topic occurrence sequences (i.e., different topic traces), are used for another set of users to answer the same question in the first task, except that the question (2) is replaced by "Which passage describes the topic more logically and smoothly?" ("more logically and smoothly" refers to better trace). In question (3) and (4) of both tasks, the readability score "5" means the simplest to read, while "1" means the hardest to read.

The evaluation was conducted through Amazon Mechanical Turk which targets at "crowdsourcing" of Human Intelligence Tasks (HIT) in large scale. Only the high-qualification turkers are used (i.e., HIT Approval Rate (%) $\geq$ 95). We filtered the data of turkers who did not answer the filtering question (i.e., question (1)) correctly, whose dwell time was less than 40 s or whose individual HIT is uncompleted. As a result, we collected the high-quality data of 20 Mechanical Turk users for each pair. For every pair of passages, we computed the average readability score for each passage (with average standard deviation 0.89), and we calculated the consistency of users' judgements on topic scope (topic trace) with average readability score in terms of *Similar Rate*. Through referring to Table 1, we derived *SimilarRate* for topic scope in Eq. (10), where n_{as}, n_{bs}, n_{at}, n_{bt} means the number of users who picked the corresponding choice. $Read(A)$ and $Read(B)$ are the average readability scores assigned by all users. In addition, we calculate it for topic trace in the same way.

Table 2. Similar rate for users' judgements

Qid	Query1	Query2	Query3	Query4	Query5	Query6	AvgSimlarRate
TS (by users)	0.70	0.10	0.85	0.25	0.40	0.80	0.52
TT (by users)	0.75	0.80	0.70	0.95	0.95	0.90	0.84

Table 3. Pearson correlation coefficient for user evaluation

	TT+SI $(x = 0, y = 1, z = 1)$	TS+TT+SI $(x = y = 0.5, z = 1)$	Yan(Scope)
Pearson	**0.63**	**0.22**	**0.18**

Table 2 summarizes the results of *Similar Rate*, in which "TT, by users" shows a good average similar rate with 0.84 among users. It means that users tend to assign higher readability score to the passage with better topic trace.

In addition, we calculated the Pearson Correlation Coefficient between average assigned readability scores and that computed by our model and Yan's, which have been shown in Table 3 with best tuned combing parameters (i.e., x, y and z in Eq. (7)). "TT+SI" (combination of Topic Trace and Surface Indicator) has the highest coefficient among all combinations, and "TS+TT+SI" (combination of all indicators) also correlates more closely with average assigned score than Yan's model, which also implies the potential of our proposed model.

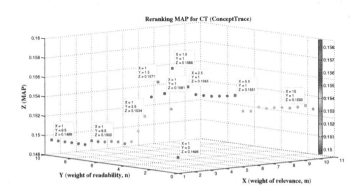

Fig. 6. Reranking MAP for CT (Concept Trace).

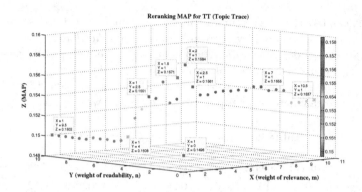

Fig. 7. Reranking MAP for TT (Topic Trace).

System-oriented Evaluation. We also conducted system experiment to examine the proposed indicators and combinations of them to rerank the top 20 documents for all 50 test queries in CLEF eHealth 2013. To explore the relative effect of readability and relevance, we tuned the weights of m and n. Parts of the tuning results have been shown in Figs. 6 and 7, through which we can infer that by integrating a certain weight of readability, i.e., n (for instance in Fig. 6, when n is around 1), we can get consistent improvement by increasing weight of relevance, i.e., m.

Specifically, we compared the reranking MAP of each indicator and some combinations of them, and picked up their best performance to do the significance test. Detailed results have been shown in Table 4, in which "CT" (*Concept Trace* that implements the idea of *Trace* by referring to MeSH), gains the highest improvement of 5.92 % that is better than Yan's model. Meanwhile, "TT" (*Topic Trace*) and "TT+SI" (combination of *Topic Trace* and *Surface Indicator*) also improve the reranking performance significantly. Compared with "CT", "TT" (*Topic Trace*) is competitive by constructing taxonomy automatically, which indicates the good potential of the idea of *Trace*.

Table 4. MAP comparisons for CLEF eHealth 2013 (symbol † means p < 0.05 with paired t-test)

	Baseline	Yan	CT	TS	TT	SI
MAP	0.1496	0.1515	0.1586	0.1504	0.1584	0.1548
(x,y,z)	-	(1,0,0)	(1,0,0)	(1,0,0)	(0,1,0)	(0,0,1)
(n,m)	-	(2.5,1)	(1.5,1)	(1,3)	(2,1)	(5,1)
(-/+)	-	+1.27 %	+5.92 % †	+0.53 %	+5.88 % †	+3.48 %
	-	-	**TS+TT**	**TT+SI**	**TS+SI**	**TS+TT+SI**
MAP	-	-	0.1571	0.1583	0.1505	0.1570
(x,y,z)	-	-	(0.5,0.5,0)	(0,1,1)	(1,0,1)	(0.5,0.5,1)
(n,m)	-	-	(1,1)	(1,1)	(3.5,1)	(1,0.5)
(-/+)	-	-	+5.01 %	+5.82 % †	+0.60 %	+4.95 %

5   Conclusions and Future Work

In this paper, we proposed a sequential latent topic-based document readability model for domain-specific information retrieval. In our model, two topical readability indicators, namely *Topic Scope* and *Topic Trace* have been developed, which can capture the overall coverage and sequential trace of the latent topics in the document, respectively. Compared with Yan's work [16], on one hand, our model does not require referencing to an existing domain taxonomy. Instead, we automatically construct a latent topic taxonomy from the data. Therefore, our approach is more general and applicable to any domains that may not have an existing taxonomy. On the other hand, we take advantage of the sequential information between adjacent latent topics. Through user-oriented evaluation, our proposed readability indicators and the re-ranking model demonstrate a good correlation with human judgments. Furthermore, our model outperforms a state of the art concept-based model.

In the future, we plan to improve topic taxonomy construction by incorporating n-grams. Meanwhile, refined algorithms and more suitable combinations of readability indicators will be tested.

Acknowledgments. This work is supported in part by Chinese National Program on Key Basic Research Project (973 Program, grant No. 2013CB329304, 2014CB744604), the Chinese 863 Program (grant No. 2015AA015403), the Natural Science Foundation of China (grant No. 61272265, 61402324), and the Research Fund for the Doctoral Program of Higher Education of China (grant No. 20130032120044).

References

1. Bendersky, M., Croft, W. B., Diao, Y.: Quality-biased ranking of web documents. In: Proceedings of the fourth ACM International Conference on Web Search and Data Mining, pp. 95–104. ACM (2011)
2. Blei, D.M., Griffiths, T.L., Jordan, M.I.: The nested chinese restaurant process and bayesian nonparametric inference of topic hierarchies. J. ACM (JACM) **57**(2), 7 (2010)
3. Chall, J.S., Dale, E., Readability Revisited: The New Dale-Chall Readability Formula. Brookline Books, Cambridge (1995)
4. Chen, Y., Yin, X., Li, Z., Hu, X., Huang, J.X.: A lda-based approach to promoting ranking diversity for genomics information retrieval. BMC genomics **13**(Suppl 3), S2 (2012)
5. Goeuriot, L., Kelly, L., Jones, G.J., Zuccon, G., Suominen, H., Hanbury, A., Müller, H., Leveling, J.: Creation of a new evaluation benchmark for information retrieval targeting patient information needs (2013)
6. Jameel, S., Lam, W., Qian, X.: Ranking text documents based on conceptual difficulty using term embedding and sequential discourse cohesion. In: Proceedings of the The 2012 IEEE/WIC/ACM International Joint Conferences on Web Intelligence and Intelligent Agent Technology, Volume 01, pp. 145–152. IEEE Computer Society (2012)

7. Jameel, S., Qian, X.: An unsupervised technical readability ranking model by building a conceptual terrain in LSI. In: 2012 Eighth International Conference on Semantics, Knowledge and Grids (SKG), pp. 39–46. IEEE (2012)
8. Jiang, J.J., Conrath, D.W.: Semantic similarity based on corpus statistics and lexical taxonomy. arXiv preprint cmp-lg/9709008 (1997)
9. Kate, R.J., Luo, X., Patwardhan, S., Franz, M., Florian, R., Mooney, R.J., Roukos, S., Welty, C.: Learning to predict readability using diverse linguistic features. In: Proceedings of the 23rd International Conference on Computational Linguistics, pp. 546–554. Association for Computational Linguistics (2010)
10. Kim, J.Y., Collins-Thompson, K., Bennett, P.N., Dumais, S.T.: Characterizing web content, user interests, and search behavior by reading level and topic. In: Proceedings of the Fifth ACM International Conference on Web Search and Data Mining, pp. 213–222. ACM (2012)
11. Senter, R., Smith, E.: Automated readability index. Technical report, DTIC Document (1967)
12. Sripairojthikoon, P., Senivongse, T.: Concept-based readability of web services descriptions. In: 2013 15th International Conference on Advanced Communication Technology (ICACT), pp. 853–858. IEEE (2013)
13. Tenenbaum, J.B., Kemp, C., Griffiths, T.L., Goodman, N.D.: How to grow a mind: statistics, structure, and abstraction. Science $331(6022)$, 1279–1285 (2011)
14. WaiLam, S.X.: N-gram fragment sequence based unsupervised domain-specific document readability (2012)
15. Yamasaki, T., Tokiwa, K.-I.: A method of readability assessment for web documents using text features and html structures. Electron. Commun. Japan $97(10)$, 1–10 (2014)
16. Yan, X., Lau, R.Y., Song, D., Li, X., Ma, J.: Toward a semantic granularity model for domain-specific information retrieval. ACM Trans. Inf. Syst. (TOIS) $29(3)$, 15 (2011)
17. Yan, X., Song, D., Li, X.: Concept-based document readability in domain specific information retrieval. In: Proceedings of the 15th ACM International Conference on Information and Knowledge Management, pp. 540–549. ACM (2006)
18. Ye, Z., Huang, J.X., Lin, H.: Finding a good queryrelated topic for boosting pseudorelevance feedback. J. Am. Soc. Inform. Sci. Technol. $62(4)$, 748–760 (2011)
19. Yilmaz, E., Verma, M., Craswell, N., Radlinski, F., Bailey, P., Relevance, effort: an analysis of document utility. In: Proceedings of the 23rd ACM International Conference on Information and Knowledge Management, pp. 91–100. ACM (2014)
20. Zhang, Y., Ahmed, A., Josifovski, V., Smola, A.: Taxonomy discovery for personalized recommendation. In: Proceedings of the 7th ACM International Conference on Web Search and Data Mining, pp. 243–252. ACM (2014)
21. Zhang, Y., Zhang, J., Lease, M., Gwizdka, J.: Multidimensional relevance modeling via psychometrics and crowdsourcing. In: Proceedings of the 37th International ACM SIGIR Conference on Research & Development in Information Retrieval, pp. 435–444. ACM (2014)
22. Zuccon, G., Koopman, B.: Integrating understandability in the evaluation of consumer health search engines. In: Proceedings of MedIR 29 (2014)

Automatic Labelling of Topic Models Using Word Vectors and Letter Trigram Vectors

Wanqiu Kou[1]([⊠]), Fang Li[1], and Timothy Baldwin[2]

[1] Department of Computer Science and Engineering, Shanghai Jiaotong University,
Shanghai, China
autumn2012@qq.com, fli@sjtu.edu.cn
[2] Department of Computing and Information Systems, The University of Melbourne,
Melbourne, Australia
tb@ldwin.net

Abstract. The native representation of LDA-style topics is a multinomial distributions over words, which can be time-consuming to interpret directly. As an alternative representation, automatic labelling has been shown to help readers interpret the topics more efficiently. We propose a novel framework for topic labelling using word vectors and letter trigram vectors. We generate labels automatically and propose automatic and human evaluations of our method. First, we use a chunk parser to generate candidate labels, then map topics and candidate labels to word vectors and letter trigram vectors in order to find which candidate label is more semantically related to that topic. A label can be found by calculating the similarity between a topic and its candidate label vectors. Experiments on three common datasets show that not only the labelling method, but also out approach to automatic evaluation is effective.

Keywords: Topic labelling · Word vectors · Letter trigram vectors

1 Introduction

Topic models have been widely used in tasks like information retrieval [1], text summarization [2], word sense induction [3] and sentiment analysis [4]. Popular topic models include mixture of unigrams [5], probabilistic latent semantic indexing [6], and latent Dirichlet allocation (LDA) [7].

Topics in topic models are usually represented as word distributions, e.g. via the top-10 words of highest probability in a given topic. For example, the multinomial word distribution ⟨*feed contaminated farms company eggs animal food dioxin authorities german*⟩ is a topic extracted from a collection of news articles. The model gives high probabilities to those words like *feed*, *contaminated*, and *farms*. This topic refers to an animal food contamination incident. Our research aims to generate topic labels to make LDA topics more readily interpretable.

A good topic label has to satisfy the following requirements: (1) it should capture the meaning of a topic; and (2) it should be easy for people to understand.

© Springer International Publishing Switzerland 2015
G. Zuccon et al. (Eds.): AIRS 2015, LNCS 9460, pp. 253–264, 2015.
DOI: 10.1007/978-3-319-28940-3_20

There are many ways to represent a topic, such as a list of words, a single word or phrase, an image, or a sentence or paragraph [8]. A word can be too general in meaning, while a sentence or a paragraph can be too detailed to capture a topic. In this research, we select phrases to represent topics.

Our method consists of three steps. First, we generate candidate topic labels, then map topics and candidate labels to vectors in a vector space. Finally by calculating and comparing the similarity between a topic and its candidate label vectors, we can find a topic label for each topic.

Our contributions in this work are: (1) the proposal of a method for generating and scoring labels for topics; and (2) the proposal of a method using two word vector models and a letter trigram vector model for topic labelling. In experiments over three pre-existing corpora, we demonstrate the effectiveness of our methods.

2   Related Work

Topics are usually represented by their top-N words. For example, Blei et al. [7] simply use words ranked by their marginal probabilities $p(w|z)$ in an LDA topic model. Lau et al. [9] use features including PMI, WordNet-derived similarities and Wikipedia features to re-rank the words in a topic, and select the top three words as their topic label. A single word can often be inadequate to capture the subtleties of a topic. Some other methods use human annotation [10,11], with obvious disadvantages: on the one hand the result is influenced by subjective factors, and on the other hand, it is not an automatic method and is hard to replicate.

Some use feature-based methods to extract phrases to use as topic labels. Lau et al. [12] proposed a method that is based on: (1) querying Wikipedia using the top-N topic words, and extracting chunks from the titles of those articles; (2) using RACO [13] to select candidate labels from title chunks; and (3) ranking candidate labels according to features like PMI and the Student's t test, and selecting the top-ranked label as the final result.

Blei and Lafferty [14] used multiword expressions to visualize topics, by first training an LDA topic model and annotating each word in corpus with its most likely topic, then running hypothesis testing over the annotated corpus to identify words in the left or right of word or phrase with a given topic. The hypothesis testing is run recursively. Topics are then represented with multiword expressions.

Recent work has applied summarization methods to generate topic labels. Cano et al. [15] proposed a novel method for topic labelling that runs summarization algorithms over documents relating to a topic. Four summarization algorithms are tested: Sum basic, Hybrid TFIDF, Maximal marginal relevance and TextRank. The method shows that summarization algorithms which are independent of the external corpus can be applied to generate good topic labels.

Vector based methods have also been applied to the topic labelling task. Mei et al. [16] developed a metric to measure the "semantic distance" between a phrase and a topic model. The method represents phrase labels as word distributions, and approaches the labelling problem as an optimization problem that minimize the distance between the topic word distribution and label word distribution.

Our technique is inspired by the vector based method of Mei et al. [16] and Aletras and Stevenson [17], and work on learning vector representations of words using neural networks [18–20]. The basic intuition is that a topic and a label are semantically related in a semantic vector space.

3 Methodology

3.1 Preliminaries

Distributional vectors can be used to model word meaning, in the form of latent vector representations [18]. In order to capture correlations between a topic and a label, we map LDA topics and candidate labels to a vector space, and calculate the similarity between pairs of topic vectors and candidate label vectors. The candidate label which has the highest similarity is chosen as the label for that topic.

Table 1. An overview of the variables used to describe our models

Symbol	Description
z	A topic
T	The number of topics
ϕ_z	The word distribution of topic z
w	A word
d	A document
θ_d	The topic distribution of d
l	A topic label
L_z	A set of candidate labels for topic z
S	A letter trigram set
D	A document set
V	Vocabulary size
Sim	A word similarity measure
y_w	A vector representation of word w
GS	A gold standard label

The framework of our method is shown in Fig. 1. Note that $l_1...l_n$ represent candidate labels of topic z and Sim represents the similarity between two vectors. The symbols used in this paper to describe the top model and the topic labelling method are detailed in Table 1.

We experiment with three kinds of vectors to label topics: letter trigram vectors from [21], and two word vectors: CBOW (continuous bag-of-words model)

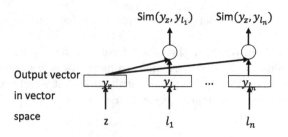

Fig. 1. Outline of our method

and Skip-gram [20]. A letter trigram vector is able to capture morphological variants of the same lemma in close proximity in a letter trigram vector space. CBOW and Skip-gram are methods for learning word embeddings based on distributional similarity. They each capture latent features from a corpus.

3.2   Candidate Label Extraction

We first identify topic-related document sets according to a topic summarization method [15]. The predominant topic of a document d can be calculated by:

$$z_d = \arg\max_z \theta_d(z) \tag{1}$$

Given a topic z, the set of documents whose predominant topic is z_d is then simply the set of documents that have z as their predominant topic. For each topic z, we then use OpenNLP[1] to full-text chunk parse each document in z_d, and extract chunks that contain at least words in the top-10 words in z, as candidate labels.

3.3   Vector Generation

CBOW Vectors. CBOW generates continuous distributed representations of words from their context of use. The model builds a log-linear classifier with bi-directional context words as input, where the training criterion is to correctly classify the current (middle) word. It captures the latent document features and has been shown to perform well over shallow syntactic relation classification tasks [22].

Recent research has extended the CBOW model to go beyond the word level to capture phrase- or sentence-level representations [22–24]. We simply apply the weighted additive method [23] to generate a phrase vector. Word vectors of candidate labels and LDA topics are generated as follows:

$$y_l^{cbow} = \sum_{w_j \in l} y_{w_j}^{cbow} \tag{2}$$

$$y_z^{cbow} = \sum_{w_j \in z} y_{w_j}^{cbow} \times \phi_z(w_j) \tag{3}$$

where $y_{w_j}^{cbow}$ is the word vector of word w_j based on CBOW.

---

[1] http://opennlp.apache.org/.

Skip-gram Vectors. The Skip-gram model [22] is similar to CBOW, but instead of predicting the current word based on bidirectional context, it uses each word as an input to a log-linear classifier with a continuous projection layer, and predicts the bidirectional context.

The Skip-gram model can capture latent document features and has been shown to perform well over semantic relation classification tasks [22].

Phrase vectors are, once again, generated using the weighted additive method [23]. Skip-gram vectors of candidate labels and LDA topics are generated in the same manner as CBOW, based on $y_{w_j}^{skip}$ (i.e. word vectors from Skip-gram).

Letter Trigram Vectors. We use the method of [21] to generate vectors for the topic and its candidate labels based on letter trigrams. Each dimension in a letter trigram vector represents a letter trigram (e.g. *abc* or *acd*). We generate a letter trigram set for each phrase l. A letter trigram set is defined as the multiset of letter trigrams from the phrase. For example, the letter trigram multiset of the phrase *stock market* is { ˆst, sto, toc, ock, ck␣, ␣ma, mar, ark, rke, ket, et$ }. For each dimension i in the letter trigram vector of phrase l, we assign an integer value based on the frequency of the corresponding letter trigram in the multiset, and normalize the counts to sum to one.

Using a similar method, we generate the letter trigram multiset of each of the top-10 LDA words, and take the union of the individual letter trigram multisets to calculate the overall letter trigram distribution for the top-10 words. We derive a vector representation for the topic based on the combined letter frequencies, and once again, normalize the counts to sum to one.

3.4 Topic Label Selection

After generating vectors for candidate labels and LDA topics, we then calculate the similarity between them based on cosine similarity.

4 Experiments

4.1 Dataset & Gold Standard

We use three corpora in our experiments: (1) NEWS, (2) TWITTER, and (3) NIPS. The NEWS and TWITTER corpora are from [15], while the NIPS corpus is a collection of NIPS abstracts from 2001 to 2010, commonly used for topic model evaluation.

The LDA training parameter α is set to $50/T$ and β is set to 0.01. We test the effect of the topic labelling method when T (the number of topics) is set to 30, 40 and 50 for each corpus. We use a within-topic entropy-based method to filter bland topics, i.e. topics where the probability distribution over the component words is relatively uniform, based on:

$$H(z) = -\sum_{i=1}^{V} \phi_z(w_i) \log_2(\phi_z(w_i)) \tag{4}$$

Table 2. The datasets used in this research

Corpus	# documents	# topics (T)	# pruned topics
NEWS-30	3743	30	2
NEWS-40	3743	40	6
NEWS-50	3743	50	11
TWITTER-30	35815	30	19
TWITTER-40	35815	40	30
TWITTER-50	35815	50	40
NIPS-30	2075	30	5
NIPS-40	2075	40	8
NIPS-50	2075	50	20

In the NEWS and TWITTER corpora, topics with an entropy higher than 0.9 were eliminated, and in the NIPS corpus, topics with an entropy higher than 1.4 were eliminated; these thresholds were set based on manual analysis of a handful of topics for each document collection. For the TWITTER corpus, we further filtered topics which lack a meaningful gold-standard topic label, based on the method described later in this section. Table 2 provides details of the datasets.

Yang [25] observed that gold standard labels from human beings suffer from inconsistency. The inter-annotator F-measure between human annotators for our task is 70–80 %. In an attempt to boost agreement, we developed an automatic method to generate gold standard labels to evaluate the proposed method: for each topic z, we extract chunks from titles in D_z, assign a weight to each chunk according to the word frequency in that chunk, and select the chunk that has the highest weight as the label ("GS") for that topic. Our underlying motivation in this is that each headline is the main focus of a document. A phrase from a title is a good representation of a document. Therefore a phrase from a title can be a good label for the predominant topic associated with that document.

Note that the NEWS and NIPS corpora have titles for each document, while the TWITTER corpus has no title information. The gold standard for the NEWS and NIPS corpora were thus generated automatically, while for the TWITTER corpus — which was collected over the same period of time as the NEWS corpus — we apply the following method, based on [15]: (1) calculate the cosine similarity between each pair of TWITTER and NEWS topics, based on their word distributions; (2) for each TWITTER topic i, select the NEWS topic j that has the highest cosine similarity with i and where the similarity score is greater than a threshold (0.3 in this paper). The label (GS) of NEWS topic j is then regarded as the gold standard (GS) label for TWITTER topic i.

4.2 Evaluation Metrics

We evaluate our results automatically and via human evaluation.

Automatic Evaluation Method. Because of the potential for semantically similar but lexically divergent labels, we can't compare the generated label directly with the GS automatically. Rather, we propose the following evaluation:

$$score_z = \frac{\sum_{w \in GS} \max_{w' \in l} Lin'(w, w') + \sum_{w' \in l} \max_{w \in GS} Lin'(w, w')}{\#words(GS) + \#words(l)} \quad (5)$$

$$Lin'(w, w') = \begin{cases} 1 & \text{if } stem(w) = stem(w') \\ Lin(w, w') & \text{otherwise} \end{cases} \quad (6)$$

where Lin is word similarity based on WordNet, in the form of the information-theoretic method of Lin [26]. GS and l represent the gold standard and the label generated for topic z, respectively. The Porter stemmer[2] is used to stem all words. The score is used to measure the semantic similarity between an automatically-generated and GS label.

Human Evaluation Method. We also had six human annotators manually score the extracted labels. Each annotator was presented with the top-10 LDA words for a given topic, the gold standard label, and a series of extracted labels using the methods described in Sect. 3. They then score each extracted label as follows: 3 for a very good label; 2 for a reasonable label, which does not completely capture the topic; 1 for a label semantically related to the topic, but which is not a good topic label; and 0 for a label which is completely inappropriate and unrelated to the topic. We average the scores from the six annotators to calculate the overall topic label score.

4.3 Baseline Methods

LDA-1. Simply select the top-ranked topic word as the topic label.

DistSim. This method was proposed by [16], and involves generating a word vector of candidate labels according to first-order cooccurrence-based PMI values in the original corpus. In this paper, the first-order vector is used in our vector-based method shown in Fig. 1.

4.4 Experimental Results

The word2vec toolbox[3] was used to train the CBOW and Skip-gram models. The window size was set to 5 for both models. We experimented with word vectors of varying dimensions; the results are shown in Fig. 2, based on automatic evaluation. When the number of dimensions is 100, the result is the best on average, and this is the size we use for both CBOW and Skip-gram throughout our experiments. The dimension of the letter trigram vector is 18252.

Figure 3 shows the automatic evaluation results for topic labelling with different numbers of topics. We can see that the results vary with the number

---

[2] http://tartarus.org/~martin/PorterStemmer/.

[3] https://github.com/NLPchina/Word2VEC_java.

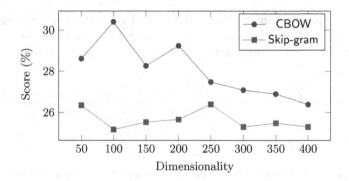

Fig. 2. Automatic evaluation results for word vectors over the NEWS corpus for different embedding dimensionalities

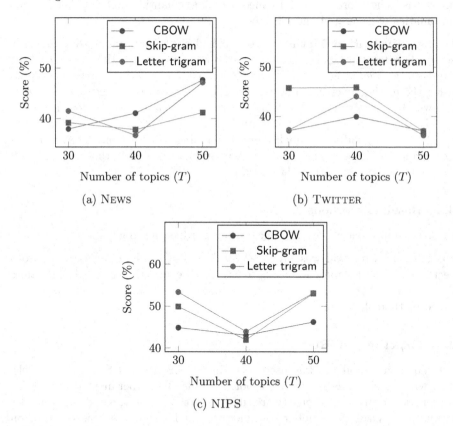

Fig. 3. Automatic evaluation results

of topics. When the topic number T is 50, the score for the NEWS and NIPS corpora is the highest; and when the topic number is 40, the score for the Twitter corpus is highest.

Table 3 shows the result between the baseline methods (LDA-1 and DistSim) and our methods, over the NEWS-50, TWITTER-40 and NIPS-50 corpora.

Table 3. Evaluation of the baselines and the proposed approaches

Method	Automatic evaluation (%)			Human evaluation		
	NEWS-50	TWITTER-40	NIPS-50	NEWS-50	TWITTER-40	NIPS-50
LDA-1	33.73	32.65	41.62	1.04	1.00	1.02
DistSim	42.77	41.90	42.74	1.60	1.60	1.70
CBOW	47.64	39.94	46.12	1.94	1.10	2.10
Skip-gram	41.15	45.90	53.05	1.85	1.60	2.07
Letter trigram	47.17	44.08	53.14	1.86	1.50	2.13

Based on the experimental results in Table 3, we summarize our findings as follows:

1. Most methods perform better over NIPS than NEWS and TWITTER. The primary reason is that we use NIPS abstracts (and not full papers) to train the LDA topic model. Abstracts are more closely related to the paper titles. This means that automatically-generated gold standard labels are more likely to score well for NIPS.
2. The Skip-gram model performs much better than CBOW over TWITTER and NIPS, while over NEWS, CBOW is better than Skip-gram; CBOW performs relatively badly over TWITTER. The reason is that Skip-gram works better over sparse corpora like TWITTER and NIPS, while CBOW works better over dense corpora. Mikolov et al. [22] show that Skip-gram performs better over semantic tasks while CBOW performs better over shallow syntactic tasks, based on which we assumed that Skip-gram should be better for topic labelling. However, our experiments indicate that results are also dependent on the genre of the corpus: NEWS topics usually refer to concrete information like the agents and details of a particular event; NIPS topics, on the other hand, usually refer to scientific concepts, while TWITTER topics are more comments on certain events, and informal and brief.
3. The letter trigram vectors perform surprisingly consistently over the three corpora in Table 3. Letter trigrams simply capture character features of a word, and the method is therefore not dependent on the corpus genre. Compared with DistSim, letter trigram vectors have reduced dimensionality, and are able to capture morphological variations of the same word to points that are close to each other in the letter trigram space.

The three methods proposed in this paper are all better than LDA-1 baseline. The reason might be that in this method, we compare the top-1 word with a phrase (GS). Our methods are also better than the DistSim baseline in most cases. Our result shows that trigram vectors are more suitable for topic labelling over different types of corpus. Skip-gram is better than CBOW for TWITTER and NIPS, while CBOW is more suitable for NEWS.

Table 4. Sample topic labels

	NEWS	TWITTER	NIPS
LDA top-10	⟨ *snow weather service heavy airport closed storm power county north*⟩	⟨ *prison guilty murder htt trial rights iran ex human jail*⟩	⟨ *motion human model visual attention range tracking body target task*⟩
GS	ice and snow hit	Amanda Knox murder appeal	human motion perception
CBOW	heavy snow and winds	convicted ex	motion and camera motion
Skip-gram	storm closed	murder trial	human visual motion perception
Letter trigram	weather service	prison sentence	human visual motion perception
DistSim	Derry airport closed	human rights violation	motion estimation

Table 3 also shows the results for human evaluation. We summarize the results as follows:

1. Similar to the automatic evaluation results, the score over NIPS is higher than the other two corpora. The score for NEWS is higher than the score for TWITTER. Under human evaluation, labels generated using vector-based methods are on average reasonable labels for NIPS, and somewhat reasonable labels for NEWS. Even for a corpus without title information like TWITTER, it can extract related topic labels.
2. Human evaluation achieves very similar results to our automatic evaluation; in fact, we calculated the Pearson correlation between the two and found it to be remarkably high at $r = 0.84$. This shows that our automatic evaluation method is effective, and can potentially save manual labor for future work on topic label evaluation.

4.5   Effectiveness of Topic Labelling Method

To show the effectiveness of our method, some sample topic labels from NEWS, TWITTER and NIPS are shown in the Table 4. Full results over the three corpora are available for download from:

http://lt-lab.sjtu.edu.cn/wordpress/wp-content/uploads/2014/
05/topic%20label%20result.zip

5   Conclusion

We have proposed a novel method for topic labelling using embeddings and letter trigrams. Experiments over three corpora indicate that all three kinds of vectors are better than two baseline methods. Based on the results for automatic and human evaluation, labels extracted using the three vector methods have reasonable utility. The results of word vector models vary across the different corpora, while the letter trigram model is less influenced by the genre of the corpus. The limitation of word vectors is that the quality of a topic label relies on the quality of the word vector representation, which in turn is influenced by the corpus size. The novelty of our work includes the use of embeddings for label ranking, the automatic method to generate gold-standard labels, and the method to automatically evaluate labels. In the future, we plan to do more experiments on different types of corpora. Letter trigram vectors do not need training, and are more suitable for different types of corpus. We also plan to do more experiments on different types of vector representations and on vector combination, and also extrinsic evaluation of the topic labels [8].

Acknowledgements. This research was supported in part by the Australian Research Council.

References

1. Xu, J., Croft, W.B.: Cluster-based language models for distributed retrieval. In: Proceedings of 22nd International ACM-SIGIR Conference on Research and Development in Information Retrieval (SIGIR 1999), pp. 254–261. Berkeley, USA (1999)
2. Haghighi, A., Vanderwende, L.: Exploring content models for multi-document summarization. In: Proceedings of the North American Chapter of the Association for Computational Linguistics - Human Language Technologies 2009 (NAACL HLT 2009), pp. 362–370. Boulder, USA (2009)
3. Lau, J.H., Cook, P., McCarthy, D., Newman, D., Baldwin, T.: Word sense induction for novel sense detection. In: Proceedings of the 13th Conference of the EACL (EACL 2012), pp. 591–601. Avignon, France (2012)
4. Lin, C., He, Y.: Joint sentiment/topic model for sentiment analysis. In: Proceedings of the 18th ACM Conference on Information and Knowledge Management (CIKM 2009), pp. 375–384. Hong Kong, China (2009)
5. Gimpel, K.: Modeling topics. Inf. Retrieval **5**, 1–23 (2006)
6. Hofmann, T.: Probabilistic latent semantic indexing. In: Proceedings of 22nd International ACM-SIGIR Conference on Research and Development in Information Retrieval (SIGIR 1999), pp. 50–57. Berkeley, USA (1999)
7. Blei, D.M., Ng, A.Y., Jordan, M.I.: Latent dirichlet allocation. J. Mach. Learn. Res. **3**, 993–1022 (2003)
8. Aletras, N., Baldwin, T., Lau, J.H., Stevenson, M.: Evaluating topic representations for exploring document collections. J. Assoc. Inf. Sci. Technol. (to appear) (2015)
9. Lau, J.H., Newman, D., Karimi, S., Baldwin, T.: Best topic word selection for topic labelling. In: Proceedings of the 23rd International Conference on Computational Linguistics (COLING 2010), Posters Volume, pp. 605–613. Beijing, China (2010)
10. Wang, X., McCallum, A.: Topics over time: a non-Markov continuous-time model of topical trends. In: Proceedings of the 12th ACM SIGKDD International Conference on Knowledge Discovery and Data Mining, pp. 424–433. Philadelphia, USA (2006)

11. Mei, Q., Liu, C., Su, H., Zhai, C.: A probabilistic approach to spatiotemporal theme pattern mining on weblogs. In: Proceedings of the 15th International Conference on the World Wide Web (WWW 2006), pp. 533–542. Edinburgh, UK (2006)

12. Lau, J.H., Grieser, K., Newman, D., Baldwin, T.: Automatic labelling of topic models. In: Proceedings of the 49th Annual Meeting of the Association for Computational Linguistics: Human Language Technologies (ACL HLT 2011), pp. 1536–1545. Portland, USA (2011)

13. Grieser, K., Baldwin, T., Bohnert, F., Sonenberg, L.: Using ontological and document similarity to estimate museum exhibit relatedness. ACM J. Comput. Cult. Heritage 3(3), 1–20 (2011)

14. Blei, D.M., Lafferty, J.D.: Visualizing topics with multi-word expressions. arXiv preprint arXiv:0907.1013 (2009)

15. Cano Basave, E.A., He, Y., Xu, R.: Automatic labelling of topic models learned from twitter by summarisation. In: Proceedings of the 52nd Annual Meeting of the Association for Computational Linguistics (ACL 2014), pp. 618–624 (2014)

16. Mei, Q., Shen, X., Zhai, C.: Automatic labeling of multinomial topic models. In: Proceedings of the 13th ACM SIGKDD International Conference on Knowledge Discovery and Data Mining (KDD 2007), pp. 490–499 (2007)

17. Aletras, N., Stevenson, M.: Measuring the similarity between automatically generated topics, pp. 22–27 (2014)

18. Bengio, Y., Ducharme, R., Vincent, P., Janvin, C.: A neural probabilistic language model. J. Mach. Learn. Res. 3, 1137–1155 (2003)

19. Collobert, R., Weston, J.: A unified architecture for natural language processing: deep neural networks with multitask learning. In: Proceedings of the 25th International Conference on Machine Learning (ICML 2008), pp. 160–167. Helsinki, Finland (2008)

20. Mikolov, T., Chen, K., Corrado, G., Dean, J.: Efficient estimation of word representations in vector space. In: Proceedings of Workshop at the International Conference on Learning Representations. Scottsdale, USA (2013)

21. Huang, P.S., He, X., Gao, J., Deng, L., Acero, A., Heck, L.: Learning deep structured semantic models for web search using clickthrough data. In: Proceedings of the 22nd ACM Conference on Information and Knowledge Management (CIKM 2013), pp. 2333–2338. San Francisco, USA (2013)

22. Mikolov, T., Sutskever, I., Chen, K., Corrado, G.S., Dean, J.: Distributed representations of words and phrases and their compositionality. In: Advances in Neural Information Processing Systems, pp. 3111–3119 (2013)

23. Mitchell, J., Lapata, M.: Composition in distributional models of semantics. Cogn. Sci. 34(8), 1388–1429 (2010)

24. Socher, R., Huval, B., Manning, C.D., Ng, A.Y.: Semantic compositionality through recursive matrix-vector spaces. In: Proceedings of the Joint Conference on Empirical Methods in Natural Language Processing and Computational Natural Language Learning 2012 (EMNLP-CoNLL 2012), pp. 1201–1211. Jeju Island, Korea (2012)

25. Yang, J., Xu, W., Tan, S.: Task and data designing of sentiment sentence analysis evaluation in COAE2014. J. Shanxi Univ. (Natural Science Edit) 1(3), 51 (2015)

26. Lin, D.: An information-theoretic definition of similarity. In: Proceedings of the International Conference on Machine Learning, pp. 296–304. Madison, USA (1998)

A Study of Collection-Based Features for Adapting the Balance Parameter in Pseudo Relevance Feedback

Ye Meng[1], Peng Zhang[1], Dawei Song[1,2] (✉), and Yuexian Hou[1]

[1] Tianjin Key Laboratory of Cognitive Computing and Application,
Tianjin University, Tianjin, China
ye.meng04@gmail.com, {pzhang,dwsong,yxhou}@tju.edu.cn
[2] The Computing Department, The Open University, Buckinghamshire, UK

Abstract. Pseudo-relevance feedback (PRF) is an effective technique to improve the ad-hoc retrieval performance. For PRF methods, how to optimize the balance parameter between the original query model and feedback model is an important but difficult problem. Traditionally, the balance parameter is often manually tested and set to a fixed value across collections and queries. However, due to the difference among collections and individual queries, this parameter should be tuned differently. Recent research has studied various query based and feedback documents based features to predict the optimal balance parameter for each query on a specific collection, through a learning approach based on logistic regression. In this paper, we hypothesize that characteristics of collections are also important for the prediction. We propose and systematically investigate a series of collection-based features for queries, feedback documents and candidate expansion terms. The experiments show that our method is competitive in improving retrieval performance and particularly for cross-collection prediction, in comparison with the state-of-the-art approaches.

Keywords: Information retrieval · Pseudo-relevance feedback · Collection characteristics

1 Introduction

Pseudo-relevance feedback (PRF) has been proven effective for improving retrieval performance. The basic idea is to assume a certain number of top-ranked documents as relevant and select expansion terms from these documents to refine the query representation [18]. A fundamental question is whether the feedback information are truly relevant to the query. Cao et al. [4] show that the expansion process indeed adds more bad terms than good ones, and the proportions of bad terms in different collections are different. This means that there is noise in the expansion terms and we can not always trust the expansion information. Thus, we need to carefully balance the original query model and the expansion model

G. Zuccon et al. (Eds.): AIRS 2015, LNCS 9460, pp. 265–276, 2015.
DOI: 10.1007/978-3-319-28940-3_21

derived from the feedback documents. If we over-trust the feedback information, the retrieval performance can be harmed due to the noise in the expansion model. If we under-trust it, we will not be able to take full advantage of the feedback information. Currently, the balance parameter is often manually tested and set to a fixed value across queries for a specific collection, to combine the original query and the expansion terms derived from the feedback documents. Due to the difference between different collections and different queries, this parameter should be set differently. Recently, Lv and Zhai [10] present a learning approach to adaptively predict the optimal weight of the original query model for different queries and collections. They explore a number of features and combine them using a regression approach for the prediction. The features they used are mostly based on the original query and feedback information, yet do not sufficiently consider features of the candidate expansion terms and the collection.

It has long been recognized in information retrieval that document collection has a great impact on the performance of a retrieval model [17]. In this paper, we propose and systematically investigate a set of collection-based features about queries, feedback documents and candidate terms, which are complementary to the features used in Lv and Zhai [10]. Specifically, three types of features are studied, including (1) Information amount of query: we suppose that a query is more reliable when it carries more information; (2) Reliability of feedback documents; (3) Reliability of candidate terms: We will trust the feedback documents and candidate terms only when they are highly reliable. The proposed features are feed into a logistic regression model to predict the feedback parameter.

2   Related Work

Pseudo-relevance feedback has been implemented in different retrieval models: e.g., vector space model, probabilistic model, and language model. In the vector space model [6], feedback is usually done by using the Rocchio algorithm, which forms a new query vector by maximizing its similarity to relevant documents and minimizing its similarity to non-relevant document. The feedback method in classical probabilistic models [3,16] is to select expanded terms primarily based on Robertson/Sparck-Jones weight. In the language modeling approaches [9,20], relevance feedback can be implemented through estimating a query language model or relevance model through exploiting a set of feedback documents. All those works used a fixed parameter to control the balance parameter between original query and feedback information.

Recently, Lv and Zhai [10] present a learning approach to adaptively predict the optimal balance parameter for each query and each collection. They leverage state-of-the-art language models for ranking documents and use logistic regression to optimize an important parameter inside the language modeling framework. Three heuristics to characterize feedback balance parameter are used, including the discrimination of query, discrimination of feedback documents and divergence between query and feedback documents. These three heuristics are then taken as a road map to explore a number of features and combined them using the logistic regression model to predict the balance parameter. The experiments show that the proposed adaptive relevance feedback is more robust and effective than the regular fixed-parameter feedback. Nevertheless there is still room to explore when the training and testing sets are different. Our work uses a similar method, but adds features based on characteristic of collection. The experiments show that our method is competitive in improving retrieval performance, in comparison with their approaches.

3   Basic Formulation

The Relevance Model (RM) [21] is a representative and state-of-the-art approach for re-estimating query language models based on PRF [9]. We will carry out our study in the RM framework.

For a given query $Q = (q_1, q_2, ..., q_m)$, based on the corresponding PRF document set F ($|F| = n$), RM estimates an expanded query model [22]:

$$P(w|\theta_F) \propto \sum_{D \in F} P(w|\theta_D)P(\theta_D) \prod_{i=1}^{m} P(q_i|\theta_D) \tag{1}$$

where $P(\theta_D)$ is a prior on documents and is often assumed to be uniform without any additional prior knowledge about the document D. Thus, the estimated relevance model is essentially a weighted combination of individual feedback document models with the query likelihood score of each document as the weight.

The estimated relevance model, $P(w|\theta_F)$, can then be interpolated with the original query model θ_Q to improve performance:

$$P(w|\theta'_Q) = \lambda P(w|\theta_Q) + (1 - \lambda)P(w|\theta_F) \tag{2}$$

where λ is a balance parameter to control the weight of the feedback information. The model in Eq. (2) is often referred to as RM3 [9]. When $\lambda = 1$, we only use the original query model (i.e., no feedback). If $\lambda = 0$, we ignore the original query and rely only on the feedback model.

4   The Proposed Collection-Based Features

As aforementioned, due to the difference of collections in document type, size and other characteristics, and the difference of query difficulties, the expansion terms selected from the feedback documents are not always good terms [4]. Accordingly, the balance parameter should be set differently for different collections and queries. In this section, we investigate three types of collection-based features about query, feedback documents and candidate terms, for adaptive setting of the balance parameter.

4.1   Information Amount of Query

Intuitively, if a query contains a sufficient amount of information about the search topic, then the expansion terms may be less important and thus more weight should be given to the original query. As the query performance is largely related to the information amount of the query, it is natural to borrow some features that have been used in query performance prediction [5]. As a step further, we also propose to look at two extra features, namely the mutual information and information entropy.

4.1.1. The Distribution of Information Amount in the Query Terms

In general, each query term t can be associated with an inverse document frequency $(idf(t))$ describing the information amount that the term carries. According to Pirkola and Järvelin [13], the difference in the discriminative power of query terms, which is reflected by the $idf(t)$ values, could affect the retrieval effectiveness. Therefore, the distribution of the $idf(t)$ over query terms, denoted DI, might be an intrinsic feature that affects the selection of balance parameter. DI is represented as:

$$DI = \sigma_{idf} \tag{3}$$

where σ_{idf} is the standard deviation of the idf values of the terms in Q. In our study, idf is defined as follows:

$$idf(t) = \frac{\log \frac{(N+0.5)}{N_t}}{\log(N+1)} \tag{4}$$

where N_t is the number of documents containing the query term t, and N is the number of documents in the collection. The higher DI score, the more dispersive the query's information amount distribution is. Then we would need to bring in more precise information from the expansion terms, and thus give more weight to the feedback/expansion model.

4.1.2. Query Scope

The notion of query scope characterizes the generality of a query. For example, the query "Chinese food" is more general than "Chinese dumplings", as the latter is about a particular Chinese food. The query scope was originally studied in [14], defined as a decay function of the number of documents containing at least one query term, and has been shown to be an important property of the query. Similarly, in this paper, we define the query scope as follows:

$$QS = -\log(\frac{n_Q}{N}) \tag{5}$$

where n_Q is the number of documents containing at least one of the query terms, and N is the number of documents in the whole collection. A larger n_Q value will result in a lower query scope. The higher QS value means clearer information contained by the query, then we should give more weight to the original query.

4.1.3. Average Inverse Collection Term Frequency

According to Kwok [8], the inverse collection term frequency $(ICTF)$ can be seen as an alternative of idf and is correlated with the quality of a query term. The average $ICTF$ $(AvICTF)$ is given by:

$$AvICTF = \frac{log\Pi_{q \in Q} \frac{|C|}{tf_q}}{|Q|} \tag{6}$$

where tf_q is the occurrence frequency of a query term in the collection; $|C|$ is the number of tokens in the collection; and $|Q|$ is the query length. $AvICTF$ measures the overall discriminative power of query terms. The higher $AvICTF$ of the query indicates that more weight may be needed for the original query while the expansion terms may not bring much extra benefit.

4.1.4. Mutual Information Among Query Terms

Mutual information (MI) [12] is used to quantify how the terms in a query are associated to each other. The MI is a quantity that measures the mutual dependence of the two discrete random variables X and Y, defined as follows:

$$MI(Q) = I(X;Y) = \sum_{y \in Y} \sum_{x \in X} P(x,y) \log \frac{P(x,y)}{P(x)P(y)} \tag{7}$$

where $P(x,y)$ is the joint probability distribution function of X and Y, $P(x)$ and $P(y)$ are the marginal probability distribution functions of X and Y respectively. In our study, they can be defined as follows:

$$\begin{aligned} P(x,y) &= df_{xy}/N \\ P(x) &= df_x/N \\ P(y) &= df_y/N \end{aligned} \tag{8}$$

where x and y are two original query terms; df_{xy} is the document frequency where terms x and y co-occur; N is the number of documents in the whole collection; df_x and df_y are document frequency of the query term x and y respectively. The higher MI score means a high correlation among query terms, and thus more coherent information is carried by the original query. In turn, less weight can be given to candidate expansion terms.

4.1.5. Information Entropy of Query

We propose to analyze the term distribution in a query using information entropy [2]. In information theory, entropy measures the average amount of information contained in a message received, thus characterizing the uncertainty of information. For a random variable X with n outcomes $\{x_1, ..., x_n\}$, the widely used Shannon entropy (denoted by $H(X)$), is defined as follows:

$$IE(Q) = H(x) = -\sum_{i=1}^{n} P(x_i) \log P(x_i) \tag{9}$$

where $P(x_i)$ is the probability mass function of outcome x_i. In this study, it is calculated as follows:

$$P(x_i) = tf_i/Ntf \tag{10}$$

where tf_i is the frequency of query term x_i, Ntf is the sum of the all tfs in the collection. The high IE score means less certainty of the query, then more weight should be given to candidate expansion terms.

4.2 Reliability of Feedback Documents

We expect that the more reliable the feedback documents are, the more weight should be given to the expansion model derived from these documents.

4.2.1. Clarity of Feedback Documents

The clarity of feedback documents is defined as follows, as also used in [10].

$$CFD = \sum_{\omega \in F} p(\omega|\theta_F) \log \frac{p(\omega|\theta_F)}{p(\omega|C)} \tag{11}$$

where F is the set of feedback documents, $p(\omega|C)$ is the collection language model, and $p(\omega|\theta_F)$ is estimated as $p(\omega|\theta_F) = \frac{c(\omega,F)}{\sum_\omega c(\omega,F)}$. The higher CFD value, the more reliable the feedback documents tend to be.

4.2.2. The Content of Least Frequent Terms in Feedback Documents

The least frequent terms (LFT) are terms appearing less than a certain number of times (e.g., 3 in our experiments) in the collection and containing non-alphabetical characters, such as "00", "1", "2d". These terms usually have little practical significance. The content of LFT in feedback documents is defined as:

$$LFT_F = \frac{N(LFT)}{|F|} \tag{12}$$

where $N(LFT)$ is the number of LFT terms in the feedback documents, and $|F|$ is the total number of terms in the feedback documents. The higher LFT_F, the less reliable the feedback documents tend to be.

4.3 Reliability of Candidate Terms

We expect that the higher reliability of candidate expansion terms, the more weight should be given to them when combining with the original query model.

4.3.1. Mutual Information Between Candidate Expansion Terms and Query

The definition of $MI(C)$ is the same as $MI(Q)$ roughly, except the different meaning of the variables in Eqs. (7) and (8). For $MI(Q)$, the X and Y represent the original query terms, but for $MI(C)$, they represent the original query and candidate terms respectively.

4.3.2. Information Entropy of Candidate Expansion Terms

Similar to the definition of $IE(Q)$, the $IE(C)$ can be calculated using Eqs. (9) and (10), with x_i representing candidate terms.

4.3.3. The Content of LFT in Candidate Expansion Terms

This can be measured in the same way as for the feedback documents in Eq. (12), and we defined it as $LFT_C = \frac{N(LFT)}{|C|}$. For candidate expansion terms, $N(LFT)$ is the number of LFT terms in the candidate terms, and $|C|$ is the total number of candidate terms.

5 The Logistic Regression Model

Logistic regression is widely used in data mining and machine learning. We use a logistic regression model to combine our features and generate a score for predicting the balance parameter, whereas the output is confined to values between 0 and 1. The method is the same as the one used in [10], defined as follows:

$$f(X) = \frac{1}{1 + exp(-X)} \tag{13}$$

where the variable $X = \overline{\omega} * \overline{x}$ represents the set of features. Specifically, $\overline{x}$ is a vector of numeric values representing the features and $\overline{\omega}$ represents a set of weights, which indicates the relative weights for each features. $f(X)$ represents the probability of a particular outcome given the set of features.

6 Experiments and Results

6.1 Experimental Setup

We used five standard benchmarking collections in our experiments: AP8890 (AP), WSJ8792 (WSJ), ROBUST2004 (ROBUST), WT10G and SJM, which are different in size and genre. The WSJ, AP and SJM collections are relatively small and consist of news articles, science and technology reports and government documents, whereas WT10G is a larger Web collection. The details of these collections are shown in Table 1.

Table 1. Information of Collections

Collection	Size	#doc(K)	Queries	#qry	Avg_dl	Dev(dl)
SJM	286 MB	90K	101-150	46	218	364
AP	728 MB	243K	151-200	50	244	244
WSJ	508 MB	173K	151-200	50	247	455
ROBUST	1.85 GB	528K	601-700	99	254	869
WT10G	10.2 GB	1692K	501-550	50	379	2941

In all the experiments, we only used the title field of the TREC queries for retrieval, because it is closer to the actual queries used in the real web search applications and relevance feedback is expected to be the most useful for short queries [19].

First, we used Indri which is part of the Lemur Toolkit [11] to index document collection. In the indexing process, all terms were stemmed using Porter's English stemmer [15], and stopwords from the standard InQuery stoplist [1] were removed. Then, we initially retrieved a document list for each query using language model with the Dirichlet prior (takes a hyper-parameter of μ applied to smooth the document language model which is better than other smoothing methods for title query.) and fixed the smoothing parameter to 1500 for all queries. This is our baseline for all pseudo relevance models in our experiments denoted as LM. After that, for each query, we used the expanded query model (Eq. (1)) to get the candidate expansion terms. In this part, we fixed the number of feedback documents to top 30, and the number of candidate expansion terms to 100 according to the settings in existing work.

To train the proposed adaptive relevance feedback, we needed to obtain the training data first. Considering the reliability and authority of training data, 90 % of the queries were selected randomly for training, resulting in a total of 41 out of queries 101-150, 45 out of queries 151-200, 89 out of queries 601-700, and 45 out of query 501-550, and the rest were taken as testing queries. In this way, we aimed to make the training data more diversified and the test results more general and reliable. It turned out that 262 queries were taken as training data of different types and 33 queries were taken as testing data.

For traditional RM3 model, as we have discussed in Sect. 3, the balance parameter λ (Eq. (2)) changed from 0 to 1 (0.1, 0.2,....., 1) on five collections to find the optimal λ (this parameter is fixed for all queries in the same collection) and we called it RM3-Manual. For our adaptive relevance feedback, we chosen the optimal λ for each query. All the above processes were the same for our training and testing data.

Finally, the effectiveness of the IR models on each collection was measured by the Mean Average Precision (MAP) [7] at the top 1000 retrieved documents.

6.2   Sensitivity of Balance Parameter

We investigated the sensitivity of balance parameter on AP8890 collection and some queries of AP8890 in relevance feedback experiments by varying λ form 0 to 1, as it is showed in Fig. 1. We could observe that the setting of λ could affect the retrieval performance significantly, and the optimal parameter for different queries on the same collection could be quite different.

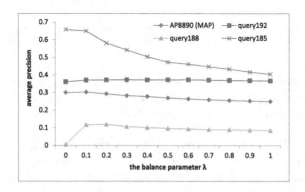

Fig. 1. Sensitivity of the balance parameter (λ) for different queries on AP8890 collection

6.3 Correlation Between Features and the Optimal Balance Parameter

We measured the correlation between features and the optimal balance parameter for each query in the training data using Pearson and Spearman methods which are common. Based on the Sect. 4, we could obtain a matrix [262,10] of query-features, each query had its own 10 feature values and optimal λ which were the base of analysis. As showing in Table 2, DI, QS, $IE(Q)$ and LFT_F are more correlated with the optimal feedback coefficient than other features. It may mean that the information of query plays an important role in predicting the balance parameter.

Table 2. Pearson and Spearman correlation coefficients between features and the optimal λ on training data

Features	Pearson	Spearman
DI	**-0.0453**	**-0.1495**
QS	**-0.133**	**-0.1022**
AvICTF	0.0531	0.0892
MI(Q)	0.0028	-0.0262
IE(Q)	**-0.0832**	**-0.1347**
CFD	0.0751	0.05
LFT_F	**0.1158**	**0.1037**
MI(C)	-0.0509	-0.0702
IE(C)	-0.0662	-0.0559
LFT_C	-0.0189	-0.0178

6.4 Prediction Models and the Results

In this part, we trained three prediction models on our training data by using three different sets of features respectively and the assessment of fit is based on significance tests for the balance parameter.: (1) all of the proposed features (ten); (2) only DI, QS, $IE(Q)$ and LFT_F; (3) five important features proposed in [10], including clarity of queries, feedback length, clarity of feedback documents and the absolute divergence between queries and feedback documents (see [10] for more details). They are called as "RM3-A", "RM3-A2" and "RM3-B" respectively. Given a new query, we could predict its feedback balance parameter directly using the formula: $f(X) = \frac{1}{1+exp(-X)}$ which was introduced in Sect. 5, and X for all ten features ($X1$) and four important features ($X2$) are showed below:

$$X1 = -0.2444 - 3.6127 * DI - 0.4249 * QS - 0.0214 * AvICTF$$
$$+184.2403 * MI(Q) - 44.8057 * IE(Q) + 3.1588 * CFD$$
$$+0.6834 * LFT_F - 0.7406 * MI(C) + 0.5376 * IE(C) - 4.7051 * LFT_C$$

$$(14)$$

$$X2 = -0.5594 - 5.5303 * DI - 0.3347 * QS - 43.2822 * IE(Q) + 0.418 * LFT_F \tag{15}$$

From the above formula, we can see that, for the distribution of information amount in queries, DI and $IE(Q)$ are correlated negatively to the λ and $MI(Q)$ shows a positive correlation. This is consistent with our expectation that more weight should be given to the original query when the query has more information. For the reliability of feedback documents and expansion terms, LFT_F, $EI(C)$ and LFT_C should be positively to the λ, while CFD and $MI(C)$ should show a negative correlation. That means high weight $(1 - \lambda)$ should be given to candidate terms when the feedback information is more reliable. However, the different behaviors of LFT_F and LFT_C in Eq. (14) could be explained as a trade-off between "credibility" and "quantity of information".

Table 3. Performance comparison of RM3-A and RM3-B on all testing data.

	SJM	AP8890	WSJ8792	ROBUST	WT10G
LM	0.2461	0.3279	0.3321	0.2258	0.2840
RM3-A	**0.3105**	0.3249	**0.3360**	**0.2597**	**0.3061**
RM3-B	0.3081	0.3309	0.3357	0.2495	0.3009
RM3-A2	0.3104	**0.3363**	0.3342	0.2485	0.2999
RM3-Manual	0.3175	0.3361	0.3293	0.2555	0.3044

The performance (MAP) on baseline, RM3-A, RM3-A2, RM3-B and RM3-Manual are demonstrated in Table 3. It shows that RM3-A, RM3-A2 and RM3-B all outperform the LM, but comparing with RM3-Manual, there is still room improvement by further optimizing the feedback parameter. RM3-A is better than RM3-B on SJM, WSJ8792, ROBUST2004 and WT10G, and only fails on

Table 4. Performance comparison of RM3-A and RM3-B on ROBUST2004.

ROBUST2004	LM	RM3-A	RM3-B	RM3-A2	RM3-Manual
query609	0.0280	**0.0404**	0.0308	0.0359	0.0443
query621	0.0812	0.0634	0.0868	0.0671	0.0634
query635	0.5471	**0.6152**	0.6067	0.6168	0.6026
query642	0.3503	**0.4014**	0.4014	0.3882	0.4014
query651	0.0220	**0.0786**	0.0175	0.0186	0.0164
query666	0.7081	**0.6815**	0.6727	0.6815	0.6935
query678	0.1509	**0.1961**	0.1933	0.1933	0.213
query683	0.0976	**0.2677**	0.2227	0.2227	0.2677
query691	0.0125	**0.0144**	0.0138	0.0144	0.0139
query700	0.2600	0.2384	0.2503	0.2469	0.2384

AP8890. RM3-A2 is better than RM3-B on AP8890. The result is encouraging, our method which explicitly takes into account the collection-based features and the multiple types of our training data make the result more robust when predicting for different type of collections. As for RM3-A and RM3-A2, the results indicate that the performance of using all features is better than some important features in generally.

Further more, we show the performance of baseline, RM3-A, RM3-B, RM3-A2 and RM3-Manual on ROBUST2004 for each testing query in Table 4. The results show that our method (RM3-A) is really effectiveness for per query when comparing with RM3-B.

7 Conclusions and Future Work

In this paper, we propose a series of collection-based features about query, feedback documents and candidate expansion terms, then combine them using a logistic regression model to adapt the balance parameter of PRF for different queries and collections (RM3-A). The experiments show that our method outperforms a state-of-art method (RM3-B) when the training and test data are of very different types. This verifies our hypothesis on incorporating collection-sensitive features will help improve the retrieval performance. On the other hand, there is still a room for further improvement when comparing with the manual setting of optimal balancing parameter. We will keep improving our work in the future by investigating other features about collection and analyzing the relationship between different features. We will also evaluate our method on different PRF methods and using different training and test data.

Acknowledgments. This work is supported in part by Chinese National Program on Key Basic Research Project (973 Program, grant No. 2013CB329304, 2014CB744604), the Chinese 863 Program (grant No. 2015AA015403), the Natural Science Foundation of China (grant No. 61272265, 61402324), and the Research Fund for the Doctoral Program of Higher Education of China (grant No. 20130032120044).

References

1. Allan, J., Connell, M.E., Croft, W.B., Feng, F.-F., Fisher, D., Li, X.: Inquery and trec-9. Technical report, DTIC Document (2000)
2. Bialynicki-Birula, I., Mycielski, J.: Uncertainty relations for information entropy in wave mechanics. Commun. Math. Phys. **44**(2), 129–132 (1975)
3. Buckley, C., Robertson, S.: Relevance feedback track overview: Trec 2008. In: Proceedings of TREC 2008 (2008)
4. Cao, G., Nie, J.-Y., Gao, J., Robertson, S.: Selecting good expansion terms for pseudo-relevance feedback. In: SIGIR, pp. 243–250. ACM (2008)
5. He, B., Ounis, I.: Query performance prediction. Inf. Syst. **31**(7), 585–594 (2006)
6. Jones, K.S.: Experiments in relevance weighting of search terms. Inf. Process. Manage. **15**(79), 133–144 (1979)

7. Kishida, K.: Property of mean average precision as performance measure in retrieval experiment. IPSJ SIG. Notes **74**, 97–104 (2001)
8. Kwok, K.-L.: A new method of weighting query terms for ad-hoc retrieval. In Proceedings of the 19th Annual International ACM SIGIR Conference on Research and Development in Information Retrieval, pp. 187–195. ACM (1996)
9. Lavrenko, V., Croft, W.B.: Relevance based language models. In Proceedings of the 24th Annual International ACM SIGIR Conference on Research and Development in Information Retrieval, pp. 120–127. ACM (2001)
10. Lv, Y., Zhai, C.: Adaptive relevance feedback in information retrieval. In Proceedings of the 18th ACM Conference on Information and Knowledge Management, pp. 255–264. ACM(2009)
11. Ogilvie, P., Callan, J.P.: Experiments using the lemur toolkit. TREC **10**, 103–108 (2001)
12. Peng, H., Long, F., Ding, C.: Feature selection based on mutual information: criteria of max-dependency, max-relevance, and min-redundancy. IEEE Trans. Pattern Anal. Mach. Intell. **27**(8), 1226–1238 (2005)
13. Pirkola, A., Järvelin, K.: Employing the resolution power of search keys. J. Am. Soc. Inform. Sci. Technol. **52**(7), 575–583 (2001)
14. Plachouras, V., Ounis, I., van Rijsbergen, C.J., Cacheda, F.: University of glasgow at the web track: dynamic application of hyperlink analysis using the query scope. TREC **3**, 636–642 (2003)
15. Porter, M.F.: An algorithm for suffix stripping. Program **14**(3), 130–137 (1980)
16. Salton, G.: Improving retrieval performance by relevance feedback. J. Am. Soc. Inf. Sci. **41**(4), 288–297 (1990)
17. Sanderson, M., Turpin, A., Zhang, Y., Scholer, F.: Differences in effectiveness across sub-collections. In: CIKM ACM Conference on Information & Knowledge Management, pp. 1965–1969 (2012)
18. Xu, J., Croft, W.B.: Query expansion using local and global document analysis. In: Proceedings of the 19th Annual International ACM SIGIR Conference on Research and Development in Information Retrieval, pp. 4–11. ACM (1996)
19. Z. Ye and J. X. Huang.: A simple term frequency transformation model for effective pseudo relevance feedback. In: Proceedings of the 37th International ACM SIGIR Conference on Research & Development in Information Retrieval, pp. 323–332. ACM (2014)
20. Zhai, C., Lafferty, J.: Model-based feedback in the language modeling approach to information retrieval. In: Proceedings of Tenth International Conference on Information & Knowledge Management, pp. 403–410 (2001)
21. Zhai, C., Lafferty, J.: A study of smoothing methods for language models applied to ad hoc information retrieval. In: Proceedings of the 24th Annual International ACM SIGIR Conference on Research and Development in Information Retrieval, pp. 334–342. ACM (2001)
22. Zhang, P., Song, D., Zhao, X., Hou, Y.: A study of document weight smoothness in pseudo relevance feedback. In: Cheng, P.-J., Kan, M.-Y., Lam, W., Nakov, P. (eds.) AIRS 2010. LNCS, vol. 6458, pp. 527–538. Springer, Heidelberg (2010)

Clustering

An MDL-Based Frequent Itemset Hierarchical Clustering Technique to Improve Query Search Results of an Individual Search Engine

Diyah Puspitaningrum[1(✉)], Fauzi[1], Boko Susilo[1],
Jeri Apriansyah Pagua[1], Aan Erlansari[1], Desi Andreswari[1],
Rusdi Efendi[1], and I.S.W.B. Prasetya[2]

[1] Department of Computer Science,
University of Bengkulu (UNIB), Bengkulu, Indonesia
diyahpuspitaningrum@gmail.com,
{fauzi.faisal,boko.susilo,jeri.apriansyahpagua,
aan.erlansari,desi.andreswari,
rusdi.efendi}@unib.ac.id
[2] Department of Information and Computing Sciences,
Utrecht University, Utrecht, The Netherlands
s.w.b.prasetya@cs.uu.nl

Abstract. In this research we propose a technique of frequent itemset hierarchical clustering (FIHC) using an MDL-based algorithm, viz KRIMP. Different from the FIHC technique, in this proposed method we define clustering as a rank sequence problem of the top-3 ranked list of each itemsets-of-keywords clusters in web documents search results of a given query to a search engine. The key idea of an MDL compression based approach is the code table. Only frequent and representative keywords as those in a KRIMP code table can be used as candidates, instead of using all important keywords from keywords extractor such as RAKE. To simulate information needs in the real world, the web documents are originated from the search results of a multi domain query. By starting in a meta-search engine environment to grab many relevant documents, we set up $k = \{50, 100, 200\}$ for k-toplist retrieved documents of each search engine to build a dataset for automatic relevance judgement. We implement a clustering technique to the best individual search engine the MDL-based FIHC algorithm with setting of $k = \{50, 100, 200\}$ for k-toplist of retrieved documents of each search engine, minimum support = 5 for itemset KRIMP compression, and minimum cluster support = 0.1 for FIHC clustering. Our results show that the MDL-based FIHC clustering can improve the relevance scores of web search results on an individual search engine significantly (until 39.2 % at precision P@10, k-toplist = 50).

Keywords: MDL-based FIHC · Frequent itemset hierarchical clustering · KRIMP · Search engine · Relevance score

© Springer International Publishing Switzerland 2015
G. Zuccon et al. (Eds.): AIRS 2015, LNCS 9460, pp. 279–291, 2015.
DOI: 10.1007/978-3-319-28940-3_22

1 Introduction

The World Wide Web contains huge amount of information from around the world. Given a query, existing search engines such as Google, Lycos, Bing, Exalead, and Ask.com return different lists of web search results, ranked by their relevance to the given query. This difference corresponds to different rank algorithms employed by those search engines. Looking for a search engine that has high relevance score then become a need. To overcome this problem one possible solution is by clustering web document to many different groups of topics so a user can match user information need by directly traversing along the intended topic. By clustering web search results of a search engine, the user will quickly find high precision documents. This is the idea behind the cluster utilization in a search engine results [1, 2, 3, 4, 5]. We also know that the major advantage of a meta-search engines is their coverage of multiple search engines [6] hence we can use this to develop a high relevance data set using the Condorcet method [7] as a gold standard for a multi domain query. Furthermore, while FIHC outperforms best existing methods in terms of both clustering accuracy and scalability [8], KRIMP, an MDL-based algorithm, models the database very well [12]. As a result one can use KRIMP to generate only keywords that best represent web documents search results of a given query, and then pass it to FIHC clusters to obtain more relevant results of the query search results.

In this paper, we present an MDL-based approach to web search result clustering (FIHC) that captures associations among keywords of sets of documents extracted from titles and snippets, given a query. Each document URL is then mapped to the most appropriate cluster and the cluster results of triples URL-title-snippet is returned.

This paper provides one main contribution: it shows that an MDL-based FIHC can be used to significantly improve the relevance score of an individual search engine.

2 Related Works

The Minimum Description Length (MDL) principle is a method for inductive inference, or better, for the model selection problem. The basic idea of MDL is to try to find regularity in data. 'Regularity' here may be identified with the 'ability to compress' or viewing learning as data compression. Learning here means finding frequent keywords among web documents that usually go together. For a given set of hypotheses H and dataset D, one should try to find the hypothesis in H that compresses D most.

MDL (Minimum Description Length) is closely related to MML (Minimum Message Length) [9] and also to Kolmogorov Complexity [10]; in fact, one could see MML as fully *Bayesian* MDL. All three embrace the slogan *Induction by Compression*. Below is a brief description of MDL principle.

Given a set of models H, the best model $H \in H$ is the one that minimises

$$L(H) + L(D|H)$$

in which L(H) is the length, in bits, of the description of H; and L(D|H) is the length, in bits, of the description of the data when encoded with H. This is called two-part MDL or *crude* MDL, as opposed to *refined* MDL, where model and data are encoded together [11]. Refined MDL has a major weakness: it cannot be computed except for some special cases [11]. Hence, for modeling a database we use *crude* MDL, that is for compression purpose. MDL finds the set of frequent itemsets that yields the best compression. In this research a database is a model of a database of frequent itemsets-of-keywords that originated from retrieved web documents search results given a query to a search engine.

Central in an MDL-based approach, viz KRIMP algorithm, is the notion of a *code table*. A code table is a simple two-column translation table that has itemsets on the left-hand side and a code for each itemset on its right-hand side. Using such a table we can encode and decode databases. This is where MDL comes in: we search for the code table that compresses the data best. For further information about code tables in KRIMP see [12].

The approach presented in this paper is an MDL-based frequent itemset hierarchical clustering that is implemented on search results of an individual search engine. It differs from TDPM for evolutionary clustering that divide data into epochs where all data points inside the same epoch are exchangeable and the temporal order is maintained across epochs [13]. Even though the number of clusters produced by an MDL-based FIHC is also unbounded like in TPDM, but we do not divide data into epochs because we are focus in the offline clustering, that is a clustering stage before go to the incremental clustering stage and then to the realtime clustering stage as suggested by Vadrevu et al. [14] to build a scalable clustering system of large collections of documents. Our approach is more inspired by FIHC [8], a robust method in hierarchical clustering, that is more suitable for clustering the web search results rather than the parametric clustering techniques that use a fixed upper bound on the number of clusters.

Our approach also different from a web search clustering technique with richer representation of snippets by discover hidden topics from a very large external data collection such as proposed by Nguyen et al. [15]. Based on Nguyen et al.'s work, there is a need to collect web pages from huge resources and the data must cover many useful topics, hence in our work the focus is more to find associations among important keywords extracted from title and snippet. The MDL-based frequent itemset mining is a suitable technique for finding the association by mining only frequent itemsets that compresses the data best. Then the output of MDL-based technique is clustered using FIHC. We will use the produced clusters to create a list of high relevance retrieved documents.

3   Problem Formalization and Algorithm

3.1   Problem Formulation

We define the general steps of web search clustering of a search engine in two steps:

1. Given a query q, a search engine is used to retrieve a list of results, database $D = (r_1, ..., r_n)$; where each r_i is a triple URL-title-snippet with n is the total number of retrieved documents.

2. A clustering $C = (C_1, ..., C_m)$ of the results in D is obtained by means of a clustering algorithm.

Consider the requirement of relevance; the problem formulation is:
Given q, produce a set of relevant documents returned by a search engine:

$$\bigcup_{j=1}^{m}\bigcup_{i=1}^{\ell} r_{i,C_j}, \forall C_j, C_{j=1,..,m} \in D$$

where r_iC_j is a triple URL-title-snippet of cluster C_j at rank i-th; ℓ is the maximum number of documents retrieved from cluster C_j; and m is the maximum number of clusters used as search results.
or we paraphrase it as:

given a query q, assume C is the MDL-based FIHC clusters formed from a database D, and D is a list of a search engine search results, then let the system presents a high relevance documents to the user by merge a list of top-ℓ search results documents from m clusters, ordered by first found cluster first out.

3.2   MDL-Based FIHC Steps for Clustering Individual Search Engine

Our MDL-based FIHC algorithm is composed of four steps:

1. *Search result fetching*: Given a query q to a search engine (e.g. Google) we build a database D of web search results returned by the search engine.
2. *Document parsing and keywords extractor*: The obtained webpages' titles and snippets are analyzed by an HTML parser and then passed to the RAKE important keyword extractor [16]. Titles and snippets are informative enough to represent most relevant contents for a given query [14]. The RAKE extractor results in keywords of one or two words; it detects array of words separated by words such as and, the, and of, as its key candidates. Several metrics are used in RAKE: (1) word frequency (*freq(w)*), (2) word degree (*deg(w)*), and (3) the ratio of the frequency and the degree (*deg(w)/freq(w)*), where w is a keyword. For further discussion on RAKE see [16].
3. *MDL-based FIHC clustering*: The database of RAKE keywords named *DB*, $(DB \subset D)$, is then compressed by KRIMP to produce a code table $CT(DB, q)$. Only keywords that exist in CT will be used for clustering, since they are frequent and compress the database DB very well, which means that the resulting CT is a good representation of the keywords of all retrieved documents in a search engine's results of a given query q. As a $CT(DB, q)$ exists we then use it to form clusters using FIHC (see more detail of MDL-based FIHC clustering in Sect. 4). All the formed clusters is written to the *final table*. For simplicity, there is no ranking within clusters.
4. *Post-processing*: Once a final table is created, the post-processing involves a merging of top-ℓ search results documents from m clusters, ordered by first found cluster first out. The merging results is the final search results of the search engine.

4 MDL-Based FIHC Clustering

The KRIMP algorithm [12] uses a code table $CT(DB, q)$ that best describes the database DB, $(DB \subset D)$. To do this, one should find the minimal coding set [15].

Let I be a set of items and let D be a dataset over I, *cover* a cover function, and F a candidate set. *Cover* is required to identify which elements of CT are used to encode keywords of a web document given a query q. The result of a cover function is a disjoint set of elements of CT that cover the keywords over q. Minimal coding set is found by seeking the smallest coding set $CS \subseteq F$ such that the corresponding code table CT has a minimal total encoded length $L(D|CT)$. The search space for seeking the optimal code table is far too large, therefore the KRIMP algorithm [12] uses a heuristic, a simple greedy search strategy:

1. Start with a standard code table ST, containing the singletons only of itemsets $X \in I$.
2. Add the itemsets from F one by one. If the resulting codes lead to a better compressed size, keep the code. Otherwise, discard the code.

See further in [12].

Figure 1 presents the MDL-based FIHC clustering algorithm in action when mining "all interesting" itemsets. The algorithm starts with mining large 1-itemsets and using a very low minimum support obtained by trial i.e. *minsup* = 5, KRIMP results in best quality itemsets. Mining stops when "all interesting" itemsets are found. All interesting frequent itemsets are then saved in a code table. We propose the MDL-based frequent itemset mining method (KRIMP) to produce inputs to the FIHC hierarchical clustering because we want to show that KRIMP is also represents the web search results' database best. KRIMP produces best quality frequent itemsets by rejecting frequent itemsets-of-keywords with minimum support lower than *minsup*.

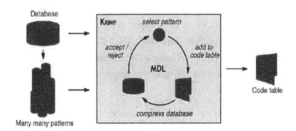

Fig. 1. An MDL-based FIHC clustering algorithm (KRIMP algorithm) [12]

The method for producing clusters, the FIHC technique [8], is "cluster-centered" in that the "cohesiveness" of a cluster is measured directly using frequent itemsets. Since the documents under a cluster contains the same topic, they are expected to share more common itemsets than those under different cluster.

Here we describe the FIHC process after KRIMP produces the code table of many frequent itemsets of important keywords produced by RAKE. Once a code table CT (DB, q) exists, we then use it to form clusters using FIHC [8]. In FIHC there are 3 main processes: clustering, tree building, and tree pruning. There are two steps to construct clusters: first, constructing initial clusters, then, second, making initial clusters disjoint. An initial cluster is constructed for each global frequent itemset to contain all documents contain the itemset. FIHC only allow any clusters that have cluster frequent items with cluster supports more than the minimum cluster support. In second step, FIHC identify the "best" initial cluster and keep a document only in the best initial cluster by measuring the goodness of a cluster for a document. In tree building process, cluster tree is built bottom-up by choose the "best" parent among such potential parents. The tree pruning criterion is based on the inter-cluster similarity between a parent and its child. A child will be pruned only if the child (the subtopic) is similar enough to its parent topic. Once final clusters are formed after tree pruning process, the next step is creating a set of *relevant* documents as final answer to the given q. This is done by merge top-ℓ search results documents from m clusters.

5 Experiments

We performed experiments in two steps: first, we build a gold standard, viz., the relevant search results; and second, we build a list of documents with high score of relevance. To do the first step, a meta-search engine environment is used to build 4 data fusions that then elected using Condorcet, an automatic relevance judgement suggested by Nuray and Can [7]. For the second step, extensive experiments of the MDL-based FIHC technique are performed on an individual search engine.

Table 1. The multi domain queries [18]

Two terms queries	Three terms queries
Database overlap	Comparative education methodology
Multilingual OPACs	Java applet programming
Programming algorithm	Indexing AND digital libraries
Road-map plan	Geographical stroke incidence
Adolescent alcoholism	Culturally responsive teaching

Table 1 shows queries taken from Mohamed [18] that we use for all of our experiments including both of the two steps above. According to Jansen et al. [19], 97 % of internet queries consist of less than 6 *terms*. Even more, the average length of those queries is 2.5 *terms*. For our experiment, we use queries with length of 2 and 3 *terms* and expand them using operator AND/OR. Therefore using, for example, a three terms query "culturally responsive teaching" will result several combinations: "culturally AND responsive AND teaching", "culturally AND responsive OR teaching", "culturally OR responsive AND teaching", and "culturally OR responsive OR teaching".

5.1   Building the Gold Standards

A meta-search engine allows us to search multiple search engines fast at once, returning more comprehensive and relevant results [20]. We use this ability to build different ideal relevant search results (the data fusions). The data fusions then are used to build gold standard for each query using an election technique namely *Condorcet* (Nuray and Can [7]). Our meta-search engine uses 3 out of 5 popular search engines: Google, Bing, AskJeeves, Lycos, and Exalead.

In the Condorcet method [7], voters rank the candidates in the order of preference. The vote counting procedure then takes into account each preference of each voter for one candidate over another. The Condorcet voting method specifies the winner as the candidate, which beats each of the other candidates in a pair wise comparison. The Condorcet method is chosen as an automatic relevance judgement technique in our experiments since for data fusion the Condorcet method provides the best performance in terms of automatic ranking compares to the random selection method or RS, and the reference count method or RC. According to [7], the mean correlation values for Condorcet is 0.560, compares to 0.506 for RC and 0.493 for RS.

For our Condorcet gold standard [21], we built the gold standard dataset using 4 data fusions. Each data fusion is built in a meta-search engine setting using all combinations of 3 out of 5 component engines using a rank fusion algorithms. The four rank fusion algorithms used are: KE [17], two variants of Weight Borda-Fuse [22, 23], and Count Function algorithm [24]. The choise of k-toplist search results ($k = 50, 100, 200$) that is used by 3 component engines will produce different ranking of retrieved documents in the data fusion. The KE, the two variants of Weight Borda-Fuse (WBF), and the Count Function algorithms act as "the voters", and their union of 10-toplist documents act as "the candidates" in Condorcet. As a gold standard dataset we only take the 10-toplist documents of the Condorcet results.

5.2   The Unclustered Web Search Result Experiments

Google is a good sample of individual search engines. From Table 2, the Google search engine in average above all queries shows best relevant search results.

Table 2. Relevance of web search results (averaged over all queries)

Search engine	Evaluation metrics					
	k-toplist	P@1	P@3	P@5	P@10	MRR
Google	50	0.6003	0.5353	0.4584	0.3203	0.3806
Bing	50	0.3418	0.2476	0.2203	0.1734	0.29
Ask.com	50	0.1591	0.245	0.2387	0.1972	0.3125
Lycos	50	0.5312	0.4403	0.3985	0.2869	0.3807
Exalead	50	0.2683	0.1343	0.1116	0.0679	0.1183

To measure the relevance of search results, we adopted the Precision at rank n (P@n) metric and the Mean Reciprocal Rank (MRR) as in [25]. Precision at rank n is defined as the proportion of retrieved documents that is relevant with the gold standard, averaged over all documents. MRR measures where in the ranking the first relevant document (with the gold standard) is returned by the system, averaged over all the web documents. This measure provides insight in the ability of the system to return a relevant document at the top of the ranking.

As an example the query is "adolescent AND alcoholism" built using Google, Bing, and Lycos. Table 3 shows an evaluation result using Condorcet (*50*-toplist), for each original Google, Bing and Lycos web search results; these are before a clustering technique is applied to the individual results of the search engines. In this example Google outperforms other search engines.

Table 3. Relevance of web search results (query "adolescent AND alcoholism", 50-toplist)

Search engine	Evaluation metrics					
	k-toplist	P@1	P@3	P@5	P@10	MRR
Google	50	1.0	0.33	0.6	0.5	1
Bing	50	0.0	0.67	0.4	0.3	0.5
Lycos	50	0.0	0.0	0.2	0.4	0.2

5.3   The MDL-Based FIHC Algorithm Experiments

In our experiments, we cluster Google's search results using *k-means* or the MDL-based FIHC with different number of clusters. For MDL-based FIHC we set $\ell = 1$ and $m = 30$, while for *k-means* $\ell = 3$ and $m = 10$. Number of clusters in *k-means* is set to 10 clusters per query search due to we have enough documents from *k*-toplist a search engine's search results $\{k = 50, 100, 200\}$. Number of clusters in FIHC is not defined by the user but it is generated automatically by the hierarchical FIHC clustering by measuring "cohesiveness" of frequent itemsets to a cluster. As an example is for a search results of query "adolescent AND alcoholism" from a search engine with *k*-toplist ($k = 200$), minimum cluster support = 0.1 and 1533 keywords, from 200 retrieved documents at initial stage of the FIHC we have 1535 clusters that then they drop to 200 clusters at final stage. From hundreds of clusters we take only the 30 first clusters of the hierarchical FIHC final result since when we observed data the number of document in each cluster almost all is 1. For relevance evaluation we test the FIHC final result using only the 10 first clusters due to the Condorcet dataset is also 10 documents. Our search engine prototype displays only 10-toplist of search results.

In final result of hierarchical FIHC we have 30 documents. In *k-means*, since the number of clusters is constant equal to 10 clusters, then we must generate the same number of documents (30 documents) as the final result by picking 3-toplist of documents from each clusters.

For all of our experiments, relevance is measured using only the 10-toplist of a final result's documents of either *k-means* or hierarchical FIHC against the Condorcet dataset as the gold standard.

Here is how we set up parameters in the MDL-based FIHC. In the MDL-based FIHC clustering stage, a parameter for KRIMP we must set up is: the candidate type of itemset collection. For any candidate type of itemset collection we set all itemsets to minimum support equal to 5 ("*all-5d-pop*"). The choice of "*all*", and not "*closed*" frequent itemsets, is because we want to mine all interesting itemsets. We use a small value for the minimum support due to our web search results database is a sparse database, contains important keywords generated from RAKE. In the FIHC clustering stage, we must define two parameters: the *minimum global support* and the *minimum cluster support*. The minimum global support is 0 (ignored) because, in general, majority of the data are itemsets with global frequent itemsets' global support is below 1 %. If for example we set the minimum global support to 1 we will lose so many global frequent itemsets data and it is not good for further clustering process. The minimum cluster support is set to 0.1. The *cluster support* of an item in C_j is the percentage of the documents in C_j that contain the item. The minimum cluster support value must be set properly depend on the data. In our case, by setting minimum cluster support to 0.1 we still have enough cluster frequent items with cluster supports ≥10 % in initial clusters while at the same time remove many unimportant frequent items with cluster supports below 10 %.

We evaluate the performance of all clustered search results of an individual search engine to the gold standard (Condorcet dataset). See Table 4 and Fig. 2.

Table 4. Relevance of web search results on Google (averaged over all queries)

Search engine	k-toplist	Evaluation metrics				
		P@1	P@3	P@5	P@10	MRR
Google	50	0.5567	0.4822	0.3993	0.2933	0.2949
k-means	50	0.4778	0.4167	0.3456	0.2656	0.6613
Hierarchical	50	0.7840	0.6605	0.5420	0.3920	0.8610
Google	100	0.5300	0.4589	0.3867	0.2863	0.3015
k-means	100	0.5278	0.3519	0.3456	0.2461	0.6682
Hierarchical	100	0.7368	0.6228	0.5237	0.3908	0.8328
Google	200	0.5200	0.4378	0.3673	0.2743	0.3116
k-*means*	200	0.3111	0.2537	0.2400	0.1794	0.5033
Hierarchical	200	0.7143	0.6667	0.4571	0.3286	0.8143

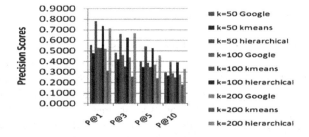

Fig. 2. Performance comparison for different web search results settings of Google. Figure 2 is taken from Table 4.

The "Google" rows show the relevance of the original (and unclustered) search results obtained directly from the Google search engine, The "*k-means*" rows show the results' relevance when the *k-means* clustering is applied, and the "hierarchical" rows show what we get when the MDL-based FIHC clustering is used.

From Fig. 2 and Table 4, we have the following observations.

1. By P@x evaluation, the MDL-based FIHC clustering outperforms other standard clustering algorithms.

Clustering web search results using *k-means* are not recommended since it performs worse than the original search results version of the search engine.

2. By P@x evaluation, the MDL-based FIHC clustering always performs better than the unclustered version, the original web search results of individual search engine.

Even at P@10 the MDL-based FIHC clustering shows significant improvement in relevance or precision than the original one with best setting *k*-toplist at $k = 50$ of web search results (improvement up to 33.65 % of original Google results). Adding more documents (*k*-toplist, $k = \{100, 200\}$) are not necessary in improving relevance.

3. By MRR evaluation, the first found relevant document rank position was significantly improved using the MDL-based FIHC clustering.

In this case the MDL-based FIHC clustering outperforms *k-means* in first relevant position is found. This also shows that there is no correlation between improvement in MRR and improvement in precisions of web documents (P@x).

Example results of the gold standard (*Condorcet*), Google, the *k-means* clustering on Google, and the MDL-based FIHC clustering on Google, can be found in the Appendix at https://www.dropbox.com/s/vfc4xobyw83yaax/Appendix.docx?dl=0.

Several statistical Welch two sample t-tests are done on the lists of P@10 values of Google, *k-means*, and hierarchical (the MDL-based FIHC). The aim is to show that there is significant difference between final search result of an individual search engine before and after it is clustered (Table 5). All tests uses confidence level = 0.95 are performed on a hierarchical clustered Google against its corresponding unclustered Google (hierarchical-vs-Google) and done also on a *k-means* clustered Google against its corresponding unclustered Google (*k-means*-vs-Google); all tests are performed 3 times (3 *k*-toplist, $k = \{50, 100, 200\}$). The detail lists of P@10 of experiments in Table 4 are not shown here due to space limitations.

See Table 5, test results of an individual search engine (clustered vs unclustered search results). Let μ_1 be the mean of a clustered individual search engine's search results and μ_2 the mean of the original unclustered search engine's search results. The hypotheses of interest are expressed as: Hypothesis 1: $\mu_1 - \mu_2 = 0$; Hypothesis 2: $\mu_1 - \mu_2 \neq 0$. The t-tests show there is significant difference between final search result of an individual search engine before and after it is clustered (Table 5, part A and B). To be more specific, by modify the Hypothesis 2 to greater or less to 0, the tests results show hierarchical-vs-Google has evidence that $\mu_1 - \mu_2 > 0$ while *k-means*-vs-Google has evidence of $\mu_1 - \mu_2 < 0$. Since Table 5 part C and part D show the *p*-value is very low, then we reject the Hypothesis 1. From Table 5 part C, for hierarchical-vs-Google the results show that there is strong evidence of a mean increase in P@10 between

Table 5. The Welch two sample t-tests on P@10 of an individual search engine

A. hierarchical-vs-Google						
Hypothesis 1: $\mu_1 - \mu_2 = 0$; Hypothesis 2: $\mu_1 - \mu_2 \neq 0$						
k-toplist	T	df	p-value	Mean(X)	Mean (Y)	|Mean (X)-Mean (Y)|
50	6.34	407	6.22e-10	0.392	0.293	0.099
100	6.69	387	7.87e-11	0.391	0.286	0.105
200	1.45	7	4.21e-13	0.329	0.274	0.055
B. k-means -vs-Google						
Hypothesis 1: $\mu_1 - \mu_2 = 0$; Hypothesis 2: $\mu_1 - \mu_2 \neq 0$						
50	-1.96	473	0.05104	0.266	0.293	|-0.027|
100	-2.64	447	0.008525	0.246	0.286	|-0.04|
200	-7.46	474	4.21e-13	0.179	0.274	|-0.095|
C. hierarchical-vs-Google						
Hypothesis 1: $\mu_1 - \mu_2 = 0$; Hypothesis 2: $\mu_1 - \mu_2 > 0$						
50	6.34	407	3.11e-10	0.392	0.293	0.099
100	6.69	387	3.94e-11	0.391	0.286	0.105
200	1.45	7	0.09514	0.329	0.274	0.055
D. k-means-vs-Google						
Hypothesis 1: $\mu_1 - \mu_2 = 0$; Hypothesis 2: $\mu_1 - \mu_2 < 0$						
50	-1.96	473	0.02552	0.266	0.293	|-0.027|
100	-2.64	447	0.004262	0.246	0.286	|-0.04|
200	-7.46	474	2.10e-13	0.179	0.274	|-0.095|

search results of the hierarchical clustered Google and the original unclustered Google. It is contrary to the fact to the *k-means*-vs-Google that show a mean decrease (Table 5, part D).

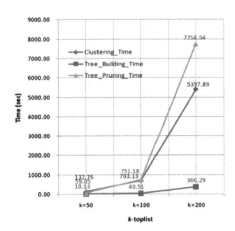

Fig. 3. Scalability of the MDL-based FIHC clustering with the scale-up *Google* web documents search results (minimum cluster support = 0.1)

Figure 3 depicts the runtime of experiments in Table 4 with respect to the number of documents (*k*-toplist, $k = \{50, 100, 200\}$). The whole process completes within 129 min tested on an Asus K45VD laptop machine with processor Intel Pentium dualcore 2.4 GHz, RAM 4 GB, and hardisk 500 GB. Implementation of the system was built in Python v2.6. The system uses very low minimum cluster support 0.1 since the system is not intended for specific domain only but multi domain that in real situation leads to many very low cluster frequent items and cluster supports (CS). As a consequence it filters them at initial clusters and processes only cluster (label) with CS >= 10 %. It demonstrates that the MDL-based FIHC clustering technique is a scalable method.

Figure 3 also shows that tree pruning and clustering are the most time consuming stages in the MDL-based FIHC. This is different with runtime of tree building that completes in 6.1 min. In the clustering stage, most time is spent on constructing initial clusters while in tree pruning there is an indication of many subtopics are very similar to their parents' topics. The tree pruning scan the tree in the bottom-up oder and calculates *Inter_Sim* similarity between the node and each of its children. If *Inter_Sim* is above 1 then the system prunes the child cluster.

Both in the tree pruning and in the clustering stages, their runtimes are linear with respect to the number of web search results that is handled (*k*-toplist, $k = \{50, 100, 200\}$ for each settings of individual search engines).

Expensive clustering time as shown in Fig. 3 is due to we performed the MDL-based FIHC clustering viz. KRIMP in sequential. If we split a database of web documents into several smaller databases and feed them into KRIMP, the problem then is about how to merge different results of frequent keywords of each databases in KRIMP code tables into best representative of frequent keywords of the original web documents' database. MapReduce technique is suitable to be implemented in this MDL-based FIHC clustering algorithm. Tree pruning time in future also can be improved by parallelizing processes of computing many *Inter_Sims*. We leave these problems for further research and discussion.

6 Conclusions

We introduced a new approach for frequent itemset-based hierarchical clustering to address the issue of improving the relevance of search results of an individual search engine. The novelty of this research is that it exploits frequent itemsets using an MDL-based algorithm, viz. KRIMP, for defining clusters to generate more relevant retrieved documents from a search engine. Evaluated in a meta-search engine environment datasets, the experimental results show that our approach outperforms other clustering algorithms in terms of relevancy as well as shows significant relevance improvement of web search results of an individual search engine.

References

1. Di Marco, A., Navigli, R.: Clustering and diversifying web search results with graph-based word sense induction. J. Comput. Linguist. **39**, 709–754 (2013)
2. Carpineto, C., Osinski, S., Romano, G., Weiss, D.: A survey of web clustering engines. J. ACM Comput. Surv. **41**, 1–38 (2009)
3. Zamir, O., Etzioni, O.: Grouper: a dynamic clustering interface to web search results. In: 8th International World Wide Web Conference (WWW8), pp. 1361–1374. Toronto (1999)

4. Hearst, M.A., Pedersen, J.O.: Reexamining the cluster hypothesis: scatter/gather on retrieval results. In: 19th Annual International ACM SIGIR Conference on Research and Development in Information Retrieval (SIGIR 1996). Zurich (1996)
5. Leuski, A., Allan, J.: Improving interactive retrieval by combining ranked lists and clustering. In: RIAO, pp. 665–681. France (2000)
6. Jadidoleslamy, H.: Search result merging and ranking strategies in meta-search engines: a survey. Int. J. Comput. Sci. **9**, 239–251 (2012)
7. Nuray, R., Can, F.: Automatic ranking of information retrieval systems using data fusion. J. Inf. Process. Manag. **42**, 595–614 (2006)
8. Fung, B.C.M., Wang, K., Ester, M.: Hierarchical document clustering using frequent itemsets. In: SIAM International Conference on Data Mining (SDM 2003), pp. 59–70 (2003)
9. Wallace, C.S.: Statistical and Inductive Inference by Minimum Message Length. Springer, New York (2005)
10. Li, M., Vitanyi, P.: An Introduction to Kolmogorov Complexity and Its Applications, 2nd edn. Springer, New York (1997)
11. Grünwald, P.D.: The Minimum Description Length Principle. MIT Press, Cambridge (2007)
12. Vreeken, J., van Leuwen, M., Siebes, A.: KRIMP: mining itemsets that compress. J. Data Min. Knowl. Discov. **23**, 169–214 (2011)
13. Ahmed, A., Xing, E.: Dynamic non-parametric mixture models and the recurrent Chinese restaurant process: with applications to evolutionary clustering. In: SDM, pp. 219–230 (2008)
14. Vadrevu, S., Teo, C.H., Rajan, S., Punera, K., Dom, B., Smola, A., Chang, Y., Zheng, Z.: Scalable clustering of news search results. In: WSDM 2011. Hong Kong (2011)
15. Nguyen, C.-T., Phan, X.-H., Horiguchi, S., Nguyen, T.-T., Ha, Q.-T.: Web search clustering and labeling with hidden topics. ACM Trans. Asian Lang. Inf. Process. **8**, 12 (2009)
16. Rose, S., Engel, D., Cramer, N., Cowley, W.: Automatic keyword extraction from individual documents. In: Berry, M.W., Kogan, J. (eds.) Text Mining: Applications and Theory, pp. 3–20. John Wiley & Sons, Chichester (2010)
17. Akritidis, L., Voutsakelis, G., Katsaros, D., Bozanis, P.: QuadSearch: a novel metasearch engine. In: 11th Panhellenic Conference on Informatics (PCI 2007), pp. 453–466. Patras (2007)
18. Mohamed, K.A-E-F.: Merging multiple search results approach for meta-search engines. Ph. D. thesis, University of Pittsburgh (2004)
19. Jansen, B.J., Spink, A., Bateman, J., Saracevic, T.: Real life information retrieval: a study of user queries on the web. ACM SIGIR Forum **32**, 5–17 (1998)
20. Meng, W.: Metasearch engines. Technical report, Department of Computer Science, State University of New York at Binghamton (2008)
21. Puspitaningrum, D., Pagua, J.A., Erlansari, A., Fauzi, F., Efendi, R., Andreswari, D., Prasetya, I.S.W.B.: The analysis of rank fusion techniques to improve query relevance. J. Telkomnika **13**(4) (2015)
22. Fagin, R., Kumar, R., Sivakumar, D.: Comparing top-k lists. SIAM J. Discrete Math. **17**, 134–160 (2003)
23. Dorn, J., Naz, T.: Structuring meta-search research by design patterns. In: The International Computer Science and Technology Conference (ICSTC). San Diego, California (2008)
24. Patel, B., Shah, D.: Ranking algorithm for meta search engine. Int. J. Adv. Eng. Res. Stud. (IJAERS) **2**, 39–40 (2012)
25. Sigurbjörnsson, B., van Zwol, R.: Flickr tag recommendation based on collective knowledge. In: WWW, pp. 327–336. Beijing (2008)

Improving Clustering Quality by Automatic Text Summarization

Mohsen Pourvali[1]([✉]), Salvatore Orlando[1], and Mehrad Gharagozloo[2]

[1] Università Ca' Foscari Venezia, Venice, Italy
{mohsen.pourvali,orlando}@unive.it
[2] Queensland University of Technology (QUT), Brisbane, Australia
mehrad.gharagozloo@qut.edu.au

Abstract. Automatic text summarization is the process of reducing the size of a text document, to create a summary that retains the most important points of the original document. It can thus be applied to summarize the original document by decreasing the importance or removing part of the content. The contribution of this paper in this field is twofold. First we show that text summarization can improve the performance of classical text clustering algorithms, in particular by reducing noise coming from long documents that can negatively affect clustering results. Moreover, the clustering quality can be used to quantitatively evaluate different summarization methods. In this regards, we propose a new graph-based summarization technique for keyphrase extraction, and use the Classic4 and BBC NEWS datasets to evaluate the improvement in clustering quality obtained using text summarization.

1 Introduction

The abundance of available electronic information is rapidly increasing with the advancements in digital processing. Furthermore, huge amounts of textual data have given rise to the need for efficient techniques that can organize the data in manageable forms. One of the common approaches for this aim is the use of clustering algorithms, in which sets of similar documents are grouped in clusters.

In text clustering, a text or document is always represented as a bag of words. This representation raises one severe problem: the high dimensionality of the feature space and the inherent data sparsity. Obviously, a single document has a sparse vector over the set of all terms [8]. The performance of clustering algorithms will decline dramatically due to the problems of high dimensionality and data sparseness [1]. Therefore it is highly desirable to reduce the feature space dimensionality. There have some works that deal with this problem by utilizing two popular approaches: feature selection [3] and feature extraction [13]. Graph-based ranking is one of the popular unsupervised approaches for extracting features (keyphrases) from texts. Specifically, TextRank [10] is one of the most well-known graph-based approaches for keyphrase extraction.

In this paper we investigate how we can improve the effectiveness of text clustering by summarizing some documents in a collection, specifically the ones that

© Springer International Publishing Switzerland 2015
G. Zuccon et al. (Eds.): AIRS 2015, LNCS 9460, pp. 292–303, 2015.
DOI: 10.1007/978-3-319-28940-3_23

are much significantly longer than the mean. Our method could be considered as one for unsupervised feature selection, because it chooses a subset from the original feature set, and consequently reduces vector space for each document. In particular, as mentioned above, it is particular effective when applied to longer documents, since these documents reduce purity of clustering. To this end, we propose a novel method in which *n-tsets* (i.e., non-contiguous sets of n terms that co-occur in a sentence) are extracted through a graph-based approach. Indeed, the proposed summarization method is a keyphrase extraction-based summarization method in which the goal is to select individual words or phrases to tag a document. We have utilized HITS algorithm [5], which is designed for web page ranking, in order to boost the chance of a node to be selected as a keyphrase of the document, although other graph-based algorithms have been proposed to summarize texts, For example, we can mention [9] in which sentences, instead of key-phrases, are extracted through undirected graphs.

The rest of the paper is organized as follows. A graph-based summarization algorithm is presented in Sect. 2. Section 3 discusses our graph-based ranking algorithm to extract *n-tsets* from documents. In Sect. 4 we present the experimental setup, and Sect. 5 discusses the experimental results obtained for two human-labelled datasets, namely BBC NEWS and Classic4, used for clustring purposes. In addition, we also use the DUC2002 dataset for evaluating the quality of text summarizations provided. Finally, Sect. 6 draws some conclusions.

2   Baseline Graph-Based Keyphrase Extraction

In this section we discuss the *baseline* used for testing. We start from this method because it is a simple form of graph-based ranking approach. In addition, we exploit it to boost the score of keyphrases to include in a text summary.

This graph-based method relies on HITS (Hyperlinked Induced Topic Search) [5] to rank terms. HITS is an iterative algorithm that was designed for ranking Web pages. HITS makes a distinction between "authorities" (pages with a large number of incoming links) and "hubs" (pages with a large number of outgoing links). Hence, for each vertex V_i, HITS produces an "authority" and a "hub" score:

$$HITS_A(V_i) = \sum_{V_j \in In(V_i)} HITS_H(V_j) \tag{1}$$

$$HITS_H(V_i) = \sum_{V_j \in Out(V_i)} HITS_A(V_j) \tag{2}$$

A similar idea can be applied to lexical or semantic graphs which are extracted from text documents in order to identify the most significant blocks (words, phrases, sentences, etc.) for building a summary [7,10]. Specifically, we applied HITS to directed graphs whose vertexes are terms, and edges represent co-occurrences of terms in a sentence. Before generating the graph, stopword removal and stemming are applied. Once computed the $HITS_A(V_i)$ and

$HITS_H(V_i)$ scores for each vertex V_i of the graph, we can rank the graph nodes by five simple functions of the two scores:

$$F_\Gamma(V_i) = \Gamma(HITS_A(V_i), HITS_H(V_i))$$

where Γ corresponds to different ways of combining the two HITS scores. Namely $avg/max/min/sum/prod$ (average/maximum/minimum/sum/product of the Hub and Authority scores). After the scoring of the nodes by F_Γ, we can rank them, and finally return the K-top ranked ones.

3   Our Summarization Technique

To create a keyphrase-based summary of a document, we devised a unsupervised technique, called N-tset Graph-based Ranking (*NG-Rank*) for which n-tset is a set of one or more terms co-occurring in a sentence.

In a document, the discussed subjects are presented in a specific order. For each document paragraph, the first sentence represents a general view of the discussed subject, which is examined in depth in the rest of the sentences. The rest sentences might be ended by a conclusion sentence, which is the final close of the discussed subject. In general, the first and last sentences likely include the main concepts of the document. Therefore, let D be a document of the collection, denoted by $D = (P_1, P_2, ..., P_n)$, where P_i is a paragraph of D. The sentences of P_i are thus partitioned as follows:

- First Sentences (FS): which are the first f consecutive sentences occurring of P_i.
- Middle Sentences (MS): which are the middle sentences of P_i.
- Last Sentences (LS): which are the last l consecutive sentences of P_i.

Once denoted the sentences of each paragraph, our algorithm preprocesses these sentences by removing stop words and applying the Porter stemmer. Suppose that after these processing step, the number of stemmed terms in a document is m. The next step of our algorithm builds an $m \times m$ (normalized) co-occurrence matrix $A_0 = (a_1, a_2, ..., a_m)$ of the terms. Specifically, each entry of matrix A_0 is given by $\frac{t_{ij}}{t_i}$, where t_{ij} indicates the number of times term i and term j co-occur within the various sentences of the documents, and t_i is the number of times term i occurs in the document. We can have:

$$a_{ij} = \begin{cases} 1 & \text{if } t_i = t_{ij} \quad \text{(I)} \\ < 1 & \text{otherwise} \quad \text{(II)} \end{cases}$$

In case $a_{ij} = a_{ji} = 1$ and $t_{ij} > 1$, then the terms i and j always co-occur for the same number of times within the various sentences of the documents. Then we merge them as a new *n-tset* term, and rebuild the matrix, by merging the ith and j^{th} rows (columns). This process is iterated, namely $A_{h+1} = merge(A_h)$, till $\nexists i, j$ such that $a_{ij} = a_{ji} = 1$ and $t_{ij} > 1$. The number of iteration is $I = N - 1$, where N is the biggest *n-tset* found in the document.

For example, consider a document with one paragraph, consisting of 5 sentences, partitioned into the sets FS, LS, and MS (First, Last, and Middle Sentences)[1], where the stemmed terms are represented as capital letters:

$$FS = \{(AB)\} \quad LS = \{(MSR)\} \quad MS = \{(ACDFG), (ACNDG), (MSN)\}$$

In the first iteration, terms C and D are merged as a new term C-D. In addition, also terms M and S are merged as a new term M-S. In the second iteration, terms C-D and G are merged as a new term C-D-G. Note that at the end of this iterative process, each row/column will correspond to n-tsets, $n \geq 1$. Without loss of generality, hereinafter we call "n-tset" both single and multiple terms (identified by our algorithm). The final sentences after the merging is thus:

$$FS = \{(A\ B)\} \quad LS = \{(M\text{-}S\ R)\} \quad MS = \{(A\ C\text{-}D\text{-}G\ F), (A\ C\text{-}D\text{-}G\ N), (M\text{-}S\ N)\}$$

Finally, the *primary score* for each n-tset (single or multiple terms), corresponding to a row a_i of the final matrix A_{last}, is defined as follows:

$$PScore(a_i) = \frac{1}{\sum_{j=0}^{m} a_{ij}} \tag{3}$$

If an n-tset appears in long sentences or appears multiple times in short sentences, its row a_i in the matrix is not so sparse, in comparison with n-tsets occurring in a few short sentences. If this property holds, this decreases the value of $PScore$.

In the next step, we use A_{last} as the adjacency matrix to generate a graph of relationships between n-tsets. Each node corresponds to an n-tset occurring in the document, and each edge models the co-occurrence of a pair of n-tsets in a sentence. Indeed, the graph is directed. If $a_i \rightarrow a_j$, then a_i occurs before a_j in one or more sentences. The graph of n-tsets for our running example is shown in Fig. 1. Note that the nodes of the graph are subdivided into three partitions: FS, LS, and MS. This means that each node associated with an n-tset must be univocally assigned to one partition. When the same n-tset occurs in more than one set of sentences – i.e., first, last, or middle sets of sentences – we must choose only one of the three partitions FS, LS, or and MS. Specifically, we assign the n-tset to a partition according to a priority order: we choose FS if the n-tset appears in some of the first sentences, then LS if the n-tset appears in some of the last sentences, MS otherwise.

We exploit this graph to boost the primary score assigned to some n-tsets. Since n-tsets in FS and LS are considered more discriminative than the others, we increase the primary scores of n-tsets whose associated nodes are in the FS or LS partition. In addition, we also boost the primary scores of nodes in MS that are connected to nodes in FS or LS, i.e., there exist a path that connects these nodes in MS to nodes in FS or LS partitions. Specifically, we use two boosting methods that exploit graph properties The first one simply exploits the in/out degree of each node:

$$Score(a_i) = PScore(a_i) + \log(\ \max(v_{in}(a_i), v_{out}(a_i))\ ) \tag{4}$$

---

[1] $f = l = 1$, where f and l are the number of sentences in FS and LS, respectively.

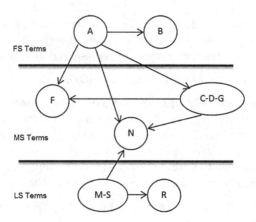

Fig. 1. Structure of graph, the nodes are n-tsets of the document, in turn partitioned into three sets. The direction of the edges corresponds to the order in which the n-tsets appear in each sentence.

The second boosting method exploits function $\Gamma = \max$ among the HITS functions discussed in Sect. 2:

$$Score(a_i) = PScore(a_i) * (1 + \max(HITS_A(a_i), HITS_H(a_i))) \qquad (5)$$

We obtained better results with $\Gamma = \max$ than with any other alternative functions Γ. It shows that the words occurred at the beginning or at the end of a paragraph are much more important candidates as keywords of the document.

It is worth noting that the nodes in MS that are not boosted maintain, however, the old primary score, i.e., $Score(a_i) = PScore(a_i)$. These nodes are still considered in the following phase.

Specifically, once all the nodes in the graph are scored by $Score(a_i)$, we rank them, and finally return the n-tsets associated with the K-top ranked ones, where the value of K depends on the length of document to be summarized. Indeed, we sort in decreasing order of $Score$ the nodes within each partition of the graph (FS, MS, or LS). After this separated reordering of each partition, we return a summary that contains the *same fraction* α, $0 < \alpha < 1$, of the top-scored n-tsets for each of the three partitions. Specifically, we return $\alpha \cdot |FS|$, $\alpha \cdot |LS|$, and $\alpha \cdot |MS|$ nodes (n-tsets) from the sets FS, MS, and LS.

Finally, the order of the terms in the generated summary is the same as the one in the original document. This step is important to evaluate the quality of the extracted summaries with respect to human-generated ones (using the DUC 2002 dataset).

Hereinafter, we call the summarization algorithm that exploits the boosting methods of Eq. (4) $NG\text{-}Rank_M$, whereas we call the ones adopting the alterative boosting method of Eq. (5) $NG\text{-}Rank_H$.

4   Experimental Setup

The principal idea of the experiments is to show the efficacy of the text summarization on clustering results through a manually predefined categorization of the corpus. In addition, in order to evaluate the absolute quality of our summarization method, we further need a standard dataset to compare our method with the baseline. We used "Classic4" and "BBC NEWS" to test the benefits of summarization on clustering quality , and "DUC 2002" for testing the quality of our summarization method.

In the following we introduce the three corpora, the preprocessing applied to them, and finally the evaluation measures used for in the experimental tests.

Datasets. The three corpora used in the experiments are described in the following:

Classic4: This dataset is often used as a benchmark for clustering and co-clustering[2]. It consists of 7095 documents classified into four classes denoted MED, CISI, CRAN and CACM.

BBC NEWS: This dataset consists of 2225 documents from the BBC news website corresponding to stories in five topical areas, which are named Business, Entertainment, Politics, Sport and Tech, from 2004–2005 [4]. We have used four classes of BBC news in our experiments. Unlike Classic4, the BBC NEWS corpus is full of names of athletes, politicians, etc. These proper names are challenging, because they could be important to be extracted as keyphrase of text. On the other hand, they could reduce the similarity between two related texts.

DUC 2002: This dataset is a collection of newswire articles provided during the Document Understanding Evaluations 2002[3]. DUC 2002 contains 567 document-summary pairs which are clustered into 59 topics. We have used the 100-words summary provided for each document.

Preprocessing. Preprocessing is an essential step in text mining. The first classing preprocessing regards stop words removal, lower case conversion, stemming[4], and finally identifying sentences and paragraphs.

In addition, we preprocess the corpora Classic4 and BBC to generate from them two new datasets. Specifically, since our aim is to evaluate the efficacy of summarizing longer documents to improve clustering, for each original dataset we generated a sub-collection of documents of different sizes: a large part of them approximatively contains the same number of terms sz, while the others are much longer than sz. Specifically, longer documents contain a number of terms not less than $3 \cdot sz$.

---

[2] http://www.dataminingresearch.com/index.php/2010/09/classic3-classic4-datasets/.
[3] http://duc.nist.gov/data.html.
[4] http://tartarus.org/martin/PorterStemmer/.

In more details, we stratified the sampling of each *original labeled dataset* as follows. Let $\mathbb{L} = \{L_1, L_2, ..., L_c\}$ be the original dataset, where L_i is the set of documents labeled with the ith class. From each L_i we thus extract a subset $\mathcal{L}_i$, thus generating the sub-collection $\mathbb{D} = \{\mathcal{L}_1, \mathcal{L}_2, ..., \mathcal{L}_c\}$. Specifically, we have:

$$\mathcal{L}_i = R_i \cup E_i \tag{6}$$

where $R_i = \{d \in L_i \mid a \le size(d) \le b\}$ and $E_i = \{d \in L_i \mid size(d) \ge 3 \cdot \mathcal{M}\}$, while $\mathcal{M}$ is the average size of the documents in R_i, i.e. $\mathcal{M} = \frac{\sum_{d \in R_i} size(d)}{|R_i|}$.

The constants a and b limit the size of documents in R_i. We tested our method on different sampled sub-collections $\mathbb{D}$, using diverse a and b. The results obtained are similar.

4.1   Evaluation Measures

For evaluating quality of summaries produced by NG-Rank, we used the ROUGE-v.1.5.5[5] evaluation toolkit. It is a method based on N-gram statistics, found to be highly correlated with human evaluations [6]. The ROUGE-N is based on n-grams and generates three scores *Recall*, *Precision*, and the usual *F-measure* for each evaluation.

$$R_n = \frac{\sum\limits_{S \in \{Ref\}} \sum\limits_{\text{n-gram} \in S} Count_{match}(\text{n-gram})}{\sum\limits_{S \in \{Ref\}} \sum\limits_{\text{n-gram} \in S} Count(\text{n-gram})} \tag{7}$$

$$P_n = \frac{\sum\limits_{S \in \{Cand\}} \sum\limits_{\text{n-gram} \in S} Count_{clip}(\text{n-gram})}{\sum\limits_{S \in \{Cand\}} \sum\limits_{\text{n-gram} \in S} Count(\text{n-gram})} \tag{8}$$

$$F = \frac{2 \times P_n \times R_n}{P_n + R_n} \tag{9}$$

R_n (recall) counts the number of overlapping n-gram pairs between the candidate summary to be evaluated and the reference summary created by humans (See [6] for more details). P_n (precision) measures how well a candidate summary overlaps with multiple human summaries using n-gram co-occurrence statistics (See [11] for more details). We used two of the ROUGE metrics in the experimental results, ROUGE-1 (unigram) and ROUGE-2 (bigram).

For evaluating the clustering results, we used *Purity* measure. The purity is a simple and transparent evaluation measure which is related to the entropy concept [12]. To compute the purity criterion, each cluster P is assigned to its majority class. Then we consider the percentage of correctly assigned documents, given the set of documents L_i in the majority class:

---

[5] http://www.berouge.com/.

$$Precision(P, L_i) = \frac{|P \bigcap L_i|}{|P|} \tag{10}$$

The final purity of the overall clustering is defined as follows:

$$Purity(\mathbb{P}, \mathbb{L}) = \sum_{P_j \in \mathbb{P}} \frac{|P_j|}{N} \arg \max_{L_i \in \mathbb{L}} Precision(P_j, L_i) \tag{11}$$

where N is the number of all documents, $\mathbb{P} = \{P_1, P_2, ..., P_k\}$ is the set of clusters and $\mathbb{L} = \{L_1, L_2, ..., L_c\}$ is the set of classes.

5   Experimental Results

As previously stated, we first evaluate NG-$Rank$ as a keyphrase extraction-based summarization method, by comparing the automatically generated summaries with human-generated ones. Then we indirectly assess the quality of the summaries, automatically extracted by our algorithm, by evaluating the clustering improvement after applying NG-$Rank$.

5.1   Assessing the Quality of the Summarization

For the former tests, we thus utilize DUC 2002, and adopt the ROUGE evaluation toolkit to measure the quality of summaries. DUC 2002 provides reference summaries of 100-words (manually produced) to be used in the evaluation process. We stemmed tokens and removed stop words from reference and extracted summaries. In our experiments, we tested both NG-$Rank_M$ and NG-$Rank_H$[6] for extracting keyphrases from documents. To compare our method with the HITS-based algorithm (our baseline), we considered the best results obtained for the possible Γ functions presented in Sect. 2. The size of the summary we have to extract for each documents should be equals to the manually produced reference summary, Since we also remove from them stop words, thus making the reference summaries smaller than the original 100-words ones, we had to choose a suitable parameter α for NG-$Rank$. Recall that α determines the percentage of top-scored graph nodes in each partition FS, LS, or MS that NG-$Rank$ returns (see Sect. 3).

We used two of the ROUGE metrics in the our comparison, ROUGE-1 (unigram) and ROUGE-2 (bigram). The obtained results are showed in Table 1. The convergence time of HITS algorithm increases the execution time of NG-$Rank_H$, but it is negligible considering the significant results obtained by NG-$Rank_H$. Due to this encouraging result, we always applied NG-$Rank_H$ to summarize long documents in our experiments on clustering.

---

[6] For the convergence of HITS, we stop iterating when for any vertex i in the graph the difference between the scores computed at two successive iterations fall below a given threshold: $\frac{|x_i^{k+1} - x_i^k|}{x_i^k} < 10^{-3}$ [10].

Table 1. *NG-Rank* vs. the baseline DUC 2002

	ROUGE-1			ROUGE-2		
	Avg-R_1	Avg-P_1	Avg-F	Avg-R_2	Avg-P_2	Avg-F
NG-Rank$_H$	**0.364**	**0.431**	**0.395**	**0.0346**	**0.0567**	**0.043**
NG-Rank$_M$	0.342	0.396	0.367	0.0341	0.0535	0.0417
Baseline	0.282	0.305	0.293	0.0084	0.0085	0.0085

5.2   Assessing the Clustering Improvement Due to Summarization

In previous experiments, we applied *NG-Rank$_H$* to summarize longer documents in our corpus, before applying a text clustering algorithm. The algorithm adopted for clustering documents was K-Means, while the vectorial representation of documents was based on a classical *tf-idf* weighting of terms, and the measure of similarity between two vector was Cosine similarity. Specifically, we utilized *RapidMiner*[7], which is an integrated environment for analytics, also providing tools for text mining.

Indeed, we tested and evaluated clustering with/without applying *NG-Rank$_H$*, to show the improvements in clustering purity due to summarization. Before reporting and examining the various results, we have first to discuss the features of the sampled corpora, which contain some longer documents. These longer documents are exactly our candidates for summarizations. As stated in Sect. 4, for each sampled corpus $\mathbb{D} = \{\mathcal{L}_1, \mathcal{L}_2, ..., \mathcal{L}_c\}$, we have $\mathcal{L}_i = R_i \cup E_i$, where E_i denotes the set of documents of the ith class that are significantly longer than the average length $\mathcal{M}$. More specifically, the documents in E_i have a size that is at least 3 times $\mathcal{M}$. In our test we used five sampled datasets $\mathbb{D}$, with different sizes of $|E_i| = \{7, 12, 18, 25\}$.

Another important remark concerns the size of the summaries extracted by *NG-Rank$_H$* from each longer document in E_i. This size is determined by the parameter α of the algorithm (see Sect. 3). For each $d \in E_i$, we chose $\alpha = \lceil \frac{\mathcal{M}}{|d|} \rceil$, where $|d|$ and $\mathcal{M}$ denote, respectively, the length of d and the average length of the shorter documents in the sampled class. Figure 2 shows the size of the documents belonging to a given class, namely the class *Sport* in a dataset sampled from the BBC corpus, before and after summarizing larger documents.

Figure 3 shows the average purity obtained by clustering documents in each sampled corpus $\mathbb{D}$, with/without summarizing longer documents. The best improvements in the average purity, due to summarization of longer documents, were about 10 %.

Table 2(a) shows the clustering results without summarizing the longer documents. The dataset used in the test were obtained from the BBC NEWS corpus, where the longer documents were added to the classes *Polit* and *Sport* only. Specifically, we have $|E_{Polit}| = 25$ and $|E_{Sport}| = 7$, while $|E_{Bus}| = 0$

---

[7] https://rapidminer.com/products/studio/.

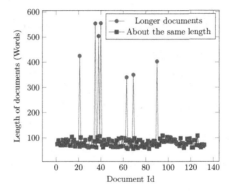

Fig. 2. Length reduction of documents in the class *Sport* (consists of 7 long documents) of a corpus sampled from BBC. After summarization, the lengths of longer documents are reduced, and all documents become of about the same length *len* (in the range $50 \leq len \leq 120$).

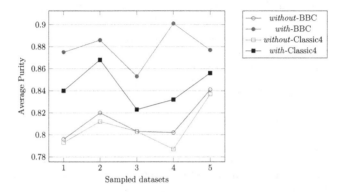

Fig. 3. Average purity of clustering, *with/without* applying $NG\text{-}Rank_H$, for five datasets sampled from the BBC and Classic4 corpora. The five sampled datasets, each corresponding to a distinct $j \in \{1, \ldots, 5\}$ on the x-axis, are characterized by different numbers of longer documents $|E_i|$, for each class i.

and $|E_{Enter}| = 0$. The size of each class before adding these longer documents was: $|R_{Bus}| = 116$, $|R_{Enter}| = 117$, $|R_{Polit}| = 75$, and $|R_{Sport}| = 125$. Table 2(b) reports the results obtained by first applying $NG\text{-}Rank_H$ to summarize the longer documents, and by then clustering all the document collection. We obtained an improvement in the average purity of about 10 %.

Table 3 reports a similar experiment conducted on a dataset sampled from Classic4. Specifically, we have $|E_{Cisi}| = 18$, $|E_{Cran}| = 12$, $|E_{Med}| = 0$, and $|E_{Cacm}| = 7$. The size of each class before adding these longer documents was: $|R_{Cisi}| = 82$, $|R_{Cran}| = 88$, $|R_{Med}| = 100$, and $|R_{Cacm}| = 93$. In this case the improvement in average purity was smaller than for the BBC dataset. However, we registered a similar behaviour, and thus summarizing longer documents by using our algorithm is always valuable.

Table 2. Clustering results: (a) original documents without any summarization; (b) after replacing longer documents with their summaries extracted by *NG-Rank_H* (**BBC Dataset**)

Cluster	Bus	Enter	Polit	Sport	Purity
Cluster 0	7	4	83	5	0.838
Cluster 1	1	61	0	0	0.984
Cluster 2	105	50	10	3	0.625
Cluster 3	3	2	7	124	0.912
Total Purity					**0.802**

(a)

Cluster	Bus	Enter	Polit	Sport	Purity
Cluster 0	10	1	93	6	0.845
Cluster 1	0	101	0	2	0.980
Cluster 2	105	15	7	2	0.814
Cluster 3	1	0	0	122	0.992
Total Purity					**0.905**

(b)

Table 3. Clustering results: (a) original documents without any summarization; (b) after replacing longer documents with their summaries extracted by *NG-Rank_H* (**Classic4 Dataset**)

Cluster	Cisi	Cran	Med	Cacm	Purity
Cluster 0	0	88	0	1	0.989
Cluster 1	3	1	0	47	0.921
Cluster 2	6	10	100	40	0.641
Cluster 3	91	1	0	12	0.875
Total Purity					**0.815**

(a)

Cluster	Cisi	Cran	Med	Cacm	Purity
Cluster 0	0	88	0	0	1
Cluster 1	4	10	7	60	0.740
Cluster 2	1	2	93	0	0.969
Cluster 3	95	0	0	40	0.703
Total Purity					**0.840**

(b)

We conclude with some final remarks about our methodology based on document summarization. When we add longer documents to a class, we likely increase the frequency of terms that are not relevant to the main topic of the class. Indeed, each document contains several topics, for each of which there are relevant terms in documents [2]. Therefore, when we increase the length of a document, we may cause the number of topics to get larger. We can think of *NG-Rank_H* as a method to remove some of these less important/relevant topics, by retaining the main topics only, hopefully those topics that are common to all the documents in a given class.

6   Conclusion

In this paper, we have presented a new graph-based algorithm for keyphrase extraction, in turn used to summarize big documents in a textual corpus, before applying a clustering algorithm. Our experiments indicate the big documents, i.e., document whose size is significantly larger than the mean size in the corpus, introduce noise that can worsen the quality of clustering result. We tested our keyphrase extraction algorithm to summarize these big documents, thus retaining only the terms that are relevant to the main topics discusses in the documents, and observed a significant improvement in clustering quality.

As a future work, we plan to combine summarization with document expansion. In particular, we plan to utilize background knowledge like WordNet to also enrich small documents, with the aim of further improving clustering quality.

References

1. Aggarwal, C.C., Yu, P.S.: Finding generalized projected clusters in high dimensional spaces, vol. 29. ACM (2000)
2. Blei, D.M.: Probabilistic topic models. Commun. ACM **55**(4), 77–84 (2012)
3. Dash, M., Koot, P.W.: Feature selection for clustering. In: Encyclopedia of Database Systems, pp. 1119–1125. Springer, New York (2009)
4. Greene, D., Cunningham, P.: Practical solutions to the problem of diagonal dominance in kernel document clustering. In: Proceedings of the 23rd international conference on Machine learning, pp. 377–384. ACM (2006)
5. Kleinberg, J.M.: Authoritative sources in a hyperlinked environment. J. ACM (JACM) **46**(5), 604–632 (1999)
6. Lin, C.-Y.: ROUGE: a package for automatic evaluation of summaries. In: Text summarization branches out, Proceedings of the ACL-04 workshop, vol. 8 (2004)
7. Litvak, M., Last, M.: Graph-based keyword extraction for single-document summarization. In: Proceedings of the workshop on Multi-source Multilingual Information Extraction and Summarization, pp. 17–24. Association for Computational Linguistics (2008)
8. Liu, T., Liu, S., Chen, Z., Ma, W.-Y.: An evaluation on feature selection for text clustering. ICML **3**, 488–495 (2003)
9. Mihalcea, R.: Graph-based ranking algorithms for sentence extraction, applied to text summarization. In: Proceedings of the ACL 2004 on Interactive poster and demonstration sessions, p. 20. Association for Computational Linguistics (2004)
10. Mihalcea, R., Tarau, P.: TextRank: Bringing order into texts. Association for Computational Linguistics (2004)
11. Papineni, K., Roukos, S., Ward, T., Zhu, W.-J.: BLEU: a method for automatic evaluation of machine translation. In: Proceedings of the 40th annual meeting on association for computational linguistics, pp. 311–318. Association for Computational Linguistics (2002)
12. Silva, J.A., Faria, E.R., Barros, R.C., Hruschka, E.R., de Carvalho, A.C., Gama, J.: Data stream clustering: a survey. ACM Comput. Surv. (CSUR) **46**(1), 13 (2013)
13. Wyse, N., Dubes, R., Jain, A.K.: A critical evaluation of intrinsic dimensionality algorithms. In: Pattern recognition in Practice, pp. 415–425 (1980)

Tweet Timeline Generation via Graph-Based Dynamic Greedy Clustering

Feifan Fan[1], Runwei Qiang[1], Chao Lv[1], Wayne Xin Zhao[2],
and Jianwu Yang[1(✉)]

[1] Institute of Computer Science and Technology,
Peking University, Beijing 100871, China
{fanff,qiangrw,lvchao,yangjw}@pku.edu.cn
[2] School of Information, Renmin University of China, Beijing, China
batmanfly@gmail.com

Abstract. When searching a query in the microblogging, a user would typically receive an archive of tweets as part of a retrospective piece on the impact of social media. For ease of understanding the retrieved tweets, it is useful to produce a summarized timeline about a given topic. However, tweet timeline generation is quite challenging due to the noisy and temporal characteristics of microblogs. In this paper, we propose a graph-based dynamic greedy clustering approach, which considers the *coverage*, *relevance* and *novelty* of the tweet timeline. First, tweet embedding representation is learned in order to construct the tweet semantic graph. Based on the graph, we estimate the *coverage* of timeline according to the graph connectivity. Furthermore, we integrate a noise tweet elimination component to remove noisy tweets with the lexical and semantic features based on *relevance* and *novelty*. Experimental results on public Text Retrieval Conference (TREC) Twitter corpora demonstrate the effectiveness of the proposed approach.

Keywords: Tweet timeline generation · Graph-based dynamic greedy clustering · Tweet embedding

1 Introduction

Microblogging has become one of the most popular social networking platforms in recent years. When users search a query in a microblogging service such as Twitter, an archive of tweets would be returned as part of a retrospective piece on the impact of social media on a specific topic. For instance, a journalist may invest a sports scandal that has been brewing for the past several weeks. She just got news of a breaking development, and turns to searching tweets to find more details. However, due to the retweeting and sharing nature of Twitter, the traditional search engine would lead to a lot of duplicates or near-duplicates tweets that contain the same or highly-similar information - a user cannot easily get an overall idea of the retrieved these tweets. Thus, it would be helpful if

© Springer International Publishing Switzerland 2015
G. Zuccon et al. (Eds.): AIRS 2015, LNCS 9460, pp. 304–316, 2015.
DOI: 10.1007/978-3-319-28940-3_24

the search system produced a "summary" timeline about the topic, which is the studied task in this paper.

In TREC 2014 Microblog track, the organizer introduced a novel pilot task named Tweet Timeline Generation (TTG) task [11]. The TTG task can be summarized as *"At time T, I have an information need expressed by query Q, and I would like a summary that captures relevant information"*. Developing effective TTG system is inherently challenging. Aside from the challenges derived from the tweet retrieval with issues from topic detection and tracking (TDT) and traditional multi-document summarization, systems should further address three additional challenges. First of all, as the length of microblog entry is limited to 140 characters and the content of tweet can be very noisy, it is hard to detect redundant tweets (or cluster similar tweets) using traditional bag-of-words representation of tweets. Generally, users are more interested with tweets which are highly relevant with the query, and a good summarized timeline should be concise and contain as few redundant tweets as possible. Secondly, topics evolve quickly in social media, people usually talk about a subtopic of a given topic in a specific time period. Hence, to measure the similarity of tweets, systems should also take the temporal information into consideration. Thirdly, different topics can attract different amount of attention, which leads to different amount of relevant tweets. As in the search scenario of TTG, users are assumed ready to consume the entire summarized tweet list (unlike a ranked list). Therefore, it is quite necessary for the system to keep the *coverage* of raw related tweet collection in the summarized timeline.

In this work, we mainly address the above challenges (beyond tweet retrieval) in TTG and propose a graph-based dynamic greedy clustering approach to characterize the *coverage*, *relevance* and *novelty* properties of the tweet timeline. The major contributions of this work are: (1) We propose to learn tweet embedding representation to characterize the similarity between tweets which considers both the semantic relatedness and time proximity. We further utilize the similarity to construct tweet semantic graph. (2) We propose a dynamic greedy clustering approach based on tweet semantic graph, where we estimate the *coverage* according to the vertex connectivity in the graph and we integrate a noise tweet elimination component based on logistic regression classifier to measure the *relevance* and *novelty* using many effective lexical and semantic features. (3) We construct extensive experiments on public Text Retrieval Conference (TREC) Twitter corpora, which demonstrate the effectiveness of the proposed approach.

The remainder of the paper is organized as follows. Section 2 introduces related work on subtopic retrieval, TDT, and timeline generation. The graph-based dynamic greedy clustering approach is presented in Sect. 3. The experimental results as well as the comparisons with the-state-of-arts are shown in Sect. 4. Finally, we conclude the paper and outline our future work in Sect. 5.

2 Related Work

Subtopic Retrieval. Zhai *et al.* [19] presented *subtopic retrieval problem*, which is concerned with finding documents that cover many different subtopics of a given topic. Subtopic retrieval is quite different from traditional retrieval problem, where the search engines just simply return the search results in general. However, the retrieved documents always contain much redundant or noisy information in reality. Agrawal *et al.* [2] proposed a systematic approach to diversify the searched results, which tries to maximize the likelihood of selecting a relevant document in the top-k positions based on the categorical information of the queries and documents. Marco and Navigli [5] employed a method relying on n-grams to cluster and diversify web search results. They constructed a co-occurrence graph based on Dice coefficient calculated over the corpus in which the senses are discovered by word sense induction algorithm. In this way, the method can better capture the similarity between the web snippets. To build a more realistic and efficient solution, which allows to label for subtopics or aspects of a given query, Wang and Zhai [17] adopted the star clustering algorithm proposed in [4].

Unlike subtopic retrieval problem, research in the area of TTG aims to obtain a sequence of documents that could describe how a topic evolves over time. In other word, the temporal information must be incorporated into the TTG system.

Topic Detection and Tracking. TDT task mainly conveys the recognition and evolution of the topics contained in text streams. Many previous works [7,9] detect topic through discovering topic bursts from a document stream. Among them, detecting the frequency peaks of topic-related phrases over time in a histogram is a common solution. Another main technique attempted to monitor the formation of a cluster from a structure perspective. Lappas *et al.* [7] presented a approach to model the burstiness of a term, using discrepancy theory concepts. They could identify the time intervals of maximum burstiness for a given term, which is an effective mechanism to address topic detection in the context of text stream. Lin *et al.* [9] proposed burst period detection based on the phenomenon that at some time point, the topic-related terms should appear more frequently than usual and should be continuously frequent around the time point. Agarwal *et al.* [1] discovered events as dense clusters in highly dynamic graphs. Following the same idea, Lee *et al.* [8] applied a clustering algorithm named DBSCAN to recognize the evolution pattern of a topic.

Though these methods have been successfully adopted in TDT task, they are not applicable to the TTG problem. The timeline generation problem represents a natural extension of traditional retrieval [11], which means the generation process is based on the documents returned by the search engines. Therefore, major techniques used in TDT such as burst period detection and dense-based clustering cannot be well applied in generating timeline since many subtopics or aspects in the timeline just contain exactly one document.

Timeline Generation. There are also several works studying the timeline generation recently. A greedy algorithm based on approximation of Minimum-Weight Dominating Set Problem (MWDS) is exploited in [9,16,21]. Among these works, Wang *et al.* [16] proposed an approach that combines image and text analysis to generate a timeline containing textual, pictorial and structural information. They first constructed a multi-view graph, in which each node contains textual and pictorial information, and then selected the representative nodes by finding a minimum dominant set on the graph. Based on the same idea, Lin et $al.$ [9] adopted the method to tweet timeline generation. Xu et $al.$ [18] proposed a novel detection approach, framing the problem of redundant tweet removal as a sequential binary decision task. Lv et $al.$ [12] applied hierarchical clustering algorithm based on Euclidean distance and adaptive relevance estimation to generate tweet timeline, which achieved the best performance of TTG task in TREC 2014 Microblog Track.

These methods had acquired decent effect on timeline generation, while they didn't well handle the unique characteristics of microblog, especially the *coverage*, *relevance* and *novelty* of the timeline. In this paper, we propose a graph-based dynamic greedy clustering approach, we first construct tweet semantic graph using tweet embedding representation, which considers both the semantic relatedness and time proximity in a more comprehensive way. Based on the graph, we estimate the *coverage* of timeline according to the vertex connectivity in the graph, and measure the *relevance* and *novelty* through noise tweet elimination component which utilizes many effective lexical and semantic features.

3   The Proposed Approach

In this section, we first introduce the problem formulation, and then present our approach. The proposed approach first constructs tweet semantic graph based on tweet embedding representation, and then apply graph-based dynamic greedy clustering algorithm to generate the summarized tweet timeline, where we integrate a detection component to eliminate noisy tweets.

3.1   Problem Formulation

We give a formal definition of TTG as follows:

Input: Given a topic query $Q = \{q_1, q_2, \cdots, q_{|Q|}\}$ from users, where q_i is a query term, we obtain a tweet collection $C = \{T_1, T_2, \cdots, T_N\}$ related to the query by traditional retrieval model, where T_i is a tweet and N is the number of retrieved tweets.

Output: A summarized tweet timeline which consists of relevant and non-redundant, chronologically ordered tweets, i.e. $R^{(Q)} = \{T_1^{(Q)}, T_2^{(Q)}, \cdots, T_K^{(Q)}\}$, where $T_i^{(Q)}$ is a relevant tweet from C for query Q, and K is the number of tweets in the timeline.

3.2   Tweet Semantic Graph Construction

Definition 1. *(TWEET SEMANTIC GRAPH): A tweet semantic graph $G = (V, E)$, where V is a set of tweet vertices and E is a set of undirected edges, which represents the semantic relatedness between tweets.*

It is infeasible and meaningless to consider all the tweets for query-specific timeline generation. We apply state-of-the-art retrieval model to derive the top 300 ranked tweets as candidate. Given a candidate tweet collection with their timestamps, we construct a tweet semantic graph by viewing the tweets as the vertices V and calculating the weights of undirected edges on the basis of both content similarity and time proximity. Let T_i and T_j be two vertices in V. We make a undirected edge between T_i and T_j if and only if the similarity between them is greater than a similarity threshold σ.

Tweet Embedding Representation. Due to the length limit and informal expression of tweets, traditional retrieval models relying on "bag-of-words" representations are faced with challenges in describing the similarity of short tweets. For example, when talking about "happy birthday" in twitter, users may use many informal words, such as "bday" and "birthdayyy". Besides, users may use different words to express same meaning, such as "word shortage" and "drought", while these words are semantically related from the context of the twitter stream.

Inspired by distributed representation methods [13], we propose to learn tweet embedding representation which can project tweets into low-dimensional semantic space. Give a tweet $T = \{w_1, w_2, \cdots, w_{|T|}\}$, the objective function is to maximize the log-likelihood, defined as

$$\mathcal{L} = \frac{1}{|T|} \sum_{i=1}^{|T|} \log Pr(w_i | w_{i-c} : w_{i+c}) \tag{1}$$

where $w_{i-c} : w_{i+c}$ is the subsequence $(w_{i-c}, \ldots, w_{i+c})$ by excluding w_i and $2 \times c$ is the window size. We model each word w_i an M-dimensional embedding vector $\boldsymbol{v}_{w_i}$. With this form of embedding representation, we further employ a multiclass classifier softmax to formulate the probability $Pr(w_i | w_{i-c} : w_{i+c})$ as follows

$$Pr(w_i | w_{i-c} : w_{i+c}) = \frac{\exp(\bar{\boldsymbol{v}}^\top \boldsymbol{v}'_{w_i})}{\sum_{w \in \mathcal{W}} \exp(\bar{\boldsymbol{v}}^\top \boldsymbol{v}'_w)} \tag{2}$$

where $\mathcal{W}$ is the word vocabulary, $\boldsymbol{v}'_{w_i}$ is the output vector representation of target word w_i and $\bar{\boldsymbol{v}}$ is the averaged vector representation of the context. We simply average all the word vectors to obtain the tweet vector. Based on the tweet vector, we can utilize classic similarity measure (i.e. cosine similarity) to estimate the similarity of tweets. Considering that tweets posted in a same time interval (e.g. the latest two hours) are more likely to talk about the same aspect of the topic, thus we combine the tweet vector with fading time factor to characterize both the semantic relatedness and time proximity in a more comprehensive way, which is defined as follows

$$sim(T_i, T_j) = \frac{|\boldsymbol{v}_i \cdot \boldsymbol{v}_j|}{|\boldsymbol{v}_i| \cdot |\boldsymbol{v}_j|} \cdot e^{-\gamma |\tau_i - \tau_j|} \tag{3}$$

where $\boldsymbol{v}_i$ and $\boldsymbol{v}_j$ are the tweet embedding representations of T_i and T_j, respectively. γ is the exponential parameter that controls the temporal influence. τ_i and τ_j are the corresponding timestamps, measured in fractions of hours.

3.3   Graph-Based Dynamic Greedy Clustering Approach

As discussed in Sect. 1, In order to obtain the summarized tweet timeline from retrieved tweets, the system must dynamically detect and eliminate the redundant or noisy tweets for distinct topics. We propose a dynamic greedy clustering approach based on tweet semantic graph, which considers both the semantic relatedness and time proximity between tweets. Considering that the returned tweets from the retrieval model can still contain much noise, we further incorporate a noise elimination component based on both lexical and semantic similarity during the clustering process.

Graph-Based Dynamic Greedy Clustering Algorithm. The proposed algorithm iteratively identify representative vertices given a query, where *relevance*, *coverage* and *novelty* have been considered. We present the overall procedure in Algorithm 1.

Given a set of retrieved tweets as the input, at Line 1 we construct the tweet semantic graph $G = (V, E)$. At the beginning of each iteration, we first (Line 3 ~ 7) compute the *coverage* score for each remaining (unmarked) vertex by using a score function $ComputeCoverageScore(v_i)$ that calculates the connectivity

---

Algorithm 1. Graph-based Dynamic Greedy Clustering.

Input: Candidate tweet collection $C = \{T_1, T_2, \cdots, T_N\}$
　　　　Query Q
　　　　Edge similarity threshold σ
　　　　The maximum number of tweets in timeline K
　　　　Exponential parameter γ in Eq. 3
Output:
　　　　Summarized tweet timeline $R^{(Q)} = \{T_1^{(Q)}, T_2^{(Q)}, \cdots, T_K^{(Q)}\}$
 1: $G(V, E) \leftarrow ConstructGraph(C, \sigma, \gamma)$
 2: **repeat**
 3: 　　**for** $v \in V$ **do**
 4: 　　　　**if** v_i is not marked **then**
 5: 　　　　　$ComputeCoverageScore(v_i)$;
 6: 　　　　**end if**
 7: 　　**end for**
 8: 　　Rank the remaining vertices by the coverage scores;
 9: 　　$v^* \leftarrow$ top ranked vertex;
10: 　　**if** $IsNoiseTweet(v^*, Q, R^{(Q)})$ **then**
11: 　　　　Mark v^* as a noise vertex;
12: 　　**else**
13: 　　　　Mark v^* as the centroid vertex;
14: 　　　　Mark the vertices connected to v^* as *visited*;
15: 　　　　$R^{(Q)} \leftarrow R^{(Q)} \cup \{v^*\}$;
16: 　　**end if**
17: **until** $\forall v \in V$ is marked or $|R^{(Q)}| = K$
18: **return** $R^{(Q)}$

---

based on unmarked vertices connected with v_i in the graph. Besides *coverage*, users would be more interested in tweets which are highly relevant with the query, and a good summarized timeline should be concise and contain as few redundant tweets as possible. In order to capture the *relevance* and *novelty* for timeline generation, we check the top ranked tweet (Line $10 \sim 16$) by using a boolean function $IsNoiseTweet(\cdot)$ based on a rich set of relevance and novelty features, which will be discussed next. We perform the update procedure for marking as follows. If a tweet is measured as *noise* by $IsNoiseTweet(\cdot)$, the corresponding vertex is marked as *noise vertex*. Otherwise, the corresponding vertex is marked as the *centroid vertex* which will be incorporated into the timeline $R^{(Q)}$. The rest neighboring vertices of the centroid vertex are marked as *visited* and no more considered for selection.

Noise Tweet Elimination. In Algorithm 1, a key component is the noise elimination function $IsNoiseTweet(\cdot)$, which aims to filter noise tweets and improve the timeline quality. To implement $IsNoiseTweet(\cdot)$, we utilize a logistic regression classifier based on a rich set of lexical and semantic features to eliminate noise tweet, which have been used to achieve the state-of-art in adaptive filtering problem of news and tweets [3,20]. These two types of features are given as follows.

- **Relevance Features** measure the relevance between tweet T and query Q. We have four relevance features in total. Based on traditional "bag-of-words" model, three lexical similarity score features are calculated by cosine similarity, Dice coefficient, and Jaccard coefficient, respectively. In addition, one semantic similarity score feature is obtained by cosine similarity measure using tweet embedding representation.
- **Novelty Features** estimate the novelty of an unprocessed tweet compared with previous generated centroid tweets of clusters. We calculate the similarity between the unprocessed tweet and each centroid tweet of each generated cluster, and choose the closest centroid tweet as comparison to estimate the novelty of the unprocessed tweet. Like *Relevance Features*, we obtain three lexical and one semantic score features based on "bag-of-words" model and tweet embedding representation, respectively.

To train the logistic regression classifier, we use the judgments on TREC2011-2012 topics as labeled data, which are released and labeled by the official TREC organizer.

In summary, *coverage* is measured by using the function $ComputeCoverage$ $Score(\cdot)$ that calculates the connectivity based on unmarked vertices in the graph, , while *relevance* and *novelty* have been considered in function $IsNoise$ $Tweet(\cdot)$ by leveraging the relevance and novelty features. Our clustering algorithm can be efficiently implemented based on the well-known Breadth-First-Search (BFS) graph traverse algorithm.

4   Experiments

In this section, we conduct extensive experiments to evaluate the effectiveness of the proposed method. In what follows, we first describe the experimental setting and then present the results.

4.1   Experimental Setting

Dataset. Two large data collections (i.e. Tweets2011 and Tweets2013 collections) are used in our experiments. TREC organizers release a streaming API to participants [10]. Using the official API[1], we crawled a set of local copies of the canonical corpora. Tweets11 collection has a sample of about 16 million tweets, ranging from January 24, 2011 to February 8, 2011 while Tweets13 collection contains about 259 million tweets, ranging from February 1, 2013 to March 31, 2013 (inclusive). Tweets11 is used for evaluating the effectiveness of the proposed Twitter TTG systems over 10 training topics in TREC 2011 and 2012. And, Tweets13 is used in evaluating the proposed TTG systems over 55 official topics in the TREC 2014 Microblog track [11]. The topics of TREC 2011-2012 are used for tuning the parameters and then we use the best parameter setting to evaluate our methods with topics for TREC 2014. Table 1 summarizes basic statistics of the two corpora.

Table 1. Data statistics of Tweets2011 and Tweets2013 test collections. K is the average number of tweets in the gold timelines.

Corpus	#queries	#tweets	#tokens	#terms	K
Tweets2011	10	4,948,137	753,021	27,180,607	130
Tweets2013	55	68,682,325	5,063,852	298,321,176	193

Evaluation Metrics. Our evaluation metrics contain the following two types of metrics.

Tweet Retrieval: As our TTG system's input is a candidate tweet collection generated by retrieval models, the performance of our system might be affected by the retrieval performance. In TREC Microblog track, tweets are judged on the basis of the defined information using a three-point scale [14]: *irrelevant* (labeled as 0), *minimally-relevant* (labeled as 1), and *highly-relevant* (labeled as 2). We use two official main metrics for the retrieval task in TREC, including Mean Average Precision (MAP) and Precision at N (P@N). Specifically, MAP for top 1000 ranked documents and P@30 with respect to *allrel* (i.e. tweet set labeled as 1 or 2).

---

[1] https://github.com/lintool/twitter-tools.

Clustering Performance as Timeline Quality: TTG results will be evaluated by two different versions of the F_1 metric, i.e., an unweighted version and a weighted version, which are used in TREC 2014 Microblog Track [11]. F_1 metric is combined by cluster precision and cluster recall. We first introduce the unweighted version as follows.

- **Cluster precision (unweighted).** Of tweets returned by the system, how many distinct semantic clusters are represented.
- **Cluster recall (unweighted).** Of the semantic clusters discovered by the assessor, how many are represented in the system's output.

For unweighted version, the system does not get "credit" for retrieving multiple tweets from the same semantic cluster. Different from unweighted F_1, the weighted F_1 (denoted as F_1^w) attempts to account for the fact that some semantic clusters are intuitively more important than others. Each cluster will be weighted by relevance grade: minimally-relevant tweets get a weight of one and highly-relevant tweets get a weight of two. These weights are then factored into the precision and recall computations. The F_1^w score is the main evaluation metric for TTG task in TREC 2014.

4.2   Methods to Compare

We consider the following methods as comparisons in our experiments.

- **TTGPKUICST2**: Hierarchical clustering algorithm based on adaptive relevance estimation and Euclidean distance, proposed in [12], which achieved the best performance in TREC 2014 Microblog Track.
- **EM50**: kNN clustering approach applied in [15], using a modified Jaccard coefficient (i.e. EM) and used top K retrieved results as candidates for clustering, which won the second place in TREC 2014 Microblog Track.
- **hltcoeTTG1**: The novel detection approach proposed by Xu *et al.* [18]. Unlike clustering methods, they framed the problem of tweet timeline generation as a sequential binary decision task. Therefore, they proposed a binary classifier to determine whether a coming tweet is novel and then compose the novel tweets as the summarized tweet timeline, which won the third place in TREC 2014 Microblog Track.
- **MWDSA**: we implement the Greedy MWDS Approximation Algorithm (denoted as **MWDSA**) that was exploited in generating storyline problem [9,16,21]. They identified the minimum-weight dominating set approximation as the most representative summary.
- **GDGC-BOW**: The proposed graph-based dynamic greedy clustering approach, which only utilizes "bag-of-words" model in both tweet graph construction and noise tweet elimination.
- **GDGC**: The proposed graph-based dynamic greedy clustering approach in Sect. 3.3.

Note that both in re-implemented systems and the proposed system, we obtain the top-ranked 300 tweets from the ranked list achieved by the retrieval models as the candidates for TTG process. We set the time factor γ as 0.01, usually around $0.005 \sim 0.02$. In tweet graph construction based on "bag-of-words" representation, we set the similarity threshold σ as 0.65, usually around $0.6 \sim 0.7$.

4.3   Results and Analysis

Recall that our TTG system's input is a candidate tweet collection generated by a retrieval model. In our study, we utilize two retrieval models for the candidate generation, namely **RTRM** and **RankSVM**. **RTRM** utilizes a two-stage pseudo-relevance feedback query expansion to estimate the query language model and expand documents with shortened URLs in microblog. In addition, **RTRM** can evaluate the temporal aspects of documents with the temporal-reranking components. **RankSVM** is a state-of-the-art pairwise learning to rank method, proposed in [6]. We follow the work of [12] and generate totally 250 features.

Table 2 presents the experimental results of TTG performance along with the candidate tweet retrieval performance. We can have the following obervations.

(1) Compared with greedy **MWDSA** algorithm, **GDGC**-based methods outperform significantly in terms of F_1 and F_1^w for both **RTRM** and **RankSVM** retrieval candidate tweet collections. Since **MWDSA** considers the *coverage* by choosing the dominating vertexs in the graph, while it does not capture the *relevance* and *novelty* of tweet timeline. Besides, from the comparisons between two retrieval candidates, (e.g. **MWDSA**$_{RTRM}$ and **MWDSA**$_{RankSVM}$), we can also observe that TTG performance will benefit from better retrieval performance, which is very reasonable since TTG utilizes retrieval tweets as input candidates.

(2) **GDGC** are consistently better than **GDGC-BOW**, which shows the importance of tweet embedding representation in constructing tweet semantic graph and estimating the *relevance* and *novelty* in noise tweet elimination component.

(3) The proposed **GDGC**$_{RankSVM}$ outperforms best three systems in TREC 2014 Microblog Track, which demonstrates the effectiveness of the proposed approach in depicting the characteristics of tweet timeline. Besides, compared with **hltcoeTTG1**, the proposed approach using weaker retrieval results (i.e. **GDGC**$_{RTRM}$) can also perform better in TTG. Specially, our method **GDGC**$_{RTRM}$ improves the F_1^w score over **EM50** and **hltcoeTTG1** by 16.7% and 20.3%, respectively; while the corresponding increments in terms of F_1 are 36.2% and 25.7%. On the other hand, when utilizing a more effective retrieval model (i.e. **RankSVM**), the graph-based dynamic greedy clustering approach will achieve more improvements in terms of F_1^w and F_1. In addition, compared with **TTGPKUICST2**, **GDGC**$_{RankSVM}$ achieves 3.91% and 3.05% further increases in terms of F_1^w and F_1, respectively.

Table 2. Performance comparisons of the proposed methods and baselines. a and b indicate that the corresponding improvements over **MWDSA**$_{RTRM}$ and **MWDSA**$_{RankSVM}$, are statistically significant ($p < 0.05$), respectively.

Method	Retrieval		TTG	
	MAP	P@30	F_1	F_1^w
TTGPKUICST2	0.5863	0.7224	0.3540	0.4575
EM50	0.5122	0.6982	0.2546	0.3815
hltcoeTTG1	0.5707	0.7121	0.2760	0.3702
MWDSA$_{RTRM}$	0.5422	0.6958	0.2581	0.3890
GDGC-BOW$_{RTRM}$	0.5422	0.6958	0.3364^a	0.4295^a
GDGC$_{RTRM}$	0.5422	0.6958	0.3468^a	0.4452^a
MWDSA$_{RankSVM}$	0.5863	0.7224	0.3143	0.4161
GDGC-BOW$_{RankSVM}$	0.5863	0.7224	0.3498^b	0.4556^b
GDGC$_{RankSVM}$	0.5863	0.7224	$\mathbf{0.3648}^b$	$\mathbf{0.4754}^b$

4.4   Parameter Tuning

Several parameters in the proposed method may affect the system performance. In this section, we analyze the parameter setting in the graph-based dynamic greedy clustering approach. All these experiments are run on TREC 2011-2012 topics.

Figure 1(a) shows the effect of tweet embedding vector in terms of metrics F_1^w and F_1. We can see that the two curves follow similar patterns, F-score increases rapidly with the increase of the embedding vector size when it is less than 300. When the vector size becomes larger, the performance changes slightly, which means that the vectors can already provide enough information to depict the tweets semantically from the contexts.

We study the effect of similarity threshold parameter σ in tweet semantic graph construction using tweet embedding representations, which is shown in Fig. 1(b). we can observe that the value of σ yields a significant effect on the evaluation metrics of TTG (i.e. F_1^w and F_1).

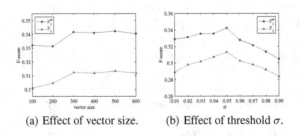

(a) Effect of vector size.   (b) Effect of threshold σ.

Fig. 1. Parameter Tuning.

Moreover, we can see that the optimal σ using tweet embedding representation is greater than that using "bag-of-words" model described in Sect. 4.2, which demonstrates the characteristics of tweet embedding representations. That is, words used in similar contexts are considered semantically similar and tend to have similar vectors, which could lead to a general high similarity score for tweets. When it comes to the "bag-of-words" representation, different words are simply regarded as irrelevant.

5   Conclusion and Future Work

In this study, we propose a graph-based dynamic greedy clustering approach. We utilize learned tweet embedding representation to construct tweet semantic graph, which considers both the semantic relatedness and time proximity in a more comprehensive way. Based on the graph, we estimate the *coverage* by highly scoring vertices with larger graph connectivity. For top ranked candidate tweet at each iteration, we measure the *relevance* and *novelty* through a logistic regression classifier which adopts many effective lexical and semantic features. Extensive experiments using public TREC Twitter collection, demonstrate the effectiveness of the proposed method.

Currently, we simply utilize the top-ranked 300 tweets from retrieval models as input candidates. In fact, the number of strong candidate tweets for distinct topics can be different due to the diverse popularity in Twitter. In the future, we will consider how to dynamically obtain candidate tweets from retrieval results.

Acknowledgments. The work reported in this paper is supported by the National Natural Science Foundation of China Grant 61370116. We thank anonymous reviewers for their beneficial comments.

References

1. Agarwal, M.K., Ramamritham, K., Bhide, M.: Real time discovery of dense clusters in highly dynamic graphs: identifying real world events in highly dynamic environments. Proc. VLDB Endowment **5**(10), 980–991 (2012)
2. Agrawal, R., Gollapudi, S., Halverson, A., Ieong, S.: Diversifying search results. In: Proceedings of the Second ACM International Conference on Web Search and Data Mining, pp. 5–14. ACM (2009)
3. Albakour, M., Macdonald, C., Ounis, I., et al.: On sparsity and drift for effective real-time filtering in microblogs. In: Proceedings of the 22nd ACM International Conference on Information & Knowledge Management, pp. 419–428. ACM (2013)
4. Aslam, J.A., Pelekhov, E., Rus, D.: The star clustering algorithm for static and dynamic information organization. J. Graph Algorithms Appl. **8**, 95–129 (2004)
5. Di Marco, A., Navigli, R.: Clustering and diversifying web search results with graph-based word sense induction. Comput. Linguistics **39**(3), 709–754 (2013)
6. Joachims, T.: Optimizing search engines using clickthrough data. In: KDD, pp. 133–142 (2002)

7. Lappas, T., Arai, B., Platakis, M., Kotsakos, D., Gunopulos, D.: On burstiness-aware search for document sequences. In: Proceedings of the 15th ACM SIGKDD International Conference on Knowledge Discovery and Data Mining, pp. 477–486. ACM (2009)
8. Lee, P., Lakshmanan, L.V., Milios, E.E.: Incremental cluster evolution tracking from highly dynamic network data. In: IEEE 30th International Conference on Data Engineering (ICDE), 2014, pp. 3–14. IEEE (2014)
9. Lin, C., Lin, C., Li, J., Wang, D., Chen, Y., Li, T.: Generating event storylines from microblogs. In: Proceedings of the 21st ACM International Conference on Information and Knowledge Management, pp. 175–184. ACM (2012)
10. Lin, J., Efron, M.: Overview of the TREC-2013 Microblog Track. In: TREC 2013 (2013)
11. Lin, J., Efron, M.: Overview of the TREC-2014 Microblog Track. In: TREC 2014 (2014)
12. Lv, C., Fan, F., Qiang, R., Fei, Y., Yang, J.: PKUICST at TREC 2014 Microblog Track: Feature Extraction for Effective Microblog Search and Adaptive Clustering Algorithms for TTG (2014)
13. Mikolov, T., Sutskever, I., Chen, K., Corrado, G.S., Dean, J.: Distributed representations of words and phrases and their compositionality. In: Advances in Neural Information Processing Systems, pp. 3111–3119 (2013)
14. Ounis, I., Macdonald, C., Lin, J., Soboroff, I.: Overview of the TREC-2011 Microblog Track. In: TREC 2011 (2012)
15. Walid, M., Wei, G., Tarek, E.: QCRI at TREC 2014: Applying the KISS Principle for TTG Task in the Microblog Track (2014)
16. Wang, D., Li, T., Ogihara, M.: Generating pictorial storylines via minimum-weight connected dominating set approximation in multi-view graphs. In: AAAI (2012)
17. Wang, X., Zhai, C.: Learn from web search logs to organize search results. In: Proceedings of the 30th Annual International ACM SIGIR Conference on Research and Development in Information Retrieval, pp. 87–94. ACM (2007)
18. Xu, T., McNamee, P., Oard, D.W.: HLTCOE at TREC 2014: Microblog and Clinical Decision Support (2014)
19. Zhai, C., Cohen, W.W., Lafferty, J.: Beyond independent relevance: methods and evaluation metrics for subtopic retrieval. In: Proceedings of the 26th Annual International ACM SIGIR Conference on Research and Development in Informaion Retrieval, pp. 10–17. ACM (2003)
20. Zhang, Y.: Using bayesian priors to combine classifiers for adaptive filtering. In: Proceedings of the 27th Annual International ACM SIGIR Conference on Research and Development in Information Retrieval, pp. 345–352. ACM (2004)
21. Zhou, W., Shen, C., Li, T., Chen, S., Xie, N., Wei, J.: Generating textual storyline to improve situation awareness in disaster management. In. In Proceedings of the 15th IEEE International Conference on Information Reuse and Integration (IRI 2014) (2014)

Evaluation

Topic Set Size Design with the Evaluation Measures for Short Text Conversation

Tetsuya Sakai[1]([✉]), Lifeng Shang[2], Zhengdong Lu[2], and Hang Li[2]

[1] Waseda University, Tokyo, Japan
tetsuyasakai@acm.org
[2] Noah's Ark Lab, Huawei, Hong Kong
{Shang.Lifeng,Lu.Zhengdong,HangLi.HL}@huawei.com

Abstract. Short Text Conversation (STC) is a new NTCIR task which tackles the following research question: given a microblog repository and a new post to that microblog, can systems reuse an old comment from the respository to satisfy the author of the new post? The official evaluation measures of STC are *normalised gain at 1* (nG@1), *normalised expected reciprocal rank at 10* (nERR@10), and P^+, all of which can be regarded as evaluation measures for navigational intents. In this study, we apply the *topic set size design* technique of Sakai to decide on the number of test topics, using variance estimates of the above evaluation measures. Our main conclusion is to create 100 test topics, but what distinguishes our work from other tasks with similar topic set sizes is that we know what this topic set size means from a statistical viewpoint for each of our evaluation measures. We also demonstrate that, under the same set of statistical requirements, the topic set sizes required by nERR@10 and P^+ are more or less the same, while nG@1 requires more than twice as many topics. To our knowledge, our task is the first among all efforts at TREC-like evaluation conferences to actually create a new test collection by using this principled approach.

1 Introduction

Short Text Conversation (STC)[1] is a new NTCIR[2] task which tackles the following research question: given a microblog repository and a new post to that microblog, can systems *reuse* an old comment from the respository to satisfy the author of the new post? For each new post, systems are expected to output a ranked list of past comments that are *coherent* with respect to the original post and *useful* from the viewpoint of the author of the post. For example, given a post "The first day in Hawaii. Watching the sunset at the balcony with a big glass of wine in hand," comments such as "Enjoy it and don't forget to share your photos!" and "How long are you going to stay there?" are coherent, and could also be considered useful to the author in Hawaii[3]. We view this as a first

---

[1] http://ntcir12.noahlab.com.hk/stc.htm.
[2] http://research.nii.ac.jp/ntcir/.
[3] Examples taken from our arxiv paper: http://arxiv.org/pdf/1408.6988.pdf.

© Springer International Publishing Switzerland 2015
G. Zuccon et al. (Eds.): AIRS 2015, LNCS 9460, pp. 319–331, 2015.
DOI: 10.1007/978-3-319-28940-3_25

Table 1. STC test collection.

(a) Repository	#posts	196,395
	#post-comment pairs	5,648,128
(b) Training data	#posts	225
	#post-comment pairs (labelled)	6,017
(c) Test data	#posts	TBD
	#post-comment pairs (labelled)	TBD

small step towards developing a system that can interact effectively with the
user in natural language; the objective of STC is to quantify how far we can go
using a purely IR-oriented approach that does not involve natural language generation. While retrieving and ranking coherent and useful comments is different
from the traditional IR task of ranking items that are *relevant* to an information
need, we expect that various wisdoms of IR such as the pooling technique and
graded relevance measures will be applicable to, and highly useful for, this task.

For the Chinese Subtask of the NTCIR-12 STC Task, a Chinese *Weibo*[4]
corpus will be used[5]. Weibo currently has over 40 million users, and is very
much like Twitter[6] in terms of user experience: just like Twitter, each Weibo
"tweet" has the length limit of 140 characters, although 140 characters in Chinese
can be significantly more informative than 140 characters in English, as the
Chinese characters are ideograms with no spaces between words[7]. Table 1 shows
the structure of the STC test collection: (a) the repository of "old" posts and
their comments; (b) labelled post-comment pairs for training; and (c) test data
that will be contructed as an outcome of the STC task. Note that the posts in our
training and test data were sampled from outside the repository to be treated
as "new" posts, while the comments in these data sets are from the repository,
which are regarded as "reused" comments. That is to say, for every labelled post-comment pair in the STC test collection, the comment was originally a response
to some other post.

The training data labels were obtained as described in the aforementioned
`arxiv` paper. Briefly, for each of our training post, we searched the repository
using three simple algorithms, and pooled the top 10 comments from each run.
The comments in the depth-10 pools were then manually assessed from multiple viewpoints to form graded "relevance" data, with relevance grades L0 (not

---

[4] http://weibo.com.

[5] A Japanese subtask using Twitter data is also in preparation.

[6] http://twitter.com.

[7] The minimum/average/maximum lengths of the 196,395 posts in the repository
are 10/32.5/140, respectively. Whereas, after translating them into English using
machine translation, the corresponding lengths are 11/115.7/724. This suggests that
a Chinese tweet can be 3–5 times as informative as an English one.

relevant), L1 (relevant) and L2 (highly relevant)[8]. In the present study, we evaluate six runs based on the training data labels in order to estimate the within-system variances of several evaluation measures and thereby determine the number of test topics (i.e., posts) in a principled way. While our training data labels are probably highly incomplete and biased, note that we are running the STC task exactly because we want to create a reliable STC test collection with a test topic set with post-comment labels obtained via a pooling of a variety of runs. See Sect. 5 for more discussions.

The official evaluation measures of STC are *normalised gain at 1* (nG@1) [15][9], *normalised expected reciprocal rank at 10* (nERR@10) [2], and P^+ [11], all of which can be regarded as evaluation measures for *navigational* intents [1]. In this study, we apply the *topic set size design* technique of Sakai [13,14] to decide on the number of test topics, using variance estimates of the above evaluation measures. Our main conclusion is to create 100 test topics, but what distinguishes our work from other tasks with similar topic set sizes is that we know what this topic set size means from a statistical viewpoint for each of our evaluation measures. We also demonstrate that, under the same set of statistical requirements, the topic set sizes required by nERR@10 and P^+ are more or less the same, while nG@1 requires more than twice as many topics. To our knowledge, our task is the first among all efforts at TREC-like evaluation conferences to actually create a new test collection by using this principled approach.

2 Related Work

2.1 Evaluation Tasks Related to STC

As the STC task requires participating systems to produce a ranked list of comments given a Weibo post, it is very similar to traditional TREC ad hoc tracks [19], in terms of input/output specifications and the test collection construction procedure. A post is like a TREC topic, and comments are like target documents; instead of retrieving relevant documents, STC systems are expected to retrieve coherent and useful comments. Just like TREC, the STC runs will be pooled, with a pool depth of 10, and graded "relevance" assessments will be conducted using multiple assessors for judging each comment.

In terms of document type, STC resembles the TREC Microblog track which uses Twitter data. At the TREC 2011 and 2012 Microblog tracks, a collection comprising 16 million tweets were used, but only tweet IDs were distributed to participating teams and each team had to download the actual data for themselves. This meant that the different downloads were not strictly identical. Whereas, from the TREC 2013 Microblog track, "Evaluation as a Service"

---

[8] While the present study uses the post-comment labels collected as described in the `arxiv` paper, we have since then revised the labelling criteria in order to clarify several different axes for labelling, including coherence and usefulness. The new labelling scheme will be used to revise the training data labels as well as to construct the official test data labels.

[9] nG@1 is sometimes referred to as nDCG@1; however, note that neither discounting ("D") nor cumulating gains ("C") is applied at rank 1.

was introduced to handle over 243 million tweets via search APIs [7], which meant that participating teams did not have direct access to the actual data. In contrast, while the STC Weibo collection is relatively small (see Table 1), the entire data set is distributed to each participating team for research purposes, in a way similar to the "TREC disks" [19].

In terms of task, STC is related to question answering (QA) tasks such as the TREC QA track [19], the NTCIR ACLIA (Advanced Crosslingual Information Access) task [8], and the NTCIR QALab task [18]. In particular, the NTCIR CQA (Community QA) task [15] is related to STC in terms of both document type and task: CQA used the Yahoo! Chiebukuro (Japanese Yahoo! Answers) data, and the task was to find the answer to a question that was selected by the questioner as the "best answer." The most important distinction between these QA-related tasks and STC is that an STC post is not necessarily a question, and therefore that each comment to the post is not necessarily an answer. For example, in the example given in Sect. 1, note that one of the *comments* is a question: "How long are you going to stay there?"[10].

2.2 Problems and Approaches Related to STC

Research on modelling human-computer dialogues started over half a century ago [21], but the recent advent of social media such as Twitter has revitalised this area using new approaches. STC is the simplest form of human-computer dialogues that deals with one post-comment pair at a time, and statistical modelling of STC and related tasks based on large scale social media corpora has become possible. For example, Ritter, Cherry and Dolan [10] utilised the Twitter data to study the feasiblity of *generating* a comment to a given post, by regarding the transformation from a post to a comment as a statistical translation problem. This is in contrast to the STC problem setting where systems are expected to *reuse* comments from a social media repository. Using Twitter and live-journal data, Jafarpour and Burges [5] tackled a problem they refer to as *learning to chat*, which is very similar to STC in that past comments are retrieved for reuse, although they mention in their paper that the retrieved comment should then be altered prior to presentation to the author of the new post. They propose a three-stage approach to ranking past comments, and also a mechanism for collecting high-quality training data from users. Higashinaka *et al.* [4] learn a conversational model from post-comment pairs (or "Two-Tweet exchanges"), and report that the learned model is comparable in effectiveness to one that utilises longer exchanges as training data.

We are hoping that many research groups that are tackling related problems such as the ones mentioned above will participate in the NTCIR-12 STC task. We shall report on the outcome of STC in our NTCIR-12 overview paper in 2016, where we hope to clarify what kind of techniques are effective for this relatively simple form of human-computer dialogue.

---

[10] Given an input remark "Men are all alike," ELIZA, the rule-based system developed in the 1960s, could respond: "IN WHAT WAY?" [21] .

2.3   Topic Set Size Design

Sakai [13,14] showed three statistically motivated methods for determining the topic set size for a test collection to be built: one based on the paired t-test, one based on one-way ANOVA and one based on confidence intervals (CIs). In the present study, we use Sakai's ANOVA-based Excel tool[11] as this method can consider comparison of $m (\geq 2)$ systems and is the most general. Sakai demonstrated that the ANOVA-based method with $m = 2$ and the t-test-based method give similar results, and also that the ANOVA-based method with $m = 10$ can be used instead of the CI-based method (see Sect. 4.2).

Sakai's ANOVA-based tool requires the following input parameters to determine the required topic set size:

α: The probability of *Type I error* (detecting a difference that does not exist).
β: The probability of *Type II error* (missing a difference that actually exists).
m: The number of systems that will be compared in one-way ANOVA ($m \geq 2$).
$minD$: The *minimum detectable range* [13,14]. That is, whenever the performance difference between the best and the worst systems is $minD$ or higher, we want to ensure a *statistical power* of $(1 - \beta)$ (i.e., the probability of detecting a difference that actually exists) given the significance level α.
$\hat{\sigma}^2$: The estimated variance of a system's performance, under the *homoscedasticity* (i.e., equal variance) assumption [9,13,14]. That is, it is assumed that the scores of the i-th system obey $N(\mu_i, \sigma^2)$, where μ_i's differ while σ^2 is common to all systems. This variance known to be heavily dependent on the evaluation measure.

Sakai [13,14] also describes simple ways to obtain $\hat{\sigma}^2$ for a particular evaluation measure, given a $n \times m$ topic-by-system matrix of scores x_{ij}, for system i and topic j. We use his variance estimation method based on one-way ANOVA: let the sample mean for system i be $\bar{x}_{i\bullet} = \frac{1}{n}\sum_{j=1}^{n} x_{ij}$; the population within-system variance can be estimated as:

$$\hat{\sigma}^2 = V_E = \frac{\sum_{i=1}^{m}\sum_{j=1}^{n}(x_{ij} - \bar{x}_{i\bullet})^2}{m(n-1)} . \tag{1}$$

3   Evaluation Measures for Short Text Conversation

The official evaluation measures of the STC task are graded-relevance IR evaluation measures for *navigational* intents [1]. This is because a human-computer conversation system that can respond naturally to a natural language post would usually require exactly one good comment. Below, we define the official measures and clarify the relationships among them. We compute these evaluation measures using the NTCIREVAL tool[12].

---

[11] http://www.f.waseda.jp/tetsuya/CIKM2014/samplesizeANOVA.xlsx.
[12] http://research.nii.ac.jp/ntcir/tools/ntcireval-en.html.

3.1   NG@1

Let $g(r)$ denote the *gain* of a document (i.e., a comment) retrieved at rank r: throughout this paper, we let $g(r) = 2^2 - 1 = 3$ if the document is L2-relevant; $g(r) = 2^1 - 1 = 1$ if it is L1-relevant; $g(r) = 0$ if it is not relevant (i.e., L0). For a given topic (i.e., a post), an *ideal ranked list* is constructed by listing up all L2-relevant documents followed by all L1-relevant ones. Let $g^*(r)$ denote the gain of a comment at rank r in the ideal list. Normalised Gain at Rank 1 is defined as follows:

$$nG@1 = \frac{g(1)}{g^*(1)} \; . \tag{2}$$

This is a crude measure, in that it only looks at the top ranked document, and that, in our setting, it only takes three values: 0, 1/3 or 1.

3.2   NERR@10

Expected Reciprocal Rank (ERR) [2] is a popular measure with a *diminishing return* property: once a relevant document is found in the list, the value of the next relevant document in the same list is guaranteed to go down. Hence, the measure is suitable for navigational intents where the user does not want redundant information. ERR assumes that the user scans a ranked list from top to bottom, and that the probability that the user is satisfied with the document at rank r is given by $p(r) = \frac{g(r)}{2^H}$, where H denotes the highest relevance level for a test collection (2 in our case). Hence, in our setting, $p(r) = 3/4$ if the document at rank r is L2-relevant; $p(r) = 1/4$ if it is L1-relevant; $p(r) = 0$ if it is not relevant. The probability that the user reaches as far as rank r and then stops scanning the list (due to satisfaction) is given by:

$$Pr_{ERR}(r) = p(r) \prod_{k=1}^{r-1} (1 - p(k)) \; , \tag{3}$$

and the *utility* of the ranked list to the user who stopped at r is computed as $1/r$ (i.e., only the final document is considered to be useful). Therefore, ERR is defined as:

$$ERR = \sum_r Pr_{ERR}(r) \frac{1}{r} \; . \tag{4}$$

ERR is known to be a member of the *Normalised Cumulative Utility* (NCU) family [16], which is defined in terms of a stopping probability distribution over ranks ($Pr_{ERR}(r)$ in this case) and the utility at a particular rank ($1/r$ in this case).

As ERR is not normalised, it may be normalised using the aforementioned ideal list. Let $p^*(r)$ denote the stopping probability at rank r in an ideal list, let $Pr^*_{ERR}(r)$ be defined in a way similar to Eq. 3. Normalised ERR at a cutoff l is given by:

$$nERR@l = \frac{\sum_{r=1}^{l} Pr_{ERR}(r)(1/r)}{\sum_{r=1}^{l} Pr^*_{ERR}(r)(1/r)} \; . \tag{5}$$

The primary measure of STC is nERR@10. Note that, when $l = 1$ in Eq. 5,

$$nERR@1 = \frac{Pr_{ERR}(1)}{Pr^*_{ERR}(1)} = \frac{p(1)}{p^*(1)} = \frac{g(1)/2^H}{g^*(1)/2^H} = \frac{g(1)}{g^*(1)} = nG@1 \ . \tag{6}$$

That is, nG@1 can alternatively be referred to as nERR@1.

3.3 P$^+$

P$^+$, proposed at AIRS 2006 [11], is another evaluation measure designed for navigational intents. Like ERR, it is a member of the NCU family. Given a ranked list, let r_p be the rank of the document that has the highest relevance level in that particular list (which may or may not be H, the highest relevance level for the entire test collection) *and* is closest to the top of the list. For example, if the ranked list has L2-relevant documents at ranks 2 and 5, and an L1-relevant document at rank 1, then $r_p = 2$; if the ranked list does not contain any L2-relevant documents but has L1-relevant document at ranks 3 and 5, then $r_p = 3$. The basic assumption behind P$^+$ is that no user will ever go beyond r_p: the *preferred rank*.

P$^+$ assumes that the distribution of users who will stop scanning the ranked list at a particular rank is uniform over all relevant documents at or above r_p. For example, if there is an L1-relevant document at rank 1 and an L2-relevant document at rank $r_p = 2$, then it is assumed that 50 % of users will stop at rank 1, and the other 50 % will stop at rank 2. More generally, let $I(r) = 0$ if the document at rank r is not relevant and $I(r) = 1$ otherwise; the stopping probability at each relevant document at or above r_p is assumed to be[13] $1/\sum_{r=1}^{r_p} I(r)$.

While ERR uses the *reciprocal rank* $(1/r)$ to measure the utility of a ranked list for users who stopped at rank r, P$^+$ employs the *blended ratio BR(r)* just like *Q-measure* [16]:

$$BR(r) = \frac{\sum_{k=1}^{r} I(k) + \sum_{k=1}^{r} g(k)}{r + \sum_{k=1}^{r} g^*(k)} \ . \tag{7}$$

Note that *precision* based on binary relevance is given by $P(r) = \sum_{k=1}^{r} I(k)/r$, while *normalised cumulative gain* [6] based on graded relevance is given by $nCG(r) = \sum_{k=1}^{r} g(k)/\sum_{k=1}^{r} g^*(k)$. $BR(r)$ combines these two measures; the r in the denominator of Eq. 7 discounts documents based on ranks.

Finally, P$^+$ is defined as follows. If the ranked list does not contain any relevant documents, let $P^+ = 0$. Otherwise,

$$P^+ = \sum_r Pr_+(r)BR(r) = \frac{1}{\sum_{r=1}^{r_p} I(r)} \sum_{r=1}^{r_p} I(r)BR(r) \ . \tag{8}$$

---

[13] Note that *Average Precision* and *Q-measure* assume a uniform distribution over *all* relevant documents, so that the stopping probability each relevant document is $1/R$, where R is the total number of relevant documents [16].

Table 2. Six pilot runs used for obtaining $\hat{\sigma}^2$'s, and their mean performances on training data.

Run name	Features used	nERR@10	P^+	nG@1
Run0	Q2P	.5839	.6050	.4015
Run1	Q2C	.6437	.6659	.4637
Run2	Q2P + Q2C	.6908	.7140	.5496
Run3	Q2P + Q2C + TransLM	.6913	.7149	.5318
Run4	Q2P + Q2C + TopicWord	.6866	.7095	.5392
Run5	Q2P + Q2C + TransLM + TopicWord	.6909	.7121	.5363

Here, $Pr_+(r)$ denotes the aforementioned uniform stopping probability distribution over relevant documents ranked at or above rank r_p.

Consider a ranked list that contains one document only. If this document is not relevant, $P^+ = 0$ by definition. If it is relevant, then $r_p = 1$ and $I(1) = 1$, and therefore

$$P^+ = \frac{1}{I(1)}I(1)BR(1) = BR(1) = \frac{I(1) + g(1)}{1 + g^*(1)} = \frac{1 + g(1)}{1 + g^*(1)} , \qquad (9)$$

which is very similar to the definition of nCG@1 (a.k.a. nERR@1). Also note that, regardless of the ranked list size, $P^+ = 1$ iff $r_p = 1$ *and* the top ranked document is one of the most relevant ones for that topic.

4   Experiments

This section reports on how we decided on the topic set size for the STC test topics (i.e., posts) using Sakai's ANOVA-based topic set size design tool [13,14], the STC repository and the training data labels described in Table 1, and the aforementioned three official evaluation measures.

4.1   Pilot Runs

As was mentioned in Sect. 2.3, topic set size design requires an estimate of the population within-system variance for a given evaluation measure. To obtain the variance estimate using Eq. 1, we created a topic-by-system matrix for each of the three evaluation measures using the $n = 225$ training topics from Table 1 and $m = 6$ pilot runs we created. Our pilot runs employ *learning-to-match* and *learning-to-rank* models as described in the aforementioned arxiv paper (see Sect. 1). Table 2 shows the combinations of features used to generate these runs, where the features used are:

Q2P. Query-post similarity based on the vector space model. Here, "query" refers to the new post as an input to an STC system, whereas "post" refers to an old post in the repository. The basic assumption is that if these two posts are similar, then their comments will likely be exchangeable.

Table 3. p-values/effect sizes (ES_{HSD}) for pairwise comparisons of the six runs. p-values smaller than $\alpha = 0.05$ are shown in bold.

(a) nERR@10	Run1	Run2	Run3	Run4	Run5
Run0	**.004**/.3392	**.000**/.6065	**.000**/.6091	**.000**/.5829	**.000**/.6070
Run1	-	**.040**/.2673	**.037**/.2699	.076/.2438	**.040**/.2678
Run2	-	-	1.000/.0026	1.000/.0236	1.000/.0005
Run3	-	-	-	1.000/.0262	1.000/.0021
Run4	-	-	-	-	1.000/.0241
(b) P$^+$	Run1	Run2	Run3	Run4	Run5
Run0	**.006**/.3450	**.000**/.6177	**.000**/.6231	**.000**/.5924	**.000**/.6073
Run1	-	.057/.2727	**.048**/.2781	.108/.2474	.075/.2622
Run2	-	-	1.000/.0054	1.000/.0253	1.000/.0104
Run3	-	-	-	.999/.0307	1.000/.0159
Run4	-	-	-	-	1.000/.0148
(c) nG@1	Run1	Run2	Run3	Run4	Run5
Run0	.194/.3528	**.000**/.8402	**.000**/.7393	**.000**/.7813	**.000**/.7645
Run1	-	**.014**/.4873	.106/.3865	.058/.4285	.071/.4117
Run2	-	-	.987/.1008	.998/.0588	.996/.0756
Run3	-	-	-	1.000/.0420	1.000/.0252
Run4	-	-	-	-	1.000/.0168

Q2C. Query-comment similarity based on the vector space model. Again, "query" refers to the new post, while "comment" refers to one from the repository. The basic assumption is that a good comment contains words that are similar to those in the new post.

TransLM. Translation-based language model for bridging the lexical gap between the query and candidate post-comment pairs, which Q2P and Q2C cannot handle. Word-to-word translation probabilities are estimated so that any word in a post or a comment can be translated with a non-zero probability into a semantically related query word.

TopicWord. Topic word model for estimating the probability that each word in a post or comment is to do with the main topic or theme. Logistic regression with features such as term frequency, inverse document frequency, whether the word is a named entity, and whether the word occurs in the first (last) sentence is employed.

Table 2 also shows the mean performances of these runs for the training data, and Table 3 shows, for each run pair, the p-value obtained with the *randomised Tukey HSD test* for multiple comparison with $B = 5000$ trials using the

Discpower tool[14], as well as the *effect size* ES_{HSD} [12][15]. However, these results should be regarded with a large grain of salt, because (a) the training data labels were contructed based on pooling only three runs and therefore may be highly incomplete and biased; and (b) the new six pilot runs have been tuned with these training data labels. The purpose of these runs in the present study is to estimate the within-system variances rather than performance comparisons. It can be observed, however, that introducing the TopicWord feature may actually hurt the mean performance (compare Run2 and Run4), and that the effect of TransLM is not statistically significant (compare Run2 and Run3, or Run4 and Run 5), even on the training data.

4.2   Topic Set Size Design Results

We created a 225×6 topic-by-system matrix for each of our evaluation measure based on NTCIREVAL, obtained the within-system variances using Eq. 1, and then used Sakai's ANOVA-based Excel tool with $(\alpha, \beta) = (0.05, 0.20)$, i.e., *Cohen's five-eighty convention* [3], which says that a Type I error is four times as serious as a Type II error. Table 4 shows the required topic set sizes given the minimum detectable range $minD = 0.05, \ldots, 0.20$ and the number of systems

Table 4. Topic Set Size Design Results for STC $(\alpha, \beta) = (0.05, 0.20)$.

$minD$	$m = 2$	$m = 5$	$m = 10$	$m = 50$	$m = 100$
	P^+ ($\hat{\sigma}^2 = .0637$)				
0.05	391	604	794	1524	2056
0.10	**98**	152	199	382	515
0.15	44	68	**89**	170	229
0.20	25	39	50	**96**	129
	nERR@10 ($\hat{\sigma}^2 = .0643$)				
0.05	395	609	802	1539	2075
0.10	**99**	153	201	385	519
0.15	45	68	**90**	172	231
0.20	26	39	51	**97**	130
	nG@1 ($\hat{\sigma}^2 = .1515$)				
0.05	928	1434	1888	3625	4889
0.10	233	359	473	907	1223
0.15	104	160	211	403	544
0.20	59	**90**	119	227	306

---

[14] http://research.nii.ac.jp/ntcir/tools/discpower-en.html.

[15] The effect size here is essentially the difference between a system pair as measured in standard deviation units, after removing the between-system and between-topic effects.

to be compared $m = 2, \ldots, 100$ for the three evaluation measures. It can be observed that the within-system variances of nERR@10 and P^+ are very similar, and therefore that the required topic set sizes are also very similar under a given set of statistical requirements $(\alpha, \beta, minD, m)$. For example, if we are to compare $m = 10$ systems using one-way ANOVA and want to guarantee $(\alpha, \beta, minD) = (0.05, 0.20, 0.15)$, that is, if we want to guarantee 80 % *statistical power* at 5 % significance level whenever there is a difference of 0.15 or more between the best and the worst systems, P^+ would require 89 topics, while nERR@10 would require 90 topics. Whereas, note that nG@1 would require as many as 211 topics under the same condition, due to the fact that it is a highly unstable measure.

Based on Table 4, we have decided to create a test set containing 100 posts for STC and release them to participating teams in November 2015. From the same table, the statistical implications of this decision under Cohen's five-eighty convention are as follows:

- If P^+ or nERR@10 is used for evaluation, this test set would achieve a minimum detectable difference of 0.10 for comparing $m = 2$ systems[16];
- If P^+ or nERR@10 is used for evaluation, this test set would achieve a minimum detectable range of 0.15 for comparing $m = 10$ systems; also, this test set would be expected to make the confidence interval width of the difference between any systems be 0.15 or smaller [13,14];
- If P^+ or nERR@10 is used for evaluation, this test set would achieve a minimum detectable range of 0.20 for comparing $m = 50$ systems;
- If nG@1 is used for evaluation, this test set would achieve a minimum detectable range of 0.20 for comparing $m = 5$ systems.

In Table 4, the topic set sizes that correspond to the above discussions are shown in bold. Topic set size design can thus provide justifications for a particular decision on the number of topics included in a new test collection.

Previous work has shown that, from a statistical viewpoint, it is more economical to have many topics with a small number of judgments than to have a small number of topics with many judgments (e.g. [13,14,17,20]). The STC task follows these recommendations and plans to rely on depth-10 pools. As of September 1, we have 29 teams that have signed up for the STC task; if each team submits five runs, we will have 145 runs in total. The pool size will therefore be $145 * 10 = 1,450$ in the worst case (although this will in fact be about several hundreds due to overlaps across runs); hence, if we have 100 test topics (posts), 145,000 comments will have to be assessed in the worst case. The STC organisers have enough budget to hire multiple assessors to judge each comment. We shall report on inter-assessor agreement in our STC overview paper in June 2016.

5   Conclusions

In this study, we applied the ANOVA-based topic set size design technique of Sakai to determine the size of the test set for the NTCIR-12 STC task. Our

---

[16] When $m = 2$, one-way ANOVA is equivalent to the unpaired t-test.

main conclusion is to create 100 test topics, but what distinguishes our work from other tasks with similar topic set sizes is that we know what this topic set size means from a statistical viewpoint for each of our evaluation measures. We also demonstrated that, under the same set of statistical requirements, the topic set sizes required by nERR@10 and P^+ are more or less the same, while nG@1 requires more than twice as many topics. To our knowledge, our task is the first among all efforts at TREC-like evaluation conferences to actually create a new test collection by using this principled approach.

There are a few limitations to the present study. First, our training data labels were devised based on pooling only three runs, which probably means that they are highly incomplete and biased. Our six runs used for estimating the within-system variances of the three evaluation measures were evaluated using the incomplete training labels. The fundamental assumption behind the present study is that the estimates of the within-system variances ($\hat{\sigma}^2$'s) are of reasonable accuracy despite the above limitations. We shall verify whether our $\hat{\sigma}^2$'s are indeed reasonably accurate once we have collected the official STC runs from participants and have completed the contruction of the test data labels. Using the new topic-by-run matrices, where the rows represent 100 new topics and the columns represent the STC participants' runs, we will obtain more accurate estimates of the $\hat{\sigma}^2$ for each evaluation measure. Using these new estimates, we can decide on the topic set sizes for the *next* round of STC. We believe that, in this way, tasks should keep trying to improve the design of their test collections in terms of statistical reliability. Our hope is that the present effort will set a good example for other tasks at TREC-like evaluation conferences.

References

1. Broder, A.: A taxonomy of web search. SIGIR Forum **36**(2), 3–10 (2002)
2. Chapelle, O., Ji, S., Liao, C., Velipasaoglu, E., Lai, L., Wu, S.L.: Intent-based diversification of web search results: metrics and algorithms. Inf. Retrieval **14**(6), 572–592 (2011)
3. Ellis, P.D.: The Essential Guide to Effect Sizes. Cambridge University Press, New York (2010)
4. Higashinaka, R., Kawamae, N., Sadamitsu, K., Minami, Y., Meguro, T., Dohsaka, K., Inagaki, H.: Building a conversational model from two-tweets. In: Proceedings of IEEE ASRU 2011 (2011)
5. Jafarpour, S., Burges, C.J.: Filter, rank and transfer the knowledge: learning to chat. Technical report, MSR-TR-2010-93 (2010)
6. Järvelin, K., Kekäläinen, J.: Cumulated gain-based evaluation of IR techniques. ACM TOIS **20**(4), 422–446 (2002)
7. Lin, J., Efron, M.: Overview of the TREC-2013 microblog track. In: Proceedings of TREC 2013 (2014)
8. Mitamura, T., Shima, H., Sakai, T., Kando, N., Mori, T., Takeda, K., Lin, C.Y., Song, R., Lin, C.J., Lee, C.W.: Overview of the NTCIR-8 ACLIA tasks: advanced cross-lingual information access. In: Proceedings of NTCIR-8, pp. 15–24 (2010)
9. Nagata, Y.: How to Design the Sample Size (in Japanese), Asakura Shoten (2003)

10. Ritter, A., Cherry, C., Dolan, W.B.: Data-driven response generation in social media. Proc. EMNLP **2011**, 583–593 (2011)
11. Sakai, T.: Bootstrap-based comparisons of IR metrics for finding one relevant document. In: Ng, H.T., Leong, M.-K., Kan, M.-Y., Ji, D. (eds.) AIRS 2006. LNCS, vol. 4182, pp. 374–389. Springer, Heidelberg (2006)
12. Sakai, T.: Statistical reform in information retrieval? SIGIR Forum **48**(1), 3–12 (2014)
13. Sakai, T.: Information Access Evaluation Methodology: For the Progress of Search Engines (in Japanese), Coronasha (2015)
14. Sakai, T.: Topic set size design. Information Retrieval Journal (submitted) (2015)
15. Sakai, T., Ishikawa, D., Kando, N., Seki, Y., Kuriyama, K., Lin, C.Y.: Using graded-relevance metrics for evaluating community QA answer selection. Proc. ACM WSDM **2011**, 187–196 (2011)
16. Sakai, T., Robertson, S.: Modelling a user population for designing information retrieval metrics. Proc. EVIA **2008**, 30–41 (2008)
17. Sanderson, M., Zobel, J.: Information retrieval system evaluation: effort, sensitivity, and reliability. Proc. ACM SIGIR **2005**, 162–169 (2005)
18. Shibuki, H., Sakamoto, K., Kano, Y., Mitamura, T., Ishioroshi, M., Itakura, K., Wang, D., Mori, T., Kando, N.: Overview of the NTCIR-11 QA-Lab task. In: Proceedings of NTCIR-11, pp. 518–529 (2014)
19. Voorhees, E.M., Harman, D.K. (eds.): TREC: Experiment and Evaluation in Information Retrieval. The MIT Press, Cambridge (2005)
20. Webber, W., Moffat, A., Zobel, J.: Statistical power in retrieval experimentation. Proc. ACM CIKM **2008**, 571–580 (2008)
21. Weizenbaum, J.: ELIZA - a computer program for the study of natural language communication between man and machine. Commun. ACM **9**(1), 36–45 (1966)

Towards Nuanced System Evaluation Based on Implicit User Expectations

Paul Thomas[1], Peter Bailey[2], Alistair Moffat[3]([⊠]), and Falk Scholer[4]

[1] CSIRO, Canberra, Australia
paul.thomas@csiro.au
[2] Microsoft, Canberra, Australia
pbailey@microsoft.com
[3] The University of Melbourne, Melbourne, Australia
ammoffat@unimelb.edu.au
[4] RMIT University, Melbourne, Australia
falk.scholer@rmit.edu.au

Abstract. Information retrieval systems are often evaluated through the use of effectiveness metrics. In the past, the metrics used have corresponded to fixed models of user behavior, presuming, for example, that the user will view a pre-determined number of items in the search engine results page, or that they have a constant probability of advancing from one item in the result page to the next. Recently, a number of proposals for models of user behavior have emerged that are parameterized in terms of the number of relevant documents (or other material) a user expects to be required to address their information need. That recent work has demonstrated that T, the user's *a priori* utility expectation, is correlated with the underlying nature of the information need; and hence that evaluation metrics should be sensitive to T. Here we examine the relationship between the query the user issues, and their anticipated T, seeking syntactic and other clues to guide the subsequent system evaluation. That is, we wish to develop mechanisms that, based on the query alone, can be used to adjust system evaluations so that the experience of the user of the system is better captured in the system's effectiveness score, and hence can be used as a more refined way of comparing systems. This paper reports on a first round of experimentation, and describes the progress (albeit modest) that we have achieved towards that goal.

Keywords: Retrieval evaluation · User behavior · Search user model

1 Introduction

Information retrieval systems underpin the considerable economic success of the web search industry. Billions of queries per day are processed, with search services possessing a seemingly uncanny ability to identify the page or pages that the user is searching for. A key component of retrieval system development is the use of evaluation processes, in order to measure the quality of the results that

© Springer International Publishing Switzerland 2015
G. Zuccon et al. (Eds.): AIRS 2015, LNCS 9460, pp. 332–344, 2015.
DOI: 10.1007/978-3-319-28940-3_26

are returned. Evaluation options include user focus groups, supervised observational trials, unsupervised trials using query and click logs, and corpus-based batch evaluations.

Central to batch evaluation is the notion of an *effectiveness metric*, a mapping from a search engine result ranking to a numeric score. For example, precision at depth k, denoted Prec@k, scores a ranking by the fraction of the first k documents in it that are deemed to be relevant to the query. Many effectiveness metrics have a corresponding *user model*. For example, if the user always examines the first k documents in the ranking, and forms an opinion of the search service according to the number of those k documents that are relevant, then their *expected utility per document inspected* exactly corresponds to Prec@k. A wide range of more gradual weighted-precision metrics – with corresponding user models – have been described. For example, in the *Rank-Biased Precision* (RBP) effectiveness metric [15], the user is assumed to always examine the first document in the ranking, and then, having viewed the document at depth d, go on to depth $d+1$ with a fixed probability p. A range of other metrics follow similar approaches, including *Expected Reciprocal Rank* (ERR) [9].

Our recent work has argued that rather than a fixed probability p, the user begins their search with an implicit goal of fetching T relevant documents and unconsciously adjusts their "continue to the next document" probability as a function of the depth d in the ranking they have reached, and of the extent of their unfulfilled expectation for relevant documents [5,14]. For example, a user who seeks $T = 1$ relevant documents will end their search more quickly than a user seeking $T = 10$ regardless of whether or not relevant documents are encountered; and will end their search even more quickly if they find a high fraction of relevant documents amongst the first ones they examine. These *adaptive* user models should – all other things being held constant – lead to more realistic system comparisons, and hence better-quality outcomes for search users.

There is, however, the question of somehow knowing what the user's expectation of required relevance is – the quantity denoted as T. In a batch evaluation setting, while curating the creation of a test collection, we can simply ask for this quantity directly from the individual at the point of providing a query in response to some information need. In other evaluation settings, interrogating the user ahead of their search activity may not be desirable, let alone possible. As a first step in resolving that uncertainty, the work presented in this paper explores the extent to which T can be established as a function of attributes associated with the query the user issues. Towards that goal, Sect. 2 describes the user models that we work with; Sect. 3 describes a crowdsourced expectation and query data-gathering exercise, and briefly summarizes the results that have already been achieved using that data; and then Sect. 4 describes the additional analysis carried out for the purposes of this work.

Our findings are mixed. Using a range of query-dependent features such as its frequency, and its length, we are able to provide better prediction of T over the aggregated data than simply taking the majority value of T and using it as a constant prediction. However, to date the gain in prediction accuracy that has been achieved is relatively small – an increase from 29 % to 33 %. While definitely significant, the gain is not substantial, and is rather less than the gain in accuracy resulting from the use of other features including the topic of the query (not normally available in a production system), and the identity of the user that issued the query (which might be available).

2   Background

This section introduces the concepts of search user models, static and adaptive effectiveness metrics, and search task complexity.

User Models: A *search user model* describes the way in which the elements of a ranked list of search results are inspected, and seeks to compute the value of a corresponding effectiveness metric, reflecting the expected rate at which a user gains utility from the system. Gain is a function of relevance, which is assigned to documents by human judges typically using an ordinal scale: for example, a four-level relevance scale might have values *not relevant, marginally relevant, relevant*, and *highly relevant* [19]. Relevance is then transformed into gain, commonly using a linear [10] or exponential function [8].

Eye tracking analysis has shown that, on average, users scan a search results page from top to bottom [11], although there is substantial additional variation and movement between individual result items [21]. As a result, the search user models behind many evaluation metrics such RBP and NDCG [10] incorporate the notion of a *discount*, where relevant items that are returned lower down a ranked list contribute smaller amounts of utility to the user. Utility gain was originally defined in absolute terms [10]; then in terms of expected utility per document inspected [15]; and most recently, in terms of expected utility per second spent searching [17]. User models where gain is based on rank position alone are called *static*, while those that additionally incorporate information about the relevance of the documents that have been seen earlier in the ranking are called *adaptive*.

Adaptive Effectiveness Metrics: We have recently introduced the notion of the user's expected search goal, quantified as a utility estimate T, hypothesizing that the value of T provides guidance as to the user's behavior while they are scanning the results list [14]. In particular, we suggest that "high T" queries involve the inspection of more documents in the result ranking than do "low T" queries, even before any relevance information is taken in to account. The model is adaptive, with searches in which the anticipated utility is accumulated quickly ending earlier than searches in which relatively few relevant documents are encountered. In followup work, we proposed an effectiveness metric "INST", a weighted-precision "expected utility per document inspected" sum defined by the assumption that the user always examines the first document in the ranking, and then continues from depth i to depth $i + 1$ with probability

$$C_{\text{INST}}(i) = \left(\frac{i + T + T_i - 1}{i + T + T_i} \right)^2$$

where T is the user's initial estimate of the number of relevant documents they will find, and T_i is the extent of the relevance found in the first i documents in the ranking. INST brings together a range of desirable attributes in a useful manner: it is adaptive, meaning that for any given value of T, the expected search length is less in rankings with many relevant documents than it is in rankings with few relevant documents; it respects patience, in that the continuation probability $C(i)$ slowly increases towards one, reflecting that a user who has invested heavily in a ranking and is already some way down a list is (on a conditional basis) more likely to continue scanning documents than a user who is still examining documents in the early part of the ranking; and it is not unbounded, since it has a finite expected search depth even on rankings that do not contain any relevant documents at all. As probabilistic limits, the expected search depth on a ranking with no relevant documents is $2T + 0.5$, and on a ranking with nothing but relevant documents, is $T + 0.25$ [5].

Search Task Complexity: Users carry out information seeking tasks for many different reasons. A key characteristic that may vary between tasks is their complexity: consider the difference between trying to find the answer to a short factoid question such as the name of the author of "The Odyssey", versus trying to obtain a deeper understanding of the cultural impact of Homer's work. Kelly et al. [12] propose a hierarchy of complexity of search tasks, based on a taxonomy of learning [4]. In the experiments described below, we consider three of Wu et al.'s cognitive complexity levels: *Remember*, tasks that primarily involve factoid-style answers, similar to recalling knowledge from long-term memory; *Understand*, tasks that involve the construction of meaning, for example through interpreting or exemplifying; and *Analyze*, tasks that involve breaking material into parts, and making overall decisions based on how these facets relate to one another [22]. Section 3 gives examples of information needs in these three categories.

Query Variability: Users typically turn to an information retrieval system with the aim of resolving an information need. A key step of their interaction with the system is to translate this information need into a query; for most users of modern search engines, this typically involves typing a small set of search terms into a text box. However, due to the expressiveness of language, many different queries could be used as instantiations of the same information need. This occurs for example when users refine an initial query as part of the same search session [13]. However, attempts to quantify the impact of query variability as a component of IR system evaluation have been limited. As part of the 1999 TREC-8 Query Track [7], participants were asked to generate alternative query strings for supplied information need statements (called search topics in the TREC framework). The track concluded that query variation can lead to substantial differences in retrieval effectiveness. In recent work, we have gathered variant queries intended to express the same underlying information needs [5]. We make use of these crowdsourced queries to investigate approaches to model a searcher's expected utility, as explained in the next section.

3   A Crowdsourced Experiment

As part of our study into the effects of query variability, we carried out a user experiment and gathered data using crowdsourcing. This section provides a brief overview of that experiment, and describes the data that was collected.

Crowdsourcing: Crowdsourcing is the process of soliciting work from a large group of people (the "crowd") in an online setting. The work is typically advertised through a crowdsourcing platform, such as Amazon Mechanical Turk or CrowdFlower. Internet users can register with the platform, search or browse through a list of available work, and choose whether to participate. The terminology of crowdsourcing tends to vary from platform to platform; in this work we refer to the people who offer their labor through a crowdsourcing platform as "workers", and each discrete unit of work that is carried out is called a "task". In the research world, crowdsourcing has become popular as a method to recruit participants for experiments involving humans responses. As an experimental practice, crowdsourcing has been criticized for reducing the level of control that researchers have over their pool of participants; conversely, proponents of crowdsourcing have highlighted that a more diverse user base is likely to be a positive feature, since prior to crowdsourcing the typical participant pool for human factors research studies consisted of university undergraduate students [1]. In the IR field, initial investigations have suggested that crowdsourcing, with appropriate controls to remove "spam" workers (people who do not take the job seriously, or the activities of automated bots intended to mimic human responses), can be a useful source of participants for user studies, including relevance judging [3,18].

Topics and Backstories: The NIST-sponsored TREC shared tasks have been generating useful search data for more than two decades. The test collections (consisting of sets of topics, documents, and judgments) that have been constructed have become invaluable resources for IR experimentation.[1] Table 1 summarizes the three different collections used in our experimentation, and gives a sample "title" query for each collection, noting that detailed "narrative" and "description" sections are also provided for the R03 and T04 topic statements.

In the case of the R03 and T04 queries, we started with TREC topic descriptions and narratives, and wrote what they called a *backstory* for each one, to personalize and motivate the information need. Backstories were also written based on the Q02 questions. For example, the backstories for the queries shown in Table 1 were:

– *You saw a Discovery Channel show that said that it takes eight minutes for the light from the sun to travel to the earth. You want to find out how far away the sun is in miles or kilometers.*
– *A workmate has been diagnosed with arthritis. You know she struggled once with Lyme disease, from a tick bite. You wonder what evidence there is to support (or refute) a connection between the two.*

---

[1] http://trec.nist.gov.

Table 1. Origins of queries used to create backstories.

Collection	Year	Topics	Example TREC query
Q02	2002	70	1876: how far from the earth is the sun?
R03	2003	60	604: lyme disease arthritis
T04	2004	50	730: gastric bypass complications

– *A surgeon has recently recommended gastric bypass surgery for your overweight uncle. He wants to lose weight, but you would like to help him make an informed decision by alerting him to the possible complications and potential dangers of gastric bypass surgery.*

These three information needs have, respectively, task complexity categories Remember, Analyze, and Understand.[2]

Experimental Process: For each search topic, study participants were first shown the backstory motivating an information need. They were then asked to provide three pieces of information: the total number of useful web pages they thought they would need to look at to answer the information need (T); how many different queries they thought they would need to issue in order to find that number of useful pages (Q); and what their first (written text) query would be when using a search engine to answer the information need. The first two responses were collected using single-selection radio buttons describing numeric ranges, and the third was a free-form text field. The interface, including the range of answer choices, is shown in Fig. 1, using one of the example backstories already introduced. Note the bands on values of T and Q used during the data collection.

The user study was carried out using the CrowdFlower platform.[3] Each task consisted of providing answers in response to five of the 180 different search topics. Users could choose to complete as many units as they wished, providing answers to anywhere between 5 and 180 topics if they wished. Since not all crowdworkers take their tasks seriously, data cleaning was carried out. If any worker entered the same "first query" string for more than one topic, all of their responses were removed from the subsequent analysis. Workers who simply pasted fragments of the topic statements that were deemed nonsensical as their "first query" were also removed. The remaining data consisted of 98 workers who provided 7,969 responses, with a median of 44 responses per topic.

4   Direct Estimates of T and Q

We now describe our detailed investigation of T and Q, making use of the data collected as part of the work described in the previous section.

---

[2] The backstories are available for reuse at DOI 10.4225/08/55D0B6A098248.
[3] http://www.crowdflower.com.

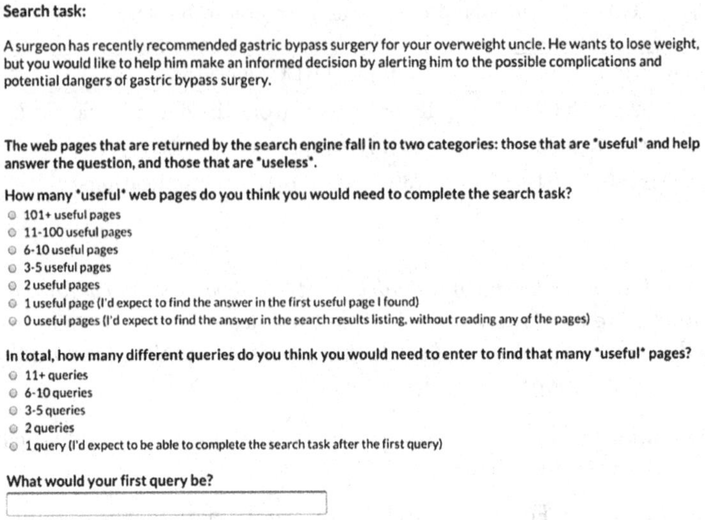

Search task:

A surgeon has recently recommended gastric bypass surgery for your overweight uncle. He wants to lose weight, but you would like to help him make an informed decision by alerting him to the possible complications and potential dangers of gastric bypass surgery.

The web pages that are returned by the search engine fall in to two categories: those that are "useful" and help answer the question, and those that are "useless".

How many "useful" web pages do you think you would need to complete the search task?
- ○ 101+ useful pages
- ○ 11-100 useful pages
- ○ 6-10 useful pages
- ○ 3-5 useful pages
- ○ 2 useful pages
- ○ 1 useful page (I'd expect to find the answer in the first useful page I found)
- ○ 0 useful pages (I'd expect to find the answer in the search results listing, without reading any of the pages)

In total, how many different queries do you think you would need to enter to find that many "useful" pages?
- ○ 11+ queries
- ○ 6-10 queries
- ○ 3-5 queries
- ○ 2 queries
- ○ 1 query (I'd expect to be able to complete the search task after the first query)

What would your first query be?
┌─────────────────────────────────┐
└─────────────────────────────────┘

Fig. 1. Screenshot of the CrowdFlower interface.

Overall Distribution: The great majority of responses were for T of one to ten – that is, people expected to read one to ten relevant documents to answer the information need. There were 436 responses (5.5 %) where $T = 0$ and people expected to answer their need from the result listing alone; 654 (8.2 %) of "11–100 useful pages"; and 62 (0.8 %) of "101+ useful pages". The most common responses were $T = 1$, with 2,329 cases (29 %), and $T = 3$–5, with 1,782 (22 %) of responses.

There was a similar skew in estimates of Q. The most common response was "one query", with 3,521 cases (44 %). At the other end of the range, there were only 600 cases (7.5 %) in the top two categories, where workers expected to need six queries or more. Across all 7,969 responses, T and Q are correlated and participants who expect to need several documents also expect to need several queries to find these documents. Figure 2 plots that correlation. As noted in the caption, this relationship is significant according to Spearman's $\rho = 0.66$, one-sided $p \ll 0.01$. A full 24 % of the workers' responses nominated $T = 1$ and $Q = 1$, indicating that they expected to need a single relevant document, and would find it by issuing a single query.

The Influence of Search Task Complexity on T and Q: To investigate the relationship between the complexity of search tasks and the number of documents and queries that a searcher believes they will need to complete the task, each of the 180 search tasks was categorized into one of three search task complexity levels identified by Wu et al. [22].

The relationships between task complexity and the user estimates of the total number of relevant documents that need to be viewed to fulfill an information need (T), and the number of queries that need to be issued to find those doc-

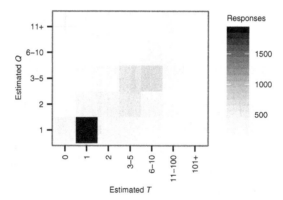

Fig. 2. Correlation between estimates of total number of useful pages (T) and queries (Q) needed to fulfill an information need. The mass along the diagonal bottom-left to top-right demonstrates a positive correlation (Spearman's $\rho = 0.66$, one-sided $p \ll 0.01$).

uments (Q), are depicted in Fig. 3. For both T and Q, a clear trend can be seen whereby a searcher expects to need fewer documents, and fewer queries to find those documents, for tasks of a lower cognitive complexity. The distinction is strongest between the Remember category on the one hand, and the two higher complexity categories on the other; with the Understand and Analyze tasks having somewhat similar distributions. Even so, the highest complexity Analyze category also has the highest overall weight allocated to requiring a larger number of documents and queries.

The demonstration that T and Q are related to the underlying complexity of the information need [5] raises an obvious question: can T and/or Q be estimated or predicted *without* asking the user? To attempt this, and to understand which factors are most relevant, we used cumulative logistic regression[4] and the crowdsourced data to model how T and Q respond to a number of potential explanatory variables. Model selection and parameter estimation were simultaneous, and all data was used to build the model.

- *Per-user and per-topic factors*: We start with two factors which are extrinsic to the query text: the identity of the user (here, CrowdFlower's worker identifier) and the information need or topic (here, the TREC topic number). The first reflects an individual's overall propensity to expect more or fewer interactions, and the second reflects characteristics such as topic complexity. Modern search engines may carry out extensive personalization, and thus can be expected to encode user identity factors relating to long term interaction patterns with documents if they prove useful, within a broader framework [6]. Simpler search engines may not carry out any personalization, and thus would not be able to encode user identity. Even with extensive contextual information modeling, in

---

[4] Cumulative logistic regression – also known as ordinal regression – used R's `ordinal::clm` and `ordinal::step.clm` functions.

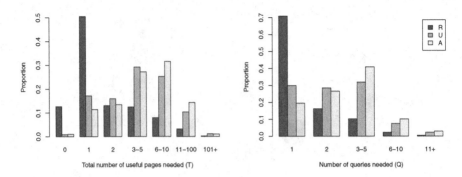

Fig. 3. Total number of useful pages (T) and queries (Q) needed to fulfill an information need of a given complexity level (*R*emember, *U*nderstand, and *A*nalyze).

the general case, an information need (as represented by a topic in our data) is not known to search engines; however topic effects have long been known to be important within test collections. We expect these two factors to explain a lot of variability.

- *Query characteristics*: This includes the number of words and the number of characters – two ways of determining length – and the mean characters per word, which is a surrogate for the complexity of each word. We also investigated the maximum inverse document frequency (IDF) and sum of IDFs assigned by Indri's Okapi BM25 similarity scoring regime, across all terms in the query, as a surrogate for specificity of the query.
- *List characteristics*: Potential explanatory factors include characteristics of the resulting ranked list, which are known after the query is processed but without any further feedback. Here we use the BM25 scores assigned by Indri to the documents ranked 1, 10, and 100, to reflect the quality of the returned list. We also use the ratio of the scores at 10 and 1, 100 and 1, and 100 and 10 (labeled "Indri@10:1", and so on), to reflect the consistency of the scores: a high ratio suggests consistent scores. We also include the number of answers if the query is treated a conjunctive Boolean query – that is, the number of documents containing all query terms.
- *Past behavior*: Finally, we consider two characteristics of past user behavior. Query frequency is drawn from the logs of the Microsoft Bing web search engine, based on usage data within the month June 2015 and is normalized to the range $(0, 1)$, where higher values represent more common queries. Relative click-through rate is drawn from the same logs and represents the ratio of click-through rate for this information need to the global average (so numbers lower than 1 represent fewer clicks). The data is aggregated by topic, and averaged. Missing query-level data is smoothed by assuming the lowest possible query frequency, and the global average click-through rate. This data, of course, is not usually available outside the large search engines.

Fitted Models: Table 2 gives the best models for T and Q, selecting from all of the listed factors except user and topic ID (which may be unknown or unknowable).

Table 2. Significant factors in fitted models for estimates of T and Q. Effect sizes >0 correspond to higher values of T or Q being more likely. All effects significant at $p < 0.05$, Wald test.

Estimating T		Estimating Q	
Factor	Effect	Factor	Effect
Query frequency	−3.46	Query frequency	−4.89
Mean chars/word	0.15	Mean chars/word	0.07
No. characters	0.03	No. characters	0.04
Indri@100:1	0.96	No. words	−0.15
Max. IDF	−0.07	Max. IDF	−0.05
Relative CTR	0.42	Indri@100	0.01
No. words	−0.10	Relative CTR	0.28
Indri@100:10	−0.86		
No. Boolean answers	0.04		
Indri@1	0.02		

Models were learned to minimize the Akaike information criterion (AIC) [2], which combines log-likelihood with a penalty for each factor in the model. Effects are given as modifiers to log-odds, and effects greater than zero mean higher odds for responses further up the scale – that is, positive effects mean higher values of T are more likely as the underlying variable increases. Factors are given in the order selected. For example, query frequency is the best single factor for predicting either T or Q.

The most predictive factor in both models is query frequency: queries which are more common predict lower values of T or Q. This is easy to interpret, since we may expect popular or authoritative pages for common queries; Teevan et al. [20] have also noted that common queries were less ambiguous. The other behavioral feature, relative click-through rate, has positive and moderately large effects in both models: as click-through rate increases, searchers expect to need more interactions. This means searchers are able to predict, at least crudely, how much interaction they will use to address a need.

Query features are the largest set in both cases. Longer query terms predict more interactions, which is possibly explained by longer terms capturing more complex information needs. Queries with more words predict lower T and Q, which is consistent with longer queries being more specific or possibly asking a question in natural language. This relationship between query length and specificity has been noted by Phan et al. [16], and again Teevan et al. noted queries with more characters were less ambiguous (although this did not hold for queries with more words). As the maximum IDF grows – meaning rarer words appear in the query – we see the same effect, with more specific language corresponding to lower T and Q.

List features are more useful predicting T than they are predicting Q. In the case of T, we see four effects at play. As the number of Boolean answers increases, T increases; T also increases as scores are more consistent from ranks 1 to 100.

Table 3. Summary statistics for trained models for T. Lower ΔAIC, and higher accuracy values, indicate better models.

Model	ΔAIC	Accuracy
Majority	9841	29 %
Query characteristics	6565	32 %
List characteristics	6949	30 %
Past behavior	6875	30 %
Best query-only model	6336	33 %
User only	3870	40 %
Topic only	4907	39 %
User and topic	74	51 %
Best model from all factors	0	51 %

The quality of the first result ("Indri@1") also correlates with T. The consistency of scores in the tail ("Indri@100:10") anticorrelates however, and it is not clear why this is the case. We hope to better understand this relationship in future work.

Models Compared: To further examine how well potential explanatory factors predict T, we built six models on the principles above. The first is a simple majority model (intercept only), and always predicts the most common response, $T = 1$. One model each was built with only query, list, and behavior factors; we also include the learned model reported in Table 2, which draws from all of these sets. Finally, we built a model which uses user and topic identity, which can adapt to per-user preferences and per-topic complexity.

Table 3 reports two measures of quality for each model. Accuracy is the number of times the model exactly predicts our users' estimate of T (recall that users chose from seven bands). ΔAIC is the difference in AIC between each model and the best model we have; lower scores are better. Note that AIC (and hence likelihood) improves dramatically over the majority baseline no matter which factors we use, but query characteristics are the most useful group as a whole with AIC improving by 3176. The combined model is better still with a further AIC improvement of 229. However, accuracy is not significantly better and we only see a 4 % improvement at best, from 29 % to 33 %. If we want to use these models to get a point estimate for T, rather than a distribution over possible values, they are not a great improvement.

If models are allowed to make use of the user's identity – here, we have used the CrowdFlower ID – and the topic behind the query, it is possible to do much better (bottom part of Table 3). Using either of these two factors, a further improvement of over 1600 points of AIC and accuracy of 39–40 % is possible; using both, over 6000 AIC points are gained compared to Table 2, and accuracy of 51 %. If all factors are allowed – that is, user and topic factors as well as query, list, and behavior factors – the best model includes user and topic identifiers, number of words, and mean characters per word, and is better by a further 74 points of AIC while still getting 51 % accuracy.

5   Conclusion

Combined with almost certainly unknowable information (the topic), we were able to achieve an accuracy of 51 % in estimating the users' selection of T from one of seven bands. Using only information from the query and documents, which can be reliably calculated by modern search engines, our best effort achieves only 33 %. Whether this degree of accuracy is sufficient to be useful in practice for evaluation of a search system operating over a query population in-the-wild is unknown. We are forced to conclude that there are other significant factors which we have not considered that contribute to the gap in accuracy for our best performing model.

At least two open issues have been identified with the work as described, which we will address as we continue with this project. First, due to choices made at the time the data was collected, the estimates we have been working with make use of bucketed bands of document and query counts, and involve an inevitable loss of accuracy. We hope to repeat the original experiment with a variation allowing users to provide more fine-grained estimates. Second, our modeling has been attempting to predict T for an individual's estimate relating to an information need. However, the current data actually encompasses a distribution over a number (median 44) of estimates for that information need. Instead of attempting to predict a single estimate, and assuming an information need-centric approach to evaluation (rather than a query-centric approach), we might instead predict a distribution. Adaptive effectiveness metrics such as INST may then require modification to encode T as a probabilistic variable rather than as a fixed value.

In the longer term, we are interested in crafting a fully explicated test collection, starting from information needs and encoding user variability, including queries and effort estimates, across a range of task complexities. We hope that this approach of capturing many sources of variability may assist in closing the gap of modeling effort expectations without explicitly needing to ask the user.

Acknowledgments. This work was supported by the Australian Research Council's *Discovery Projects* Scheme (projects DP110101934 and DP140102655). We thank Xiaolu Lu for assistance with the data collection and Bodo von Billerbeck for assistance with query log mining.

References

1. The roar of the crowd. The Economist (2012)
2. Akaike, H.: A new look at the statistical model identification. IEEE Trans. Autom. Control **19**(6), 716–723 (1974)
3. Alonso, O., Mizzaro, S.: Can we get rid of TREC assessors? Using Mechanical Turk for relevance assessment. In: Proceedings of the SIGIR Workshop. Future IR Evaluation, pp. 15–16 (2009)
4. Anderson, L.W., Krathwohl, D.A.: A Taxonomy for Learning, Teaching and Assessing: A Revision of Bloom's Taxonomy of Educational Objectives. Longman, New York (2001)

5. Bailey, P., Moffat, A., Scholer, F., Thomas, P.: User variability and IR system evaluation. In: Proceedings of SIGIR, pp. 625–634 (2015)
6. Bennett, P.N., White, R.W., Chu, W., Dumais, S.T., Bailey, P., Borisyuk, F., Cui, X.: Modeling the impact of short-and long-term behavior on search personalization. In: Proceedings of SIGIR, pp. 185–194 (2012)
7. Buckley, C., Walz, J.: The TREC-8 query track. In: Proceedings of TREC 1999. NIST Special Publication 500–246 (1999)
8. Burges, C., Shaked, T., Renshaw, E., Lazier, A., Deeds, M., Hamilton, N., Hullender, G.: Learning to rank using gradient descent. In: Proceedings of CIKM, pp. 89–96 (2005)
9. Chapelle, O., Metzler, D., Zhang, Y., Grinspan, P.: Expected reciprocal rank for graded relevance. In: Proceedings of CIKM, pp. 621–630 (2009)
10. Järvelin, K., Kekäläinen, J.: Cumulated gain-based evaluation of IR techniques. ACM Trans. Inf. Syst. **20**(4), 422–446 (2002)
11. Joachims, T., Granka, L., Pan, B., Hembrooke, H., Gay, G.: Accurately interpreting clickthrough data as implicit feedback. In: Proceedings of SIGIR, pp. 154–161 (2005)
12. Kelly, D., Arguello, J., Edwards, A., Wu, W.C.: Development and evaluation of search tasks for IIR experiments using a cognitive complexity framework. In: Proceeding of ICTIR (2015)
13. Lin, S.J., Belkin, N.: Validation of a model of information seeking over multiple search sessions. J. Am. Soc. Inf. Sci. Technol. **56**(4), 393–415 (2005)
14. Moffat, A., Thomas, P., Scholer, F.: Users versus models: what observation tells us about effectiveness metrics. In: Proceedings of CIKM, pp. 659–668 (2013)
15. Moffat, A., Zobel, J.: Rank-biased precision for measurement of retrieval effectiveness. ACM Trans. Inf. Syst. **27**(1), 2:1–2:27 (2008)
16. Phan, N., Bailey, P., Wilkinson, R.: Understanding the relationship of information need specificity to search query length. In: Proceedings of SIGIR, pp. 709–710 (2007)
17. Smucker, M.D., Clarke, C.L.A.: Time-based calibration of effectiveness measures. In: Proceedings of SIGIR, pp. 95–104 (2012)
18. Smucker, M., Kazai, G., Lease, M.: The TREC-12 crowdsourcing track. In: Proceedings of TREC 2012. NIST Special Publication 500–298 (2012)
19. Sormunen, E.: Liberal relevance criteria of TREC: counting on negligible documents? In: Proceedings of SIGIR, pp. 324–330 (2002)
20. Teevan, J., Dumais, S.T., Liebling, D.J.: To personalize or not to personalize: modeling queries with variation in user intent. In: Proceedings of SIGIR, pp. 163–170 (2008)
21. Thomas, P., Scholer, F., Moffat, A.: What users do: the eyes have it. In: Banchs, R.E., Silvestri, F., Liu, T.-Y., Zhang, M., Gao, S., Lang, J. (eds.) AIRS 2013. LNCS, vol. 8281, pp. 416–427. Springer, Heidelberg (2013)
22. Wu, W.C., Kelly, D., Edwards, A., Arguello, J.: Grannies, tanning beds, tattoos and NASCAR: evaluation of search tasks with varying levels of cognitive complexity. In: Proceedings of IIiX, pp. 254–257 (2012)

Social Media and Recommendation

A Study of Visual and Semantic Similarity for Social Image Search Recommendation

Yangjie Yao and Aixin Sun[✉]

School of Computer Engineering, Nanyang Technological University,
Singapore, Singapore
yyao002@e.ntu.edu.sg, axsun@ntu.edu.sg

Abstract. Partially due to the short and ambiguous keyword queries, many image search engines group search results into conceptual image clusters to minimize the chance of completely missing user search intent. Very often, a small subset of image clusters in search is relevant to user's search intent. However, existing search engines do not support further exploration once a user has located the image cluster(s) that interest her. Similar to the problem of finding similar images of a given image, in this paper, we study the problem of "finding similar image clusters of a given image cluster". We study this problem in the context of socially annotated images (*e.g.*, images annotated with tags in Flickr). Each image cluster is then represented in two feature spaces: the *visual feature space* to describe the visual characteristics of the images in the image clusters; and the *semantic feature space* to describe an image cluster based on the tags of its member images. Two measures named *relatedness* and *diversity* are proposed to evaluate the effectiveness of the visual and semantic similarities in image cluster recommendation. Our experimental results show that both visual and semantic similarities should be considered in image cluster recommendation to support search result exploration. We also note that using visual similarity leads to more diversified recommendations while the semantic similarity recommends conceptually more related image clusters.

Keywords: Image search · Image cluster · Social image · Image concepts · Concept relevance · Flickr · Tag

1 Introduction

The popularity of social image sharing platforms (*e.g.*, Flickr and instagram) leads to a large number of images accessible online. More importantly, many of these images are annotated by their owners or viewers using freely chosen keywords (also known as tags) for sharing or self-referencing among many other purposes. Many tags refer to high level concepts (*e.g.*, scene, object, and opinion) which bridges the semantic gap between the low-level visual content of an image and the high-level semantic meaning perceived by human being from the image. The availability of such socially tagged images makes it possible to search for

© Springer International Publishing Switzerland 2015
G. Zuccon et al. (Eds.): AIRS 2015, LNCS 9460, pp. 347–357, 2015.
DOI: 10.1007/978-3-319-28940-3_27

(a) Egyptian pyramids   (b) Louvre Pyramid

(c) Rock stone   (d) Rock music

Fig. 1. Example concepts of "pyramid" and "rock"

images that best match keyword queries through an interface similar to general Web search engines (*i.e.,* Tag-based Image Retrieval or TAGIR for short).

TAGIR is challenging and has attracted attention from both academic and industry. On the one hand, the tags are provided by common users from uncontrolled vocabulary for different purposes of tagging and with different understandings of the relevance between an image and a tag, or even different understandings of the semantic meaning of a tag. On the other hand, the keyword queries given by users for image search are usually very short with average 2.2 tags for each search [6]. One approach to partially addressing the challenge is to group the search results into conceptual clusters. The grouping may be based on the knowledge embedded in the social tagging or other knowledge sources. For example, Fig. 1 gives example outputs for two keyword queries "pyramid" and "rock" respectively. For the former, image results for Egyptian pyramid[1] and image results for Louvre Pyramid[2] are grouped into two concepts to best match the possible search intent. For the latter, image results are grouped into the concept of rock stone and rock music respectively because of the different semantics of the word "rock". The search results are obtained from a Concept-Aware Social Image Search system named CASIS [7]. Other example image search systems that summarize search results into conceptual clusters include Flickr tag cluster, Google Image search by subject, and Bing image search. Figure 2 captures the search results of query "rock" from Bing image search[3]. Observe that the top row of the search results enumerates the (sub-)concepts of rock such as

---

[1] http://en.wikipedia.org/wiki/Egyptian_pyramids.

[2] http://en.wikipedia.org/wiki/Louvre_Pyramid.

[3] http://www.bing.com/images/search?q=rock.

Fig. 2. Search result of "rock" from Bing image

"rocks and minerals" and "rock music", and the column on the right lists the "related topics" of the search query.

The cluster-based image search result presentation enables quick selection of images matching user's search intent. However, existing search engines do not support further exploration of search results following user's search intent (which becomes apparent once user clicks the image cluster of her choice). Given a user is interested in an image cluster, for example "rock music", what other image clusters shall be recommended to the user to view?

In this paper, we are interested in this "what to browse next?" problem. This problem is analogous to the problem of finding/recommending similar images for a given image. However, an image cluster contains much richer information than any single image. The key research issues include: (i) generating candidate image clusters for recommendation, (ii) representing image clusters, (iii) computing similarity or relatedness between two image concept clusters, and (iv) ranking and presenting the recommended image clusters. Among these research issues, we mainly focus on image cluster representation and similarity computation between image clusters. Note that, this problem is different from the problem of finding "related topics" of the original query (see the right column of Fig. 2). In the latter, the user has not reveal her search intent and the related topics are recommended purely based on the search query (*e.g.,* "rock" in Fig. 2). In our problem definition, the image clusters to be recommended are based on the image cluster from the search results that is selected by the user.

An image concept cluster is a set of images each of which can be represented by its low-level visual features and is annotated by tags. Therefore, an image cluster can be represented in two feature spaces, one for low-level visual features and the other for textual keywords. For the former, we first represent image concepts using a vector of concept-specific visual representativeness values. For the latter, an image cluster is represented by the tags derived from all its member images. We further consider the overlaps between two image clusters (*i.e.,* images that are contained in both clusters) in image similarity computation. Maximal Marginal Relevance (MMR) algorithm is adopted in the ranking process. In our experiments conducted on NUS-WIDE dataset, we evaluated the impact of using visual and semantic features in image concept cluster recommendation and analyzed the results.

We believe our findings will pave better understanding on the problem of concept-based image search. Besides social image search, the concept of using multiple similarities for search result recommendation can be applied to searching other types of domain-specific resources (*e.g.*, questions and answers in healthcare).

The rest of the paper is organized as follows. In Sect. 2, we survey the related studies. The image cluster recommendation is detailed in Sect. 3. Section 4 reports the experimental setting, evaluation criteria and the experimental results. Finally, Sect. 5 concludes this paper. Next, we review the related work in concept cluster generation and the study in visual representativeness.

2   Related Work

Clustering Image Search Results. There are two approaches for grouping images into conceptual clusters when both visual content and tag are considered in the clustering process, namely *early fusion* [9] and *late fusion* [4]. The former exploits tags and visual content of images jointly in the clustering process and the latter considers one feature space and then the other. There are also approaches which utilize one type of the features only, *e.g.*, tag. In [7], the Concept-Aware Social Image Search (CASIS) system detects concepts based on tag co-occurrences using graph clustering algorithm. More specifically, given a query, CASIS retrieves the best matching images from the image collection and then constructs tag co-occurrence graph based on the frequent tags associated with images in the result set. Graph clustering algorithm (*e.g.*, BorderFlow) is then employed to partition the graph into concepts, each of which is represented by a small number of representative tags. The tags describing the two concepts shown in Fig. 1(a) and (b) are {*cairo, egypt, giza, sphinx*} and {*architecture, france, louvre, night, paris*} respectively. Images are then retrieved using the representative tags for each concept as shown in Fig. 1. In this study, we use CASIS as our image search engine to get concepts and the corresponding image clusters as shown in Fig. 1.

Visual-Representativeness. To recommend visually related image clusters, the best way is to find the visual characteristics of the image cluster, *e.g.*, in terms of color, shape, texture, or others. For instance, the two image clusters in Fig. 1(a) and (b) share architectures in similar shapes. However, extracting effective and representative low-level visual features remains a challenging research problem [3]. In [5], the notion of *visual-representativeness* is proposed to evaluate the effectiveness of a tag in describing the visual content of its annotated images. The main idea is to consider whether the set of images associated with a tag expresses visual cohesiveness compared to a similar sized set of images randomly drawn from a large image collection. In the following, we briefly review the centroid-based separation and cohesion measures.

Let $D_c = \{x_1, x_2, \ldots, x_n\}$ be a cluster of n images and $\mathcal{D}$ be the entire image collection. The centroid-based separation measure is defined by the following equation:

$$\Psi_{cent}(D_c, \mathcal{D}) = dist(Cent(D_c), Cent(\mathcal{D})) \tag{1}$$

where $Cent(D_c)$ is the centroid of D_c and $dist(\cdot,\cdot)$ is a distance function between the two centroids represented by visual feature vectors. A larger centroid-based separation means that the image cluster has its visual feature vector more distinguished from the image collection. The centroid-based cohesion measure, on the other hand, reflects whether the images in a cluster are visually cohesive, defined in the following equation.

$$\Phi_{cent}(D_c, D_c) = \frac{1}{|D_c|} \sum_{x_i \in D_c} dist(x_i, Cent(D_c)) \tag{2}$$

Each image cluster in [5] is defined based on a tag (*i.e.*, the set of images annotated with the tag) so as to evaluate the visual-representativeness of a tag.

In this paper, we extend the idea of tag visual-representativeness to evaluate whether an image cluster has certain visual characteristics by computing the separation (and cohesion) measures using different types of visual features including color, textual, edge and other features.

3 Image Cluster Recommendation

In this section, we detail image cluster recommendation. Given an image cluster $D_c = \{x_1, x_2, \ldots x_n\}$ resulted from a keyword query search, our task is to recommend image clusters ranked by their similarity (or relatedness) to D_c in descending order. Note that, an image cluster D_r could be recommended because images in D_r demonstrate similar visual characteristics as D_c in *any* of the visual feature space include color, shape, texture, contained object or others, *e.g.*, image cluster "sunset" or "beach" can be recommended to the query cluster "sunrise". A cluster may also be recommended because of semantic relation, *e.g.*, image cluster of "apple" or "strawberry" maybe recommended to query cluster "banana" for all being fruits. Therefore, both visual similarity and semantic similarity need to be considered in the evaluating the similarity between two image clusters.

3.1 Similarity Between Image Clusters

We consider two types of similarities, namely *visual similarity* and *semantic similarity*.

- **Visual Similarity.** We consider 6 types of commonly used low-level visual features and compute centroid-based separation and centroid-based cohesion on each type of visual features. As the result, a 12-dimensional feature vector is obtained to describe each image cluster. The 6 types of features include 64-D color histogram (LAB), 144-D color auto-correlogram (HSV), 73-D edge direction histogram, 128-D wavelet texture, 225-D block-wise color moments (LAB), and 500-D bag of visual words (SIFT) features[4]. Manhattan distance is used in Eqs. 1 and 2 as the distance function. The visual similarity between

---

[4] See [2] for detailed description of these 6 types of low-level features.

two image clusters is the cosine similarity calculated on the two 12-dimensional feature vectors.

Note that, we use the visual-representativeness measures instead of using directly these 6 low-level features for capturing the visual characteristics of image clusters. For instance, if one cluster demonstrates strong appearance of red color (*e.g.,* "apple") and another cluster demonstrates strong appearance of yellow color (*e.g.,* "banana"), we consider these two clusters both demonstrate strong characteristics in terms of color. Using visual-representativeness will therefore be able to recommend this kind of image clusters. On the other hand, two clusters having similar low-level visual features will have similar visual-representativeness values by definition.

– **Semantic Similarity.** The semantic similarity between two image clusters is computed based on the tag distributions of the image clusters. More specifically, we consider all tags associated with all images in an image cluster. A tag feature vector is built consisting of all unique tags used to annotate any image in the cluster and each tag is weighted by the number of times it is used to annotate the images in the cluster. The semantic similarity between two image clusters is the cosine similarity of the tag feature vectors.

With visual similarity and semantic similarity, we are able to get the image clusters that are most similar to the query cluster for recommendation. However, it is observed that two image clusters may share a large number of images, hence the corresponding similarity between them becomes very large regardless of in visual or in semantic feature space. The recommendation becomes meaningless if redundancy is not considered among image clusters. Thus, we reduce the similarity between concepts if they share a large number of identical images. The ratio of shared identical images is calculated using Jaccard coefficient between the two image clusters.

To summarize, the similarity between image clusters D_c and D'_c is computed by the following equation:

$$Sim(D_c, D'_c) = \sigma \Upsilon_v(D_c, D'_c) + (1 - \sigma)\Upsilon_s(D_c, D'_c) - J(D_c, D'_c) \qquad (3)$$

In this equation, $\Upsilon_v(\cdot, \cdot)$ and $\Upsilon_s(\cdot, \cdot)$ denote visual similarity and semantic similarity respectively; σ is a parameter control the weight of two similarity; and $J(\cdot, \cdot)$ denotes the Jaccard coefficient.

3.2   Recommendation

The most straightforward approach for recommendation for image cluster D_c is to rank the similar image clusters in descending order based on the similarity defined in Eq. 3. Then top ranked clusters are recommended depending on the number of clusters needed in recommendation. However, the recommended image clusters may contain many overlaps with image duplications[5]. We employ

---

[5] Note that in Eq. 3, the duplicated images are taken care of between a cluster to be recommended and the query image cluster. However, the image overlap between two recommended clusters are not considered.

Maximal Marginal Relevance (MMR) that is commonly used in summarization generation to avoid such redundancy [1]. MMR takes both relevance and novelty into account. An image cluster is recommended if it is similar to the query image cluster and novel with respect to the existing clusters that have already been recommended.

$$Score(D'_c) = \lambda(Sim(D_c, D'_c) - (1 - \lambda)max_{D''_c \in S}Sim(D'_c, D''_c)) \qquad (4)$$

More specifically, let D_c be the query image cluster and S be the set of clusters that have already been recommended. The score of the next image cluster D'_c to be recommended is computed using Eq. 4 where $Sim(D_c, D'_c)$ evaluates the similarity between D'_c and D_c and $max_{D''_c \in S}Sim(D'_c, D''_c)$ is the maximum similarity between D'_c with an image cluster that has been recommended. Parameter λ balances the two similarity values.

4 Experiments

4.1 Dataset

The NUS-WIDE dataset [2] is used in our experiment. The dataset consists of 269,648 socially tagged images from Flickr, annotated by more than 425,000 distinct tags.

Query and Candidate Image Clusters. In this dataset, 81 concepts (each of which corresponds to a tag) have been manually annotated, including 31 concepts for *object* and 33 concepts for *scene*. We randomly selected 30 tags as queries whose image clusters are query clusters for image cluster recommendation. To get the candidate image clusters for recommendation, we used the 2569 popular tags as queries to get all their image clusters as candidate clusters. Each of the 2569 tags has been used to annotate at least 1 % (or 270) images in the dataset. Next, we describe how to obtain image clusters using CASIS.

Discussed in Sect. 2, given a query, CASIS constructs tag co-occurrence graph based on the frequent tags associated with resultant images of a keyword query. The tag co-occurrence graph is then cluster into tag concepts. Each tag concept is a collection of tags, *e.g.*, {*cairo, egypt, giza, sphinx*} for the Egyptian pyramid concept in Fig. 1(a). For the 30 tags to be used in testing, 59 tag concepts are obtained; and for the 2569 popular tags, 4660 tag concepts are obtained.

Table 1. List of 30 randomly selected tags

Category	Randomly selected tags as queries
Object	Animal, bear, birds, boats, book, cars, cat, computer, coral, cow, dog, elk, fish, flags, flowers
Scene	Airport, beach, bridge, castle, cityscape, clouds, fire, frost, garden, glacier, grass, house, lake, moon, mountains

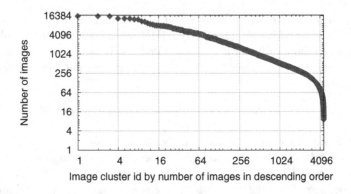

Fig. 3. Number of images in the image clusters

We then retrieve the images for each tag concept $T_c = \{t_1, t_2, ..., t_m\}$ where t_i is a tag. It is reported that for multi-tag queries (*e.g.*, a query contains multiple tags), binary match gives good search results [6]. We therefore retrieve images that are annotated by all tags in T_c. However, depending on the number of tags in a tag concept, there might be limited number of images containing all its tags. We then take images annotated with any combination of $m - 1$ tags among m tags in T_c. This process continues till there is no tags left in each subset. As the result, the retrieved images are ranked by their matching scores to the tag concepts based on the number of matching tags in descending order. Figure 3 reports the number of images in the resultant image clusters for the 4660 tag concepts from the popular tags. In our experiments, maximum top-500 images in each image cluster are considered and the image clusters with fewer than 100 images are ignored.

4.2   Evaluation Criteria

We consider two aspects in evaluating the effectiveness of a recommendation, namely *Relatedness* and *Diversity*.

Relatedness is the number of the recommended image clusters that are related to the query image cluster. For each query cluster, the top-5 recommended image clusters are manually evaluated. The range of relatedness measure is therefore 0 to 5 depending on the number of related image clusters towards the query image cluster.

Diversity reflects the differences among the recommended image clusters, which is the number of distinct visual topics contained in the recommended image sets against the query image cluster. Similarly, for each query cluster, the value for diversity ranges from 0 to 5. If the value is 0, then all the 5 recommended image clusters are redundant with respect to the query image cluster. A value of 5 means that the five recommended image clusters cover different topics and are distinct from the query image cluster.

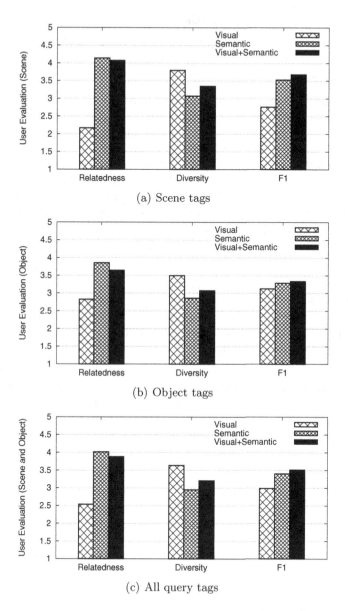

(a) Scene tags

(b) Object tags

(c) All query tags

Fig. 4. Effectiveness of using visual and semantic similarity on scene query tags, object query tags, and all query tags

Because relatedness and diversity cover two different aspects, we adopt F-measure as the overall recommendation effectiveness for a query image cluster D_c.

$$F_1(D_c) = \frac{2 * Relatedness(D_c) * Diversity(D_c)}{Relatedness(D_c) + Diversity(D_c)} \tag{5}$$

4.3 Evaluation Results

For each of the 59 query image clusters generated by the 30 randomly selected query tags (see Table 1), we evaluate the top-5 image clusters recommended from the more than 4000 candidate image clusters. Throughout the evaluation parameter λ is set to 0.7 in Eq. 4. We evaluate three versions of Eq. 3 by setting σ to be 1, 0, and 0.5 respectively. That is, we evaluate the impact of considering visual similarity, semantic similarity, and a combined visual and semantic similarity with equal importance. The evaluation results of relatedness, diversity, and F_1 on the tags of *scene* category, *object* category, and both categories are reported in Fig. 4(a), (b) and (c) respectively.

Relatedness: Shown in Fig. 4, visual similarity alone performs poorly on tags of scene category, and slightly better on tags of object category. One possible reason is that images in scene category tend to be more diverse visually, leading to less obvious visual characteristics than images in object category. Semantic similarity achieves much better values for relatedness measure. Combining both visual and semantic similarity does not give better results than semantic similarity alone, although the difference is marginal.

Diversity: Considering semantic similarity alone, however, results in poorer diversity, show in Fig. 4, for tags in both scene and object categories. One possible reason is that users tend to choose general and ambiguous tags when annotating images in order to minimize their efforts in choosing appropriate words [8]. The general and ambiguous tags make the semantic similarity less effective in identifying different image clusters, based on the noisy tag annotation alone. As expected, combining both visual and semantic similarity, better diversity is achieved.

F-Measure: Reflected by the F-measure which considers both relatedness and diversity with equal importance, both visual and semantic similarities are essential in image cluster recommendation. Overall, the poorer performance by using visual similarity alone against semantic similarity probably has its root in the ineffectiveness of existing visual features in representing the visual content of the images [3].

5   Conclusion

Grouping image search results into conceptual clusters may greatly improves user experiences in image search. However, existing image search engines do not support concept-based image exploration to let users browse the related image clusters of any image cluster that matches user search intent. In this paper, we take the first step to evaluate two different similarities between image clusters from visual aspect and semantic aspect. We conducted manual evaluation using 59 image clusters as query concepts to search for the most relevant image clusters from more than 4000 candidates. Using the two proposed measures, relatedness and diversity, we show that both visual and semantic similarities are essential for recommending image clusters. Our results also show that visual similarity is relatively weak in finding related image clusters. This calls for more study on better ways of representing image clusters in visual space.

Acknowledgements. This work was partially supported by Singapore MOE AcRF Tier-1 Grant RG142/14.

References

1. Carbonell, J., Goldstein, J.: The use of MMR, diversity-based reranking for reordering documents and producing summaries. In: SIGIR, pp. 335–336. ACM (1998)
2. Chua, T.-S., Tang, J., Hong, R., Li, H., Luo, Z., Zheng, Y.: NUS-WIDE: a real-world web image database from national university of singapore. In: CIVR, pp. 48:1–48:9. ACM (2009)
3. Jain, R., Sinha, P.: Content without context is meaningless. In: Multimedia, pp. 1259–1268. ACM (2010)
4. Moëllic, P.-A., Haugeard, J.-E., Pitel, G.: Image clustering based on a shared nearest neighbors approach for tagged collections. In: CIVR, pp. 269–278. ACM (2008)
5. Sun, A., Bhowmick, S.S.: Quantifying tag representativeness of visual content of social images. In: Multimedia, pp. 471–480. ACM (2010)
6. Sun, A., Bhowmick, S.S., Nguyen, K.T.N., Bai, G.: Tag-based social image retrieval: an empirical evaluation. J. Am. Soc. Inf. Sci. Technol. (JASIST) **62**(12), 2364–2381 (2011)
7. Truong, B.Q., Sun, A., Bhowmick, S.S.: Casis: a system for concept-aware social image search. In: WWW, pp. 425–428. ACM (2012)
8. Wu, L., Jin, R., Jain, A.K.: Tag completion for image retrieval. IEEE PAMI **35**, 716–727 (2013)
9. Xu, H., Wang, J., Hua, X.-S., Li, S.: Hybrid image summarization. In: Multimedia, pp. 1217–1220. ACM (2011)

Company Name Disambiguation in Tweets:
A Two-Step Filtering Approach

M. Atif Qureshi[1,2]([⊠]), Arjumand Younus[1,2], Colm O'Riordan[1],
and Gabriella Pasi[2]

[1] Computational Intelligence Research Group, Information Technology,
National University of Ireland, Galway, Ireland
{muhammad.qureshi,arjumand.younus,colm.oriordan}@nuigalway.ie
[2] Information Retrieval Lab, Informatics, Systems and Communication,
University of Milan Bicocca, Milan, Italy
pasi@disco.unimib.it

Abstract. Using Twitter as an effective marketing tool has become a
gold mine for companies interested in their online reputation. A quite significant research challenge related to the above issue is to disambiguate
tweets with respect to company names. In fact, finding if a particular
tweet is relevant or irrelevant to a company is an important task not
satisfactorily solved yet; to address this issue in this paper we propose a
Wikipedia-based two-step filtering algorithm. As opposed to most other
methods, the proposed approach is fully automatic and does not rely on
hand-coded rules. The first step is a precision-oriented pass that uses
Wikipedia as an external knowledge source to extract pertinent terms
and phrases from certain parts of company Wikipedia pages, and use
these as weighted filters to identify tweets about a given company. The
second pass expands the first to increase recall by including more terms
from URLs in tweets, Twitter user profile information and hashtags.
The approach is evaluated on a CLEF lab dataset, showing good performance - especially for English tweets.

1 Introduction

Online reputation management is a task that basically involves "monitoring and
handling the public image of entities (people, products, organizations, companies
or brands) on the Web" [7]. On today's World Wide Web Twitter has assumed
the role of an effective marketing platform with almost all the major companies maintaining their Twitter accounts. On the other hand, Twitter users often
express their opinions about companies via short Twitter messages (around 140-
character long) called tweets. Companies are highly interested in monitoring
their online reputation; this, however involves the significant challenge of disam-
biguating company names in texts. This task becomes even more challenging if
company names have to be identified in tweets due to their short length, to the
huge amount of noise in tweets, and to the lack of context that could help in
company name disambiguation [6].

© Springer International Publishing Switzerland 2015
G. Zuccon et al. (Eds.): AIRS 2015, LNCS 9460, pp. 358–365, 2015.
DOI: 10.1007/978-3-319-28940-3_28

This paper presents and discusses a technique that addresses the above challenges related to company name disambiguation in tweets; while this technique has recently been evaluated in the context of the RepLab task at CLEF 2012, its formal definition is presented for the first time in this paper. The task provided a set of companies and for each company a set of tweets, a subset of which was relevant to the company, and a subset was irrelevant.

The proposed technique consists of a two-step filtering algorithm that makes use of Wikipedia as an external knowledge resource in the first step, and of a concept term score propagation mechanism in the second step. The first step is precision-oriented where the aim is to reduce as much as possible the noisy tweets to a minimum. The second step has been defined to enhance recall via a score propagation technique, and through the utilization of other knowledge resources. Our technique has shown high accuracy figures over the given dataset, and it was classified as the second best algorithm.

2 Related Work

There has been an increasing interest in research on applying natural language processing techniques to tweets over the past few years. However, in spite of the immense significance of extracting commercially useful information from tweets, the amount of research dedicated to company name disambiguation in tweets is very limited. The only two serious efforts which have been undertaken to stimulate this research task are represented by the WePS online reputation management evaluation campaign at CLEF 2010 [1], and by the RepLab online reputation management evaluation campaign at CLEF 2012 [2]. The best two teams in the WePS online reputation management evaluation campaign were LSIR-EPFL [9] and ITC-UT [10]. The LSIR-EPFL system builds profiles for each company relying on external resources such as WordNet or the company homepage in addition to a manual list of keywords for the company and the most frequent unrelated senses for the company name. The profiles are then used for extraction of tweet-specific features for use in an SVM classifier. The ITC-UT system is based on a two-step algorithm. In the first step, the algorithm categorizes queries by predicting the class of each company ("organization-like names" or "general-word like names") using a Naive Bayes classifier with six binary features (for example, is the query an acronym?, is the query an entry of a dictionary? etc.). They use thresholds manually set by looking at the training data results for this categorization. The second step consists in categorizing the tweets using a set of heuristics.

Despite showing promising results, the two systems LSIR-EPFL and ITC-UT indicate heavy reliance on manual selection of both terms and thresholds for the company name disambiguation task. Moreover, their reliance on a large amount of training data makes them infeasible for real-world "company name disambiguation" tasks, and this is particularly true for tweets where obtaining training data is extremely hard. This was particularly evidenced during the RepLab 2012 online reputation management evaluation campaign, where even the best performing team relied on hand-coded rules [2] for the filtering task.

Instead of relying on semi-automatic and supervised methods, we have defined a completely automatic two-step algorithm for this task that relies on Wikipedia as an external knowledge resource of evidence for the first step, and on a score propagation mechanism for the second step.

3   Methodology

This section describes the proposed filtering method in detail. We first explain how we make use of Wikipedia as an external knowledge resource, and how only portions of a company's Wikipedia page are used. Next we explain the two steps of our algorithm.

3.1   Wikipedia as an External Knowledge Resource

Recently researchers have begun to make use of Wikipedia as an external knowledge resource for understanding social media content [5], the work by Meij et al. [6] is particularly significant with respect to linking tweets with concepts[1]. As the authors point out, a simple matching between terms in tweets and in Wikipedia text would produce a significant amount of irrelevant and noisy concept terms, and that this noise can be removed on either the Wikipedia or on the textual side. In line with this argument and on the basis of the intuition that the Wikipedia page of a company contains significant information about the company in certain portions of the Wikipedia article (i.e., concept terms exist in some portions of text), we perform the cleaning on the Wikipedia side as follows:

- The text inside the category information within a Wikipedia article contains significant information about a company. We make use of this text by splitting it into single terms, and by using these single terms as concept terms of a company to be matched with the tweets' dataset.
- The information inside Wikipedia infoboxes is highly significant with respect to any Wikipedia entity [4,8], and hence we make use of this information. For example, let's say on the Wikipedia page of the company 'Apple Inc.', we have the following information in the infobox:
  1. Founder(s): Steve Jobs, Steve Wozniak, Ronald Wayne
  2. Industry: Computer hardware, Computer software, Consumer electronics, Digital distribution

We extract Steve Jobs, Steve Wozniak, Ronald Wayne, Computer hardware, Computer software, Consumer electronics and Digital distribution from the Wikipedia infoboxes. These are then split into single terms and used as concept terms of a company to be matched with the tweets' dataset.

---

[1] Meij et al. in [6] take a concept to be any item that has a unique and unambiguous entry in Wikipedia.

– We parse the paragraphs in Wikipedia and apply POS tagging to these paragraphs. After the application of POS tagging, we extract significant unigrams, bigrams and trigrams for proper nouns (NNP and NNPS). It is important to note that a bigram or trigram will only be extracted if all of its terms are tagged as a proper noun e.g., on the Wikipedia page of the company 'Apple Inc.', the bigram "Tim Cook" is extracted while the bigram "unveiled iPad" is not extracted (this is because POS tag of unveiled does not correspond to that of a proper noun but to that of a verb). After extraction, the significant unigrams, bigrams and trigrams are split into single terms to serve as concept terms for linking with the tweets.

3.2   First Pass

We collect all concept terms extracted from Wikipedia (through the method outlined in Sect. 3.1) adding them to the corpus of *wiki_terms_to_check*. Furthermore, corresponding to every term in *wiki_terms_to_check* we store the inverse-document frequency of that term (note that each phrase that we extract from Wikipedia is treated as a separate document when computing the idf scores; as an example the phrases "computer hardware" and "computer software" from Sect. 3.1 would be treated as two separate documents with the term "computer" having a document frequency of two). The computed idf score is referred to as weight of a term. We check for the occurrence of these concept terms in the tweets and number of occurrences per term is multiplied by the weight of that particular concept term to constitute a score for the tweet. Tweets that have score above a certain threshold are considered to be relevant.

3.3   Second Pass

The second pass makes use of the idea of concept term score propagation in order to discover more tweets relevant to a particular company i.e., to increase the recall. The score propagation technique rests upon the intuition that terms co-located with significant concept terms may have some relevance to that concept. The scores for concept terms in a relevant tweet obtained from the first pass are redistributed among co-located terms as follows.

$$Redistribution\_Score = \frac{Tweet\_Score}{|W_r|} \tag{1}$$

$$S(w \mid w \in W_n) = \frac{Redistribution\_Score}{|W_n|} \tag{2}$$

Here, W_r denotes the set of concept terms found in a tweet, W_n denotes those set of terms in a tweet that do not fall under the definition of concept terms (based on the discovery process from Wikipedia that we described previously) and finally, S denotes the score assigned to a term w in W_n. We further illustrate the score propagation mechanism with the help of the example in Table 1. Consider the two tweets in Table 1 with first tweet having t_1, t_3 and t_7 as concept terms and second

Table 1. Illustrated example to explain score propagation mechanism

Tweet	Concept terms	Non-concept terms
t_1 t_2 t_3 t_4 t_5 t_6 t_7	t_1, t_3, t_7	t_2, t_4, t_5, t_6
t_1 t_3 t_5 t_6 t_{11} t_{12} t_{13}	t_1, t_3, t_{11}	t_5, t_6, t_{12}, t_{13}

tweet having an additional t_{11} as concept term (t_1 and t_3 are also concept terms in the second tweet since they were already discovered as concept terms for the first tweet). The tweet score for the first and second tweet (computed as described in Sect. 3.2) is first decomposed between each tweet's discovered concept terms to constitute a *Redistribution_Score* as shown in Eq. 1. This *Redistribution_Score* is then distributed among the terms in W_n as shown in Eq. 2. After the score redistribution, the second pass computes a new score for each tweet which now takes into account the scores of non-concept terms as well.

Moreover, the second phase also makes use of additional sources of evidence, these are:

- POS tag of the company name occurring within the tweets: We apply POS tagging to each tweet after which the POS tag corresponding to the company name occurring within a tweet is checked and if it corresponds to a proper noun (i.e., POS tag being NNP) we perform an increment in the tweet score.
- URL occurring within the tweets: We get the extended URL[2] for tweets which contain a URL and if the company name occurs in extended URL within a tweet we perform an increment in the tweet score[3].
- Twitter username occurring within the tweets: Using the Twitter API we extract the description field of the tweet author[4] and if the company name occurs in that description field we perform an increment in the tweet score.
- Hashtag occurring within the tweets: We extract all hashtags in each tweet and if the hashtag occurring within a tweet contains company name we perform an increment in the tweet score.

Note that the score increment values for each of the mentioned additional sources of evidence are set empirically. Each of the above-mentioned sources of evidence then contribute towards computing the final score of a tweet. At the end of the second pass, the score propagation technique along with the extra sources of evidence enable extraction of more tweets relevant to the company, thus increasing the recall. We show more specific results for precision and recall in the next section.

---

[2] This information is provided by lab organizers.

[3] e.g., http://www.cincodias.com/articulo/empresas/telefonica-gana-contrato-millo nes-britanica-serco/20120423cdscdsemp_19/ contains company name 'Telefonica'.

[4] Twitter enables every user to add a brief biographical description which can be extracted using the Twitter API.

Table 2. Precision and recall scores for two passes of the algorithm

	First pass	Second pass
Precision	0.84827	0.81129
Recall	0.16307	0.76229

Table 3. Per entity results at RepLab 2012

Entity	Percentage English tweets	Accuracy filtering	R filtering	S filtering	F(R,S) filtering
Bing	98 %	0.77	0.63	0.57	0.6
Intl. consolidated airlines group	87 %	0.84	0.7	0.63	0.66
Volkswagen	96 %	0.65	0.46	0.45	0.45

4 Experimental Evaluations

As mentioned in Sect. 2 the task comprises binary classification and hence we report the effectiveness of our algorithm through the standard evaluation metrics of precision and recall. We were given a very small trial dataset (six companies) and a considerably larger test dataset (31 companies). For each company a few hundred tweets were provided with the language of the tweets being English and Spanish; furthermore, most of the tweets are in Spanish. It is important to note that we translate a tweet that is not written in English into the English Language by using the Bing Translation API[5].

We report our results for the test dataset. For this purpose, we make use of the gold standard provided by the lab organizers which comprised of individual tweet messages tagged for each company. Table 2 shows the precision and recall figures after the application of the first pass, and after the application of the second pass of our algorithm.

As Table 2 shows the first pass yields a high precision but an extremely low recall. The application of the second pass increases the recall by a large degree, while not overly reducing the precision. The significantly large increase in recall proves the effectiveness of the score-propagation technique combined with the use of multiple sources of evidence.

The RepLab 2012 filtering task used the measures of Reliability and Sensitivity for evaluation purposes, these are described in detail in [3]. Table 4 presents a snapshot of the official results for the Filtering task of RepLab 2012, where CIRGIRDISCO is the name of our team. Table 4 shows that our algorithm performed competitively. It is the second best among the submitted systems and fourth best among the submitted runs. The baseline system marked all tweets as relevant; note that the measure of *Accuracy Filtering* does not reveal true

---

[5] http://www.microsofttranslator.com/.

performance which is why lab organizers used Reliability and Sensitivity [3] for the purpose of comparing team performances. It is also important to note that all the runs submitted by "Daedalus" relied on manually crafted rules whereas our system is completely unsupervised and automatic (i.e., does not require any training data or manual efforts).

Table 4. Results of RepLab 2012

Team	Accuracy filtering	R filtering	S filtering	F(R,S) filtering
Daedalus_2	0.72	0.24	0.43	0.26
Daedalus_3	0.70	0.24	0.42	0.25
Daedalus_1	0.72	0.22	0.40	0.25
CIRGIRDISCO	0.70	0.22	0.34	0.23
Baseline	0.71	0.0	0.0	0.0

Another interesting observation we made with respect to per entity (company) evaluation results is that our algorithm suffers due to translation issues[6]. This is particularly obvious for entities with high percentage of English tweets. Table 3 shows results for three entities with high percentage of English tweets: the Reliability and Sensitivity scores exhibited by our algorithm are considerably high. In fact for the entities Bing and Volkswagen our algorithm showed the best results (out of all submitted runs by all teams). This demonstrates the high performance of our algorithm specially for English tweets.

5   Conclusion

We proposed a two-pass algorithm for company name disambiguation in tweets. Our algorithm makes use of Wikipedia as a primary knowledge resource in the first pass of the algorithm, and the tweets are matched across Wikipedia terms. The matched terms are then used for score propagation in the second pass of the algorithm that also makes use of multiple sources of evidence. Our algorithm showed competitive performance and demonstrates the effectiveness of the techniques employed in the two passes. We believe that the two-step filtering approach may open a promising new dimension to address the problem of entity name disambiguation in tweets as can be evidenced by the evaluation results.

References

1. Amigó, E., Artiles, J., Gonzalo, J., Spina, D., Liu, B., Corujo, A.: WePS3 evaluation campaign: overview of the on-line reputation management task. In: 2nd Web People Search Evaluation Workshop (WePS 2010), CLEF 2010 Conference, Padova (2010)

---

[6] The Bing Translate API does not correctly translate all Spanish tweets into English e.g., it translates name of company Telefonica into telephone.

2. Amigó, E., Corujo, A., Gonzalo, J., Meij, E., de Rijke, M.: Overview of RepLab 2012: evaluating online reputation managementsystems. In: CLEF (Online Working Notes/Labs/Workshop) (2012)
3. Amigo, E., Gonzalo, J., Verdejo, F.: Reliability and sensitivity: generic evaluation measures for document organization tasks. UNED, Madrid, Technical report (2012)
4. Bizer, C., Lehmann, J., Kobilarov, G., Auer, S., Becker, C., Cyganiak, R., Hellmann, S.: DBpedia - a crystallization point for the web of data. Web Semant. **7**(3), 154–165 (2009)
5. Bontcheva, K., Rout, D.: Making sense of social media streams through semantics: a survey. Semant. Web J. **1**, 1–31 (2012)
6. Meij, E., Weerkamp, W., de Rijke, M.: Adding semantics to microblog posts. In: Proceedings of the Fifth ACM International Conference on Web Search and Data Mining, WSDM 2012, pp. 563–572. ACM, New York (2012)
7. Spina, D., Meij, E., de Rijke, M., Oghina, A., Bui, M.T., Breuss, M.: Identifying entity aspects in microblog posts. In: Proceedings of the 35th International ACM SIGIR Conference on Research and Development in Information Retrieval, SIGIR 2012, pp. 1089–1090. ACM, New York (2012)
8. Suchanek, F.M., Kasneci, G., Weikum, G.: Yago: a core of semantic knowledge. In: Proceedings of the 16th International Conference on WorldWide Web, WWW 2007, pp. 697–706. ACM, New York (2007)
9. Yerva, S.R., Miklós, Z., Aberer, K.: What have fruits to do with technology?: the case of Orange, Blackberry and Apple. In: Proceedings of the International Conference on WebIntelligence, Mining and Semantics, WIMS 2011, pp. 48:1–48:10. ACM, New York (2011)
10. Yoshida, M., Matsushima, S., Ono, S., Sato, I.: ITC-UT: tweet categorization by query categorization for on-linereputation management. In: CLEF, vol. 170 (2010)

Utilizing Word Embeddings for Result Diversification in Tweet Search

Kezban Dilek Onal, Ismail Sengor Altingovde[✉], and Pinar Karagoz

Middle East Technical University, Ankara, Turkey
{dilek,altingovde,karagoz}@ceng.metu.edu.tr

Abstract. The performance of result diversification for tweet search suffers from the well-known vocabulary mismatch problem, as tweets are too short and usually informal. As a remedy, we propose to adopt a query and tweet expansion strategy that utilizes automatically-generated word embeddings. Our experiments using state-of-the-art diversification methods on the Tweets2013 corpus reveal encouraging results for expanding queries and/or tweets based on the word embeddings to improve the diversification performance in tweet search. We further show that the expansions based on the word embeddings may serve as useful as those based on a manually constructed knowledge base, namely, ConceptNet.

Keywords: Diversification · Expansion · Tweet search · Word embedding

1 Introduction

In the last few years, microblogging –and being its most prominent platform, Twitter– has gained tremendous popularity. Users of Twitter do not only post tweets and browse those in their own feed, but also submit queries to discover the posts about events or people (i.e., mostly celebrities) of their interest [26]. According to a recent study, more than 2 billion queries are submitted to Twitter per day [5], making microblog retrieval an important and active research area.

Diversifying tweet search results is an emerging post-retrieval optimization that aims to cover different aspects of a given query in the top-ranked results, and hence provide the user a better overview of the searched topic [18]. Recent studies show that the vocabulary mismatch problem that has been identified for microblog retrieval also affects the diversification performance adversely. The vocabulary mismatch problem arises due to the short (i.e., typically 140 characters) and informal nature of tweets (e.g., see [1,20,21]), and causes highly relevant query-tweet (or, tweet-tweet) pairs yield very low similarity scores. In the context of result diversification, Ozsoy et al. [18] have shown that explicit methods with prior knowledge of the aspects of a given query are not as successful at diversifying tweet search results as they are at diversifying web search results, since the terms describing the query aspects do not usually occur in the tweet content.

© Springer International Publishing Switzerland 2015
G. Zuccon et al. (Eds.): AIRS 2015, LNCS 9460, pp. 366–378, 2015.
DOI: 10.1007/978-3-319-28940-3_29

In this paper, we address the vocabulary mismatch problem in result diversification for tweet search, and as a remedy, we propose to expand queries and/or tweets using *word embeddings* as an external knowledge base. The vocabulary of a language is mapped into a vector space such that words that occur in similar contexts have similar vector representations, so-called word embeddings [2]. Word embedding space is a convenient source as an external knowledge base because of the following reasons. First, word embeddings are obtained by unsupervised learning on raw textual data; and recent methods in the literature enable learning embedding vectors from a corpus of billion words in an hour on a personal computer. In contrast, most of the other knowledge bases are constructed by some sort of human intervention [3,12]. Second, word embeddings learned from raw textual data can encode both the linguistic knowledge that can be obtained from resources like WordNet and real-life knowledge such as the relationships between entities. Owing to this property, it can be more useful in searching highly dynamic and informal datasets, like microblogs.

Our work presents a KNN-based strategy to expand queries and tweets using word embeddings. To the best of our knowledge, while query expansion is applied for microblog retrieval in some recent works (using pseudo-random feedback [1,21], or external resources such as a community-curated knowledge base, namely, Freebase [20]), we are the first to adopt both query and tweet expansion based on the automatically-generated word embeddings for the result diversification problem in tweet search.

In our experiments, we use the Tweets2013 corpus and employ three representative diversification methods from the literature; namely, Sy and MMR as implicit and xQuAD as explicit diversification methods. We find that query and tweet expansion using word embeddings is a promising idea and can yield some moderate improvements in most cases. As a further experiment, we employ ConceptNet, a crowd-sourced knowledge base, as an external resource for query/tweet expansion; and show that the performance is comparable to that obtained via word embeddings, while the latter requires no human effort for its construction. This last finding again implies the potential of utilizing word embeddings in tweet diversification in a realistic scenario.

The rest of the paper is organized as follows. In the next section, we review the underlying theory for the word embeddings and discuss related work for retrieving and diversifying tweets. Section 3 presents our query and tweet expansion approach based on the word embeddings. Section 4 is devoted to the experimental setup and evaluation. Finally, we conclude and point to future research directions in Sect. 5.

2   Background and Related Work

2.1   Word Embeddings

Word embeddings are real-valued vector representations for words. The idea behind word embeddings is to map the whole vocabulary of a language into a vector space such that syntactically and semantically close words are represented with similar vectors. The similarity of two words can be quantified by the similarity of their embedding vectors. The concept of word embeddings is employed within the context of the *Neural Network Language Model (NNLM)* by Bengio et al. [2]. The NNLM is trained to predict the next word given a sequence of words. The model is designed to learn word embeddings and parameters of the language model that predicts the next word given a sequence, simultaneously.

Following the latter study, several other neural network models for learning word embeddings have been proposed. The very recent models Skip-Gram [16] and Glove [19] lean on simpler and scalable models that enable learning word embeddings from a corpus of a billion words on a personal computer in an hour. In this work, we utilized the Glove model for embedding learning model since it exploits global context besides the local context in training, differently from other embedding learning models.

2.2 Microblog Retrieval and Diversification

Microblog retrieval has become an active research area in the last few years as it is addressed in several works as well as the TREC Microblog Track organized since 2011 to date. A number of these previous works identify the vocabulary mismatch problem for microblog retrieval, and propose solutions based on the query and/or tweet expansion. Massoudi et al. [14] employ a traditional pseudo-relevance feedback (PRF) strategy, where the expansion terms are scored based on their temporal closeness to the submission time of the query. Miyanishi et al. [17] again employs PRF by taking into account the temporal evidence, however their model is preceded by an initial PRF stage where a single relevant tweet for a given query has to be manually picked by the query poser. Rodriguez Perez et al. [21] apply a clustering of tweets based on the named entities appearing in the tweet body and utilize these clusters for PRF. In contrast to all the latter approaches, some other works rely on the external corpora for query or tweet expansion. In the work of Bandyopadhyay et al. [1], the expansion terms for a query are based on the result document titles obtained via Google search API for each query. Gurini et al. [8] and Qiang et al. [20] employ Wikipedia and Freebase as an external resource for query expansion, respectively. Finally, there are also works that make tweet expansion, e.g., using hyperlinks in the tweets [9,11,15] or using the tweets themselves as pseudo-queries over the collection [7]. Clearly, all of these earlier works aim to enhance the retrieval performance; whereas our goal in this paper is to improve the performance of the result diversification in tweet search. To this end, we employ representative diversification methods that may require query expansion, tweet expansion, or both; and investigate to what extent our expansion strategies based on the word embeddings can help.

Search result diversification is an emerging trend that is actively investigated in various areas such as the web search engines and recommender systems. Especially for the result diversification in web search, a large number of methods are proposed, including those that utilize query expansion techniques. For instance, in [4], the authors employ ConceptNet to obtain diverse expansion terms for a given query. In a follow up work, a vector embeddings space is used to discover the diverse aspects of the query [13]. While inspiring for us, these works are not directly related as they do not address the microblog diversification. Furthermore, their approaches expand the original query with diverse terms (i.e., diversification is applied over the expansion *terms*) and then execute the expanded query to obtain the final result; whereas we apply the diversification over the *results* once the queries and/or tweets are expanded.

In the context of tweet search, there a few earlier studies that focuses on the diversity of the top-ranked tweets. In [22], an approach based on the MMR method [6] and clustering of tweets is proposed. However, they do not evaluate the proposed approach using diversity-aware evaluation metrics, as we do here. In another study, Tao et al. introduce a data set, *Tweets 2013 Corpus*, for evaluating microblog diversification, and they also propose a diversification method, *Sy* [24]. Koniaris et al. [10] consider the tweet diversification with respect to several criteria, such as the sentiment, time, location, and readability of the tweets. Finally, Ozsoy et al. [18] investigate the performance of various diversification methods on the *Tweets 2013 Corpus*. This study concluded that explicit methods perform worse than the implicit methods, a finding that contradicts to the web search literature. The failure of the explicit methods is attributed to the vocabulary gap between the query aspects and tweets. In our work, we aim to remedy the vocabulary mismatch problem using expansion strategies for both implicit and explicit result diversification methods.

3   Query and Tweet Expansion

3.1   Preliminaries of the Result Diversification for Tweet Search

The result diversification problem for tweet search is formally defined as follows [18]. Let's assume a query q that retrieves a candidate ranking C (where $|C| = N$) of tweets over a collection. The aim of the result diversification is obtaining a top-k ranking S (where $k < N$) that maximizes the total relevance and diversity among all possible size-k rankings S' of C.

In the web search literature, result diversification methods are broadly categorized as *implicit* and *explicit*, based on the source of knowledge exploited for the diversification purposes. The methods in the former category rely only on the candidate ranking C, and attempt to construct the ranking S based on some intrinsic features of the tweets (such as their pairwise similarities) in the candidate set. In contrast, the explicit methods assume the prior knowledge of the query aspects that is obtained from some external resource (e.g., from a query log in [23]) and try to diversify the candidate ranking accordingly. In this study, we employ MMR and Sy as representatives of the implicit methods, and xQuAD as a representative of the explicit methods. In what follows, we briefly review these three methods as they are adopted in [18].

MMR. The MMR [6] method employs a greedy strategy to iteratively construct a diversified ranking S. In the first iteration, the tweet with the highest relevance to the query is inserted into the set S. In the following iterations, the score of each remaining tweet t_i is computed by taking into account its relevance to q, which is denoted as $rel(t_i, q)$, and its similarity to already selected tweets in S (as shown in Eq. 1); and the tweet with the highest score is removed from C and added to S. In Eq. 1, the trade-off parameter λ is used to balance the relevance against diversity in the final result set.

$$f_{MMR}(t_i) = \lambda\, rel(t_i, q) - (1 - \lambda) \max_{t_j \in S} sim(t_i, t_j) \tag{1}$$

Sy. This is a simple yet effective method described in a framework for detecting duplicate and near-duplicate tweets in [24]. In a nutshell, Sy makes a scan of the candidate tweet ranking C in a top-down fashion. For a given tweet t_i, its near duplicates (i.e., those that have a similarity score greater than a pre-defined threshold) succeeding t_i in the ranking C (i.e., appearing at a lower rank position) are removed from C.

xQuAD. xQuAD [23] assumes the prior knowledge of the query aspects and constructs a diversified ranking by a greedy approach that selects the tweet that essentially covers the aspects that are least covered in the current set S. More formally, in each iteration, this method selects the tweet that maximizes Eq. 2, where $P(t_i|q)$ denotes the relevance of t_i to query q, $P(q_i|q)$ denotes the likelihood of the aspect q_i for the query q, $P(t_i|q_i)$ denotes the relevance of t_i to the query aspect q_i, and the product term represents the probability of the aspect q_i not being satisfied by the tweets that are already selected into S.

$$f_{xQuAD}(t_i) = (1 - \lambda)P(t_i|q) + \lambda \sum_{q_i} \left[P(q_i|q)P(t_i|q_i) \prod_{t_j \in S} (1 - P(t_j|q_i)) \right] \tag{2}$$

Our work includes xQuAD as a representative explicit diversification method since it has been the top performer in the Diversification Track of the TREC evaluation campaigns between 2009 and 2012. For the implicit methods, we include MMR for being a traditional and one of the earliest methods, and Sy for being the best-performing approach for tweet diversification in [18].

3.2   Expansion with K-Nearest Neighbours (KNN)

The aforementioned diversification methods can operate on three different types of textual units, namely, queries, query aspects or tweets. In particular, the Sy method considers only tweet-tweet similarities for constructing the diversified ranking, while MMR computes both the query-tweet and tweet-tweet similarities (see Eq. 1), and xQuAD needs to compute the query-tweet and aspect-tweet similarities as shown in Eq. 2. Hence, for different methods, it is possible to expand some or all types of these textual units.

In this work, we adopt the K-Nearest Neighbour (KNN) expansion method for expanding any textual unit. KNN follows the simple idea of collecting K-neighbours of each term from an external knowledge base. A textual unit is viewed as a set of terms, and the expansion term set is composed by merging the K-Nearest neighbour sets of unique terms in the textual unit. Formally, given a textual unit (i.e., a query, aspect or tweet) u that contains m terms, $u = \{w_1, w_2, ..w_m\}$, the expanded unit u_e with the *K-Nearest Neighbours* expansion method is computed as follows:

$$u_e = u \cup (\cup_{i=1}^{m} E(w_i, K)) \tag{3}$$

where $E(w_i, K)$ is the set of K nearest neighbours of the term w_i in the external knowledge base. The computation strategy of $E(w_i, K)$ depends on the external knowledge base and discussed in depth in the following section.

3.3   Computing the K-Nearest Neighbours

We utilize two external resources to compute $E(w_i, K)$ as follows.

Word Embeddings: In the word embedding space, each word is associated with a real-valued D dimensional vector. Since semantically similar words are likely to have similar embedding vectors, we treated K-nearest neighbours of a term w in the embedding space as the set of terms that maximize semantic similarity to w. In other words, the set of K-Nearest neighbours of a term w is determined by ranking all the words in the embedding space with respect to their similarity to w. The similarity of two words in the embedding space is computed by the Cosine similarity of their embedding vectors.

As an example, in Table 1, we provide five nearest neighbours of some words (extracted from our query set) in the Glove embedding space. Each column lists the neighbors of the target term given in the header row. Remarkably, the neighbor terms are quite relevant to the target ones.

Table 1. Five-nearest neighbours of some example terms according to the word embeddings published at the Glove Project Website.

Resignation	Hillary	Bomb	Budget	Museum
Resign	Rodham	Bombs	Spending	Art
Resigned	Barack	Exploded	Fiscal	Gallery
Appointment	Obama	Explosives	Cuts	Library
Dismissal	Mccain	Blast	Package	Exhibition
Resigning	Clinton	Detonated	Budgets	Museums

ConceptNet: The ConceptNet API[1] provides the top-K similar concepts to a given concept. A crucial point to note about ConceptNet is that concepts in ConceptNet may be sentences or phrases in which the words are separated

---

[1] http://conceptnet5.media.mit.edu/data/5.3/.

by underscores. For these types of concepts, although the concept cannot be matched to a textual unit exactly, one of the words in the concept can be found in the textual unit and provide an important information for the similarity computation. Therefore, we split such concepts and expanded the corresponding textual unit by the words instead of the whole concept. For instance, assume that *head_of_state* is the concept at hand, which is actually the second nearest neighbor to the query term *president*. In this case, we add three terms *head, of, state* to the expanded query for this single concept.

4   Experiments

4.1   Experimental Setup

Dataset. For our evaluations, we use the Tweets2013 corpus [25] that is specifically built for the tweet search result diversification problem. The dataset includes forty seven query topics and each topic has, on the average, 9 subtopics. The tweets in the data set are dated between Feb. 1, 2013 and March 31, 2013.

The owners of the Tweets2013 corpus only share the tweet identification numbers since the Twitter API licence does not allow users to share the content of the tweets. We attempted to obtain the top-100 tweets per query but failed to do so, since some of these tweets were erased or their sharing status were changed. Consequently, we ended up with 81 tweets per topic on the average. Furthermore, following the practice in the earlier works, we decided to remove the topics that have at most 2 relevant tweets among their top-100 results (with ids 5, 9, 22, 28, 7, 8, 14, 43, 46 and 47) from our query set, as they would deviate our measurements.

Preprocessing. Prior to the application of the diversification methods, we removed all the mentions and URLs from the tweets and reduced each tweet to a set of terms and a set of hashtags. We performed tokenization and stemming with the Stanford CoreNLP library (version 3.5.1). At last, we removed all the terms that appear in the default stopword list of the Indri search tool (http://www.lemurproject.org/).

Knowledge Bases. We used the 50-dimensional word embeddings obtained with the Glove algorithm from a corpus of 6 billion tokens that includes Wikipedia and Gigaword. The embeddings are published in http://nlp.stanford.edu/projects/glove/. We also employed ConceptNet, a crowd-sourced knowledge base, as a further baseline.

Evaluation Metrics. For evaluation, we used the ndeval software (http://trec.nist.gov/data/web10.html) employed in the TREC Diversity Tasks. The results are reported using three widely-used effectiveness metrics, namely, α-nDCG, Precision-IA, and Subtopic-Recall at the cut-off values of 10 and 20.

Parameters of the Diversification Methods. We measure the performance of our expansion strategy using the word embeddings and ConceptNet on three different diversification methods, namely, Sy, MMR, xQuAD. In all the diversification methods, we utilized the KNN expansion method. Depending on the method, we applied expansion to different textual unit types included in diversification. For the Sy method, we performed only tweet expansion since the query is neglected by this method. For MMR, we performed two sets of experiments. In the first set, we expanded only queries. In the second set of MMR experiments, we expanded both queries and tweets. Finally, the xQuAD diversification method relies on the queries, query aspects and tweets for diversification. We performed two sets of experiments for xQuAD. In the first set of experiments, we only expanded the queries and query aspects, whilst in the second set, we expanded queries, aspects and tweets. Note that, following the practice in the literature (e.g., [18,23]), we used the official query aspects provided in the dataset to represent an ideal scenario. For all diversification methods, we report the results for the best-performing λ value.

The diversification methods require metrics for computing the query-tweet relevance and/or tweet-tweet similarity scores. In this work, we employed Jaccard Similarity metric for both purposes. Each textual unit is viewed as a set of words and the similarity between two textual units T_1 and T_2 is computed by

$$Sim(T_1, T_2) = (T_1 \cap T_2)/(T_1 \cup T_2) \tag{4}$$

Finally, we obtain a diversified ranking of size 30 from a candidate ranking of size 100.

4.2 Experimental Results

In the results, we compare the performance of diversification methods with KNN-based expansion to two baselines. The **No-div Baseline** is the performance over the initial retrieval results (i.e., without any diversification) obtained by a system employing the query-likelihood (QL) retrieval model. These initial retrieval results were provided in the Tweets2013 corpus. However, since it is impossible to retrieve exactly 100 tweets per query due to API issues (as discussed in the previous section), we re-computed the evaluation metrics for this initial ranking based on only those tweets that could be retrieved in top-100, for the sake of fair comparison. Consequently, the effectiveness of the baseline QL run slightly differs from what is reported in [25]. As a second baseline, for each diversification method, we also provide the effectiveness scores without doing any expansion. In the following results, we denote this baseline as **No-exp Baseline**, and those results that exceed the latter baseline are shown in bold.

Results for Sy. In Table 2, we present results of the tweet expansion experiments for the Sy method. The Sy method was found to outperform all the other explicit and implicit diversification methods experimented in [18] when used with a hybrid metric that combines the Jaccard Similarity for content and the timestamp similarity. In this work, we focused only on content similarity and ignored the time attribute of tweets. The field K in the table denotes the number of neighbors in the KNN strategy. The λ parameter represents the similarity threshold used in the Sy method.

Table 2. Diversification performance of the Sy method with tweet expansion.

Method	λ	K	α-NDCG		Prec-IA		St-Recall	
			@10	@20	@10	@20	@10	@20
No-div Baseline	–	–	0.303	0.347	0.065	0.059	0.358	0.506
No-exp Baseline	0.6	–	0.345	0.385	0.080	0.069	0.416	0.547
ConceptNet	0.6	1	**0.346**	0.381	**0.081**	0.069	0.416	0.532
Glove	0.6	1	0.343	0.381	0.079	0.067	0.410	0.536
ConceptNet	0.5	3	**0.350**	0.382	0.080	0.066	**0.425**	0.538
Glove	0.5	3	**0.346**	0.382	0.080	0.068	**0.420**	0.543
ConceptNet	0.5	5	**0.349**	0.383	0.080	0.066	**0.422**	0.543
Glove	0.6	5	0.344	0.381	**0.081**	0.069	0.416	0.536
ConceptNet	0.5	8	**0.348**	0.383	0.079	0.067	**0.417**	0.537
Glove	0.6	8	**0.347**	0.384	**0.083**	**0.070**	0.416	0.547
ConceptNet	0.6	10	**0.349**	0.383	0.080	0.067	**0.429**	0.541
Glove	0.6	10	**0.350**	**0.385**	**0.083**	**0.070**	**0.423**	0.541

Our findings reveal that the performance of Sy can be improved via tweet expansion using both types of knowledge bases. The improvements are relatively higher at the cut-off value of 10 and for the ST-Recall metric; and they tend to increase with the larger number of expansion terms. This is a moderate yet encouraging finding, as Sy is reported to be the best-performing method in [18,23], and the expansion method has the potential to improve even this case.

Results for MMR. In Table 3, we present the results of the query expansion experiments for MMR. Surprisingly, MMR fails to diversify the results, as even No-exp Baseline performs worse than the No-div baseline. Effectiveness improves with increasing number of neighbors using both knowledge bases, and expansion with Glove word embeddings seems to be better than ConceptNet, especially for certain metrics, such as ST-Recall@20. Actually, the latter metric is the only one for which the diversification performance can exceed both of the baselines.

In Table 4, we present the results of the experiments in which both queries and tweets are expanded. The columns KQ and KT denote the number of neighbours used for query and tweet expansion, respectively. The trends are similar to the previous case, but the actual effectiveness scores are even worse, implying that tweet expansion does not help improving the diversification performance for MMR.

Results for xQuAD. In Table 5, we present the results of xQuAD experiments with query and aspect expansion. In this case, expansions based on both types of the external resources are again useful, and in certain cases, they can yield an absolute improvement of more than 1 %. ConceptNet is slightly better than the word embeddings. Finally, in Table 6, we present the results of the xQuAD experiments where we expand the queries, aspects and tweets. We see that expanding

Table 3. Diversification performance of MMR with query expansion.

Method	λ	K	α-NDCG		Prec-IA		St-Recall	
			@10	@20	@10	@20	@10	@20
No-div Baseline	–	–	0.303	0.347	0.065	0.059	0.358	0.506
No-exp Baseline	0.0	-	0.254	0.289	0.042	0.038	0.298	0.438
ConceptNet	0.0	1	0.254	0.289	0.042	0.038	0.298	0.438
Glove	0.0	1	0.254	0.289	0.042	0.038	0.298	0.438
ConceptNet	0.0	3	0.254	0.289	0.042	0.038	0.298	0.438
Glove	0.8	3	0.233	0.291	**0.046**	**0.048**	**0.326**	**0.503**
ConceptNet	0.8	5	0.254	**0.302**	**0.054**	**0.048**	**0.374**	**0.504**
Glove	0.8	5	**0.269**	**0.318**	**0.055**	**0.049**	**0.384**	**0.537**
ConceptNet	0.8	8	**0.268**	**0.310**	**0.056**	**0.048**	**0.365**	**0.511**
Glove	0.8	8	**0.285**	**0.320**	**0.058**	**0.050**	**0.394**	**0.514**
ConceptNet	0.8	10	**0.265**	**0.311**	**0.053**	**0.048**	**0.348**	**0.513**
Glove	0.8	10	**0.283**	**0.322**	**0.055**	**0.049**	**0.365**	**0.515**

Table 4. Diversification performance of MMR with query and tweet expansion.

Method	λ	KQ	KT	α-NDCG		Prec-IA		St-Recall	
				@10	@20	@10	@20	@10	@20
No-div Baseline	–	–	–	0.303	0.347	0.065	0.059	0.358	0.506
No-exp Baseline	0	–	–	0.254	0.289	0.042	0.038	0.298	0.438
ConceptNet	0	1	1	**0.264**	**0.298**	**0.043**	0.040	**0.338**	**0.464**
Glove	0	1	1	**0.267**	**0.307**	**0.044**	0.043	**0.331**	**0.481**
ConceptNet	0	3	3	0.249	0.291	0.037	0.039	0.292	**0.439**
Glove	0	3	3	0.243	0.294	0.039	**0.041**	0.298	**0.466**
ConceptNet	0	5	5	0.252	0.291	0.038	0.038	0.303	**0.449**
Glove	0	5	5	0.244	0.286	0.040	**0.039**	**0.309**	**0.459**
ConceptNet	0	8	8	**0.257**	**0.301**	0.040	**0.040**	**0.309**	**0.467**
Glove	0	8	8	0.236	**0.290**	0.038	**0.040**	0.284	**0.468**
ConceptNet	0	10	10	**0.255**	**0.300**	0.040	**0.040**	**0.307**	**0.468**
Glove	0	10	10	0.242	**0.292**	0.039	**0.041**	0.296	**0.475**

tweets further improve the results (cf. Table 5) especially for the cut-off value of 10 and using a few (i.e., up to 3) expansion terms. In this case, expansions based on the word embeddings are slightly more useful than those based on the ConceptNet.

Table 5. Diversification performance of xQuAD with query and aspect expansion.

Method	λ	KQ	α-NDCG		Prec-IA		St-Recall	
			@10	@20	@10	@20	@10	@20
No-div Baseline	–	–	0.303	0.347	0.065	0.059	0.358	0.506
No-exp Baseline	1	–	0.325	0.350	0.065	0.053	0.407	0.510
ConceptNet	1	1	0.306	0.335	0.065	**0.055**	0.389	0.496
Glove	1	1	0.322	0.341	**0.067**	0.053	**0.416**	0.510
ConceptNet	1	3	0.326	**0.356**	0.066	**0.057**	**0.420**	**0.517**
Glove	1	3	0.314	0.338	0.064	0.052	0.407	0.504
ConceptNet	1	5	0.310	0.346	0.065	**0.058**	0.409	**0.526**
Glove	1	5	0.287	0.325	0.058	0.051	0.380	0.502
ConceptNet	1	8	0.334	0.367	**0.068**	**0.060**	**0.414**	**0.524**
Glove	1	8	0.294	0.335	0.057	0.051	0.375	0.503
ConceptNet	1	10	0.322	**0.352**	**0.067**	**0.057**	0.402	0.510
Glove	1	10	0.266	0.309	0.049	0.049	0.350	0.495

Table 6. Diversification performance of xQuAD with query, aspect & tweet expansion.

Method	λ	KQ	KT	α-NDCG		Prec-IA		St-Recall	
				@10	@20	@10	@20	@10	@20
No-div Baseline	–	–	–	0.303	0.347	0.065	0.059	0.358	0.506
No-exp Baseline	1	–	–	0.325	0.350	0.065	0.053	0.407	0.510
ConceptNet	1	1	1	**0.330**	0.345	**0.066**	0.052	**0.408**	0.490
Glove	1	1	1	**0.337**	0.363	0.066	0.054	0.428	0.540
ConceptNet	1	3	3	**0.338**	0.360	0.067	0.054	0.420	0.512
Glove	1	3	3	0.311	0.336	0.064	0.054	0.416	0.515
ConceptNet	1	5	5	0.275	0.313	0.062	0.057	0.372	0.503
Glove	1	5	5	0.296	0.330	0.060	0.051	0.392	0.516
ConceptNet	1	8	8	0.297	0.326	0.061	0.056	0.357	0.467
Glove	1	8	8	0.291	0.328	0.058	0.052	0.391	0.506
ConceptNet	1	10	10	0.266	0.296	0.057	0.050	0.355	0.457
Glove	1	10	10	0.280	0.315	0.055	0.049	0.381	0.490

5   Conclusion and Future Work

In this work, we addressed the vocabulary mismatch problem in result diversification for microblog search. For the first time in the literature, we employed query and tweet expansion based on the automatically-generated word embeddings to improve the diversification performance. We evaluated the adopted expansion strategy using three diversification methods, namely, MMR and Sy for implicit and xQuAD for explicit diversification. Our findings revealed that while MMR is not a competitive method for diversification in this context; both xQuAD and Sy benefit from the expansion strategy, though the improvements are moderate. We further showed that the expansions based on the automatically-generated word embeddings may serve as useful as those based on the ConceptNet knowledge base, of which construction requires considerable manual effort. This is an encouraging finding for improving diversification by leveraging word embeddings in real life microblog retrieval scenarios. In our future work, we plan to generate word embeddings on a tweet corpus for further experimentation. We also aim to extend our experimental framework to analyze the impact of the other features, such as URLs, extracted from the tweets in addition to their textual content.

Acknowledgement. This work is partially funded by METU under the grant number BAP-08-11-2013-055, and The Scientific and Technological Research Council of Turkey (TÜBİTAK) under the grant numbers 113E065 and 112E275.

References

1. Bandyopadhyay, A., Mitra, M., Majumder, P.: Query expansion for microblog retrieval. In: Proceedings of TREC 2011 (2011)
2. Bengio, Y., Ducharme, R., Vincent, P., Janvin, C.: A neural probabilistic language model. J. Mach. Learn. Res. **3**, 1137–1155 (2003)
3. Bollacker, K., Evans, C., Paritosh, P., Sturge, T., Taylor, J.: Freebase: a collaboratively created graph database for structuring human knowledge. In: Proceedings of SIGMOD 2008, pp. 1247–1250 (2008)
4. Bouchoucha, A., He, J., Nie, J.: Diversified query expansion using conceptnet. In: Proceedings of CIKM 2013, pp. 1861–1864 (2013)
5. Busch, M., Gade, K., Larson, B., Lok, P., Luckenbill, S., Lin, J.: Earlybird: real-time search at twitter. In: Proceedings of ICDE 2012, pp. 1360–1369 (2012)
6. Carbonell, J., Goldstein, J.: The use of MMR, diversity-based reranking for reordering documents and producing summaries. In: Proceedings of SIGIR 1998, pp. 335–336 (1998)
7. Efron, M., Organisciak, P., Fenlon, K.: Improving retrieval of short texts through document expansion. In: Proceedings of SIGIR 2012, pp. 911–920 (2012)
8. Gurini, D.F., Gasparetti, F.: TREC microblog, : track: real-time ranking algorithm for microblog ranking systems. In: Proceedings of TREC 2012 (2012)
9. Kim, Y., Yeniterzi, R., Callan, J.: Overcoming vocabulary limitations in twitter microblogs. In: Proceedings of TREC 2012 (2012)
10. Vasileiou, Y., Sellis, T., Giannopoulos, G., Koniaris, M.: Diversifying microblog posts. In: Benatallah, B., Bestavros, A., Manolopoulos, Y., Vakali, A., Zhang, Y. (eds.) WISE 2014, Part II. LNCS, vol. 8787, pp. 189–198. Springer, Heidelberg (2014)
11. Liang, F., Qiang, R., Yang, J.: Exploiting real-time information retrieval in the microblogosphere. In: Proceedings of JCDL 2012, pp. 267–276 (2012)
12. Liu, H., Singh, P.: ConceptNet: a practical commonsense reasoning tool-kit. BT Technol. J. **22**(4), 211–226 (2004)
13. Liu, X., Bouchoucha, A., Sordoni, A., Nie, J.: Compact aspect embedding for diversified query expansions. In: Proceedings of AAAI 2014, pp. 115–121 (2014)

14. Massoudi, K., Tsagkias, M., de Rijke, M., Weerkamp, W.: Incorporating query expansion and quality indicators in searching microblog posts. In: Clough, P., Foley, C., Gurrin, C., Jones, G.J.F., Kraaij, W., Lee, H., Mudoch, V. (eds.) ECIR 2011. LNCS, vol. 6611, pp. 362–367. Springer, Heidelberg (2011)
15. McCreadie, R., Macdonald, C.: Relevance in microblogs: enhancing tweet retrieval using hyperlinked documents. In: Proceedings of OAIR 2013, pp. 189–196 (2013)
16. Mikolov, T., Chen, K., Corrado, G., Dean, J.: Efficient estimation of word representations in vector space. CoRR, arxiv.org/pdf/1301.37811301.3781 (2013)
17. Miyanishi, T., Seki, K., Uehara, K.: Improving pseudo-relevance feedback via tweet selection. In: Proceedings of CIKM 2013, pp. 439–448 (2013)
18. Ozsoy, M.G., Onal, K.D., Altingovde, I.S.: Result diversification for tweet search. In: Benatallah, B., Bestavros, A., Manolopoulos, Y., Vakali, A., Zhang, Y. (eds.) WISE 2014, Part II. LNCS, vol. 8787, pp. 78–89. Springer, Heidelberg (2014)
19. Pennington, J., Socher, R., Manning, C.D.: Glove: Global vectors for word representation. In: Proceedings of EMNLP 2014, pp. 1532–1543 (2014)
20. Qiang, R., Fan, F., Lv, C., Yang, J.: Knowledge-based query expansion in real-time microblog search (2015). CoRR, arXiv:1503.03961
21. Rodriguez Perez, J.A., McMinn, A.J., Jose, J.M.: University of glasgow (uog_twteam) at TREC microblog 2013. In: Proceedings of TREC 2013. (2013)
22. Rodriguez Perez, J.A., Moshfeghi, Y., Jose, J.M.: On using inter-document relations in microblog retrieval. In: Proceedings of WWW 2013, pp. 75–76 (2013)
23. Santos, R.L., Macdonald, C., Ounis, I.: Exploiting query reformulations for web search result diversification. In: Proceedings of WWW 2010, pp. 881–890 (2010)
24. Tao, K., Abel, F., Hauff, C., Houben, G.-J., Gadiraju, U.: Groundhog day: near-duplicate detection on twitter. In: Proceedings of WWW 2013, pp. 1273–1284 (2013)
25. Tao, K., Hauff, C., Houben, G.-J.: Building a microblog corpus for search result diversification. In: Banchs, R.E., Silvestri, F., Liu, T.-Y., Zhang, M., Gao, S., Lang, J. (eds.) AIRS 2013. LNCS, vol. 8281, pp. 251–262. Springer, Heidelberg (2013)
26. Teevan, J., Ramage, D., Morris, M.R.: #twittersearch: a comparison of microblog search and web search. In: Proceedings of WSDM 2011, pp. 35–44 (2011)

Short Papers

Displaying People with Old Addresses on a Map

Gang Zhang$^{(\boxtimes)}$ and Harumi Murakami

Graduate School for Creative Cities, Osaka City University,
3-3-138, Sugimoto, Sumiyoshi, Osaka 558-8585, Japan
springsimb@gmail.com

Abstract. This paper proposes a method of converting old addresses to current addresses for geocoding, with the aim of displaying on a map people who have such old addresses. Existing geocoding services often fail to handle old addresses since the names of towns, cities, or prefectures can be different from those of current addresses. To solve this geocoding problem, we focus on postal codes, extracting them from Web search result snippets using the query "prefecture name AND important place name AND *postal code.*" The frequency of postal codes and the edit distance between the old address and the addresses obtained using the postal codes are used to judge the most suitable postal code and thus the corresponding current address. The effectiveness of the proposed method is evaluated in an experiment using a relative dataset. A prototype system was implemented in which users could display people using their birthdate and birthplace addresses on a map chronologically with an associated history chart.

Keywords: People search · Old address · Postal code · Edit distance · Map display · Information extraction

1 Introduction

In our private lives, it is very important to learn about our ancestors and relatives and organize their information. Marking their positions on a map is a useful and feasible approach. The aim of this research is to develop a map interface to access people from the past to the present. Geocoding services are available to display people using their addresses. We want to obtain addresses from documents like census registers or other relevant documents. However, existing geocoding services often fail to obtain location information from old addresses. Most of the addresses contained in documents like family registers follow an older addressing system out of use today. It is difficult to convert old addresses to currently used addresses because house numbers, region names, and even basic town names change over time.

To solve the geocoding problems caused by old addresses, we focus on postal codes found on the Web. We assume that useful information for finding current addresses exists on the Web, and postal codes are good clues since they are easy to extract and easy to convert to current addresses using data provided by Japan Post [1].

G. Zuccon et al. (Eds.): AIRS 2015, LNCS 9460, pp. 381–386, 2015.
DOI: 10.1007/978-3-319-28940-3_30

The frequency of postal codes and the edit distance between old addresses and addresses obtained from the postal codes are used to identify the most suitable postal code and thus the corresponding current address. The effectiveness of the proposed method is confirmed in an experiment using a relative dataset that contains addresses from 1905 to 2006. A prototype system was implemented in which users could display people using their birthdate and addresses on a map chronologically with an associated history chart.

The structure of this paper is as follows. In Sect. 2 we explain our method. Experimental results are described in Sect. 3. The prototype is shown in Sect. 4. Related work and future work are described in Sect. 5. The examples presented in this paper were translated from Japanese into English for publication.

2   Approach

2.1   Overview

An overview of the proposed method consists of three stages:

(a) postal code collection: collects postal codes relevant to an old address from the Web; (b) best postal code judgment: calculates scores of postal codes using the frequency of each postal code and the edit distance between the old address and an address associated with the postal code obtained from Japan Post postal code data [1], and then judges the best one; and (c) address refinement: refines the address using Ministry of Land, Infrastructure, Transport and Tourism (MLIT) positioning reference data [2].

Table 1 presents example results from using the proposed method. We arbitrarily set house numbers as "99" in order to protect personal information. In this paper, we discuss example number 1.

Table 1. Example results of proposed method

No	Address in family register	Output address by proposed method
1	99-banchi, 1-chome, Uchiandojimachidori, Minami-ku, Osaka-shi	1-chome, Andojimachi, Chuo-ku, Osaka-shi, Osaka-fu
2	99-banchi, Kitanokofukacho, Kita-ku, Osaka-shi	1-chome, Shibata, Kita-ku, Osaka-shi, Osaka-fu
3	99-ban-yashiki, Takagi, Kokufu-mura, Ashida-gun, Hiroshima-ken	Takagi-cho, Fuchu-shi, Hiroshima-ken

2.2   Extracting Postal Codes

Addresses change with time. We focus on postal codes since they can be easily converted to the current addresses. To search for the postal code of an address, we use Web search engines; in this paper, we use Google. The combination of important place name and prefecture name were assumed to be a good choice for the query sent to the search engines, and thus the postal codes included in the snippets of the results were collected.

Extracting Important Place Name. We assume the most important type of place name that "lives long" is town names. To extract a Japanese town name, we remove the character strings focusing on numerical positions and such positional keywords as aza, mura, cho (town), gun, ku (ward), shi (city), to, do, fu, or ken (prefecture). For example, "Uchiandojimachidori" is extracted as an important place name.

Extracting Prefecture Names from the Web. To obtain a prefecture name, the old address without house numbers (e.g. 99-banchi, 99-ban-yashiki) is used for the query to the search engine. When the query "*prefecture* AND old address without house numbers" is given, our method obtains five snippets of results from Google. Prefecture names are extracted by matching a prefecture dictionary to the snippets and thus obtaining the most frequent prefecture name. For example, when the query "*prefecture* 1-chome Uchiandojimajidori Minami-ku Osaka-shi" is given, the result "Osaka-fu" is obtained.

Extracting Postal Codes from the Web. When the query "prefecture name AND important place name AND *postal code*" is given to Google, our method obtains one result snippet, and it extracts postal codes by using the following regular expression from the snippet: \d{3}-\d{4}.

When there is no result, our method repeats the above process for the other four result snippets. When no result is found even after five snippets, our method ends and *none* becomes the answer. For example, when the query "Osaka-fu AND Uchiandojimachidori AND *postal code*" is given, such postal codes as 542-0061, 542-0067 and 539-0000 are obtained from the snippet of the top page.

2.3 Judging Best Postal Code

We assume that the more frequently appearing postal codes are related to the current address, and the more similar character strings are related to the current address. Therefore, we calculate scores of postal codes by using the frequency of postal codes and the edit distance between addresses to judge the most suitable postal code. We use Japan Post postal code data to find an address associated with a postal code.

Edit distance is a way of quantifying how dissimilar two strings are to one another by counting the minimum number of operations required to transform one string into the other. We use Levenshtein distance as the edit distance.

The score is calculated as follows:

$$score = \frac{f}{d} \tag{1}$$

Here, f is the frequency of postal codes and d is the edit distance between an old address string without house numbers and an address associated with the postal code.

We select one postal code having the highest score. For example, the frequency of postal code 542-0061 is 3 and its edit distance is 11 (in Japanese), and thus its score (3/11=0.273) becomes the highest among the collected postal codes. Consequently, the address associated with the postal code 542-0061 is obtained as "Andojimachi, Chuo-ku, Osaka-shi, Osaka-fu."

2.4   Address Refinement

Most of the current official Japanese town names consist of a "chome" (district of a town) and a "town". The "chome" information is seldom contained in Japan Post data. Very old addresses do not have the "chome" designation either. We add a "chome" to the address obtained in Sect. 2.3 to complete the official town name: the available chome information is used when the old address includes it; otherwise, "1-chome" is used tentatively. Next, we check MLIT position reference data provided by the Ministry of Land, Infrastructure, Transport and Tourism, which contain current official addresses, and if the address exactly matches an entry, it becomes the final answer; if not, the "chome" is deleted. For example, "1-chome, Andojimachi, Chuo-ku, Osaka-shi, Osaka-fu" becomes the answer.

2.5   Geocoding

We obtain latitude/longitude information of the addresses obtained in Sect. 2.4 by using Google Maps Geocoding API v3 [3].

3   Experiment

3.1   Method

We use a relative dataset created in a previous work [4]. The oldest year with an address (birthplace) was 1905, and the newest year with an address (death place) was 2006. Since this is a dataset for Japanese relatives from the modern era, many addresses are identical. Thirty-two different addresses of birth or death places are used in the experiment.

The proposed method was compared with a method that gives old addresses without house numbers directly to Google Geocoding API v3.

We investigated current addresses for the old addresses without house numbers, in other words, town (including chome) levels at each public office corresponding to the old address. We evaluated the answers (output by the two methods) at prefecture, city, and town (including chome) levels. Here, traditional precision and recall are used for evaluation:

$$precision = \frac{r}{n} \tag{2}$$

$$recall = \frac{r}{c} \tag{3}$$

where r is the number of obtained correct answers, n is the number of answers output by the methods, and c is the total number of addresses in the dataset.

Table 2. Results of experiment

	Proposed method				Google geocoding			
	Precision		Recall		Precision		Recall	
Prefecture	100 %	(31/31)	97 %	(31/32)	96 %	(24/25)	75 %	(24/32)
City	87 %	(27/31)	84 %	(27/32)	80 %	(20/25)	63 %	(20/32)
Town	64 %	(16/25)	64 %	(16/25)	60 %	(9/15)	32 %	(8/25)

3.2 Results

Table 2 shows the results of the experiment. Our method outperformed Google
Geocoding API, particularly in recall.

4 Prototype

We have developed a prototype system that enables users to display people on a
map using Google Maps JavaScript API v3 [5] through a timeline. Figure 1 shows
living people in 1943 on a map with birthplace addresses. This application has
the following functions: (a) users can designate the year to display by inputting
it or by using a slider tool; and (b) users can browse historical events (extracted
from a history book [6]) before and after the year in a history chart, and they
can also change the year by selecting a historical event in the chart.

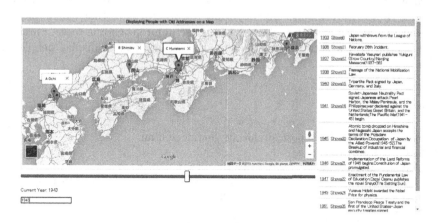

Fig. 1. Prototype system

5 Related Work and Discussion

FamilySearch [7] enables users to display people on a map based on manual
input of their prefecture-level location. Yamamoto et al. [8] proposed techniques for achieving the geographical navigation of historical events described in

Web pages as a Virtual History Tour. They extracted historical event sentences from the Web and identified one representative place name for each event. The place names were mainly at the prefecture or city level. However, we converted addresses in family registers to current addresses and focused on the town level.

Many Web sites provide a facility that converts addresses or landmarks into latitude/longitude, such as geocoding services. In general, however, they cannot handle old addresses. We aim to handle both old and current addresses in geocoding tasks.

The experimental results here reveal that our method outperformed Google Geocoding, thus demonstrating our method's usefulness. In addition, we successfully built a prototype interface that displays relatives on a map.

Future work includes the following. First, we need to improve our method to work effectively in earlier eras. Second, we need to evaluate other kinds of places like permanent addresses. Third, the map interface still has much room for improvement, such as adding the ability to display the relationships of relatives.

6   Conclusions

We proposed a method of converting old addresses to current addresses in order to use geocoding to display people on a map. The approach of our method follows three stages: (1) collecting postal codes relevant to an old address from the Web, (2) judging the best postal code by considering the frequency of each obtained postal code and the edit distance between the old address and an address associated with a postal code, and (3) refining the address by using location data provided by the government. The experimental results show that our method is superior to Google Geocoding API and thus useful for geocoding tasks. Furthermore, a prototype system was implemented as an interface to display people on a map.

References

1. http://www.post.japanpost.jp/zipcode/download.html
2. http://nlftp.mlit.go.jp/isj/
3. https://developers.google.com/maps/documentation/geocoding/
4. Murakami, H., Zheng, N.: Visualizing Family Trees: development of a dynamic family retrieval system. In: CSCW 2012 5th Workshop on Personal Information Management, PIM 2012 (2012)
5. https://developers.google.com/maps/web/
6. Kodansha International Ltd.: Japanese History: 11 Experts Reflect on the Past Large Print Edition. Kodansha International Ltd., Tokyo (2003)
7. https://familysearch.org/
8. Yamamoto, M., Takahashi, Y., Iwasaki, H., Oyama, S., Ohshima, H., Tanaka, K.: Extraction and geographical navigation of important historical events in the web. In: Kim, K.-S., Fröhlich, P., Tanaka, K. (eds.) W2GIS 2011. LNCS, vol. 6574, pp. 21–35. Springer, Heidelberg (2010)

A Query Expansion Approach Using Entity Distribution Based on Markov Random Fields

Rui Li[1], Linxue Hao[1], Xiaozhao Zhao[1], Peng Zhang[1],
Dawei Song[1,2(✉)], and Yuexian Hou[1]

[1] Tianjin Key Laboratory of Cognitive Computing and Application,
Tianjin University, Tianjin, China
{sig_lirui,haolinxue,0.25eye,darcyzzj,dawei.song2010,krete1941}@gmail.com
[2] The Computing Department, The Open University, Buckinghamshire, UK

Abstract. The development of knowledge graph construction has prompted more and more commercial engines to improve the retrieval performance by using knowledge graphs as the basic semantic web. Knowledge graph is often used for knowledge inference and entity search, however, the potential ability of its entities and properties for better improving search performance in query expansion remains to be further excavated. In this paper, we propose a novel query expansion technique with knowledge graph (KG) based on the Markov random fields (MRF) model to enhance retrieval performance. This technique, called MRF-KG, models the joint distribution of original query terms, documents and two expanded variants, i.e. entities and properties. We conduct experiments on two TREC collections, WT10G and ClueWeb12B, annotated with Freebase entities. Experiment results demonstrate that MRF-KG outperforms traditional graph-based models.

Keywords: Knowledge graph · Entity · MRF · Query expansion

1 Introduction

With the development of the semantic web, knowledge graph has been constructed for recording different entities in the world and connecting them with different links. Many commercial search engines (e.g., Google, Bing, and Baidu) have been constructing their own knowledge graphs, due to linked data in knowledge graph supplies their users with comprehensive knowledge which is helpful for improving retrieval performance and users' satisfaction. Traditional way of utilizing knowledge graph in information retrieval task is regarding it as external expansion source [1,4,8]. Diaz and Metzler has demonstrated that external expansion outperforms simulated relevance feedback in their research [2].

In this paper, we are interested in exploring Freebase for query expansion. Freebase is a large structured knowledge graph which consists of multi-relational data. Each node in it represents a specific entity (e.g., person/place/event) and a set of properties are attached to it, these properties establish relationships

© Springer International Publishing Switzerland 2015
G. Zuccon et al. (Eds.): AIRS 2015, LNCS 9460, pp. 387–393, 2015.
DOI: 10.1007/978-3-319-28940-3_31

between nodes and link them together as a graph. It now contains 46 million entities that covered massive domains, such as music, film and book. Therefore Freebase has been thought as a good external resource for information retrieval.

Previous works [8], processed only query entities in Freebase into single terms and combined with relevance model for query expansion. However, in general, keyword-based queries contain few entities and may not be enough to represent users' information needs. Considering the positive impact of entities in top-ranked pseudo-relevance feedback documents, Dalton et al. explored entities in both queries and documents for query expansion in their resent work [1]. Different from these works, we propose using the global and local importance of entities to discover expansion Freebase entities that related to query and the top-ranked pseudo-relevance feedback documents. Inspired by the idea of using MRF model for modeling dependence between query and document [6], we build MRF-KG model and improve it by adding two variants: expansion entities and associate properties to enhance retrieval performance.

Three tasks are involved in the study to achieve our goal: (a) how to model query-related entities and associate properties into MRF model, (b) when estimating the distribution of entities, how to balance the global importance of them and the local importance of entities in the context of the query entities, (c) how to filter and weight entity property terms as expansion terms.

2   Model

2.1   Overview of MRF-KG

The MRF for IR is a graph model used for model the joint distribution of one or more query terms with each document in [6], we make it as the baseline model. Based on the three cliques mentioned in MRF, including full independence, sequential dependence, and full dependence of query terms (shown in Fig. 1(a)), we add another clique to it, which is constructed by expansion entity set EN, the associate entity property set EP, documents D and queries Q ($\{EN, EP, D, Q\}$), as shown in Fig. 1(b). With the expansion graph H, a set of potentials ψ, and a parameter vector Λ, the joint distribution $P_{H,\Lambda}(EN, EP, D, Q)$ can be defined as follows:

$$P_{H,\Lambda}(EN, EP, Q, D) = \frac{1}{Z_\Lambda} \prod_{c \in C(H)} \psi(c; \Lambda) \tag{1}$$

where $Z_\Lambda = \sum_{EN,EP,Q,D} \prod_{c \in C(H)} \psi(c; \Lambda)$ is a normalizing constant. $C(H)$ is the set of cliques in H. Each $\psi(c; \Lambda)$ is a non-negative potential function over clique configurations parameterized by Λ. For ranking purpose, the document score is computed based on the posterior:

$$
\begin{aligned}
P_{H,\Lambda}(D|EN, EP, Q) &= P_{H,\Lambda}(EN, EP, Q, D) / P_{H,\Lambda}(EN, EP, Q) \\
&\overset{rank}{=} log P_{H,\Lambda}(EN, EP, Q, D) - log P_{H,\Lambda}(EN, EP, Q) \\
&\overset{rank}{=} \sum_{c \in C(H)} log \psi(c; \Lambda)
\end{aligned}
\tag{2}
$$

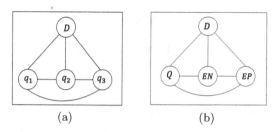

Fig. 1. (a)MRF model constructing the joint distribution of one or more query terms with the document; (b)MRF-KG model integrating two other variants entities EN and properties EP into MRF.

We denote the potential functions of cliques that are composed of document D and query terms q_i as $\psi_{MRF}(c; \Lambda)$, which is detailedly described in [6]. So the ranking function can be simplified in the following form:

$$P_{H,\Lambda}(D|EN, EP, Q) \stackrel{rank}{=} (1 - \lambda)\psi_{MRF}(c; \Lambda) + \lambda \sum_{c \in C(P)} f(EN, EP, Q, D) \quad (3)$$

where $C(P)$ is the set of cliques on $\{EN, EP, Q, D\}$, $f(EN, EP, Q, D)$ is the feature function on $C(P)$. This feature function describes how well query-related entities and properties match the documents. Hypothesizing that entities en in EN are independent to each other, much the same as property terms ep in EP, we can approximate the computation formulation as:

$$
\begin{aligned}
f(EN, EP, Q, D) &\propto \prod_{ep \in EP} P(ep|D) \cdot weight(ep) \\
&\approx \prod_{ep \in EP} P(ep|D) \sum_{en \in E_R} P(ep|en) \cdot weight(en) \\
&\approx \prod_{ep \in EP} P(ep|D) \sum_{en \in E_R} \sum_{d_i \in D_R} P(ep|en)P(en|Q, d_i) \cdot weight(d_i) \\
&\approx \prod_{ep \in EP} P(ep|D) \sum_{en \in E_R} \sum_{d_i \in D_R} P(ep|en)P(en|Q, d_i)P(d_i|Q)
\end{aligned}
\quad (4)
$$

where E_R denotes the set of top-ranked expansion entities, D_R denotes the top ranked documents in first-round retrieval. We get $P(ep|D)$ with the traditional language model. $P(en|Q, d_i)$ is estimated as the distribution of expansion entities; $P(ep|en)$ is estimated as the distribution of expansion property terms. We will illustrate these two distributions in the following parts.

2.2 Estimating Entity Distribution

With the constructed graph H, we will select expansion entities first and use it as sources for expanding property terms. Entity distribution is influenced by two factors, i.e., the global and local importance, showing as follows:

$$P(en|Q,d) = \alpha GI(en) + (1-\alpha)LI(en) \tag{5}$$

where $GI(en)$ implies the global importance of candidate entities, and $LI(en)$ implies the local importance of candidate entities, α is a coefficient to balance the global and local importance.

The global importance measures the influence of queries Q and all documents D over candidate entities. Limited by available annotated resources, we approximate the joint distribution $P(Q,D)$ to $P(D_R)$. Therefore, global importance can be estimated as:

$$GI(en) = P(en|Q,D) \propto P(en|D_R) \propto \sum_{d_i \in D_R} P(en|d_i)P(d_i|Q) \tag{6}$$

where $P(d_i|Q)$ indicates the retrieval score of d_i on Q, and $P(en|d_i)$ is entity frequency over document d_i's all annotated entities.

The local importance reflects the positive influence of contextual entities in d_i received from entities E_Q (i.e., the set of entities that explicitly appears in query). Each document has an entity sequence which is built by the position order of entities of it. According to the intuition that entities closer to E_Q are more relevant to query topic [5], we assign more weights to contextual entities which are closer to E_Q. The estimation of local importance is as follows:

$$LI(en_i) \propto \sum_{en_j \in E_Q} \sum_{d \in D_R} \frac{P(en_j|d)}{2^{|i-j|}} P(d|Q) \tag{7}$$

where i and j are the index of entity in the sequence. In order to fully capture contextual information in D_R, we regard all entities in d as context snippet.

2.3 Estimating Property Distribution

For the distribution of query-related property terms, we adopt the less-noisy properties (i.e., description and article) of entities in Freebase, which contain succinct description about entities. The probability of a property term ep in entity en is estimated by the term frequency (TF) of ep in the property values of en.

$$P(ep|en) = \frac{\#ep \; in \; entity \; property}{\#terms \; in \; entity \; property} \tag{8}$$

After estimating expansion entities and property terms's distribution above, we can directly derive the feature function $f(EN, EP, Q, D)$ in Formulate (4).

3   Experiments and Results

3.1 Experimental Setting

Two web collections ClueWeb12B and WT10G are employed in our experiment. The Clueweb12B dataset contains 52,343,021 documents (1.95T) with title queries from 201 to 250, whose entity annotation coverage is 48/50. And the WT10G dataset contains 1,692,096 documents (11G) with title queries from 501 to 550, whose entity annotation coverage is 40/50. Indri is used to implement our proposed model and baseline model, i.e. MRF and latent concept expansion (LCE) [7]. For both datasets, the processing of documents and queries includes stemming with Porter stemmer and stopwords removal using the standard Stopwords list. We use four metrics (i.e., MAP, P@20, NDCG@20 and ERR@20) to evaluate the retrieval effectiveness.

3.2 Entity Annotation

Queries and documents are annotated with Freebase entities in our experiment. We split each query into several fragments, then search them in Freebase knowledge graph using Freebase's API. The matched entities are regarded as annotations of entities, which is weighted by the match score provided by Freebase's API. Notice that there are 2 in topic 201-250 and 10 in topic 501-550 that have no entity been annotated, so we only employ global importance on it. Different methods are applied to annotate entities on different datasets. We use Google FACC1 [3] data for ClueWeb12, which automatically annotate ClueWeb12 documents with Freebase entities and the annotations are of generally high quality. Because no publicly available entity annotations exist for WT10G, we leverage the Alchemy API to annotate entities for WT10G documents, which can provide an entity identifier in Freebase for the annotated entity.

3.3 Parameter Setups and Experimental Results

We adopt the ranking function in Formulate (3) to the retrieval task, where our proposed feature function is estimated in Sect. 2.2, and the MRF potential function with its parameters are the same with [6]. In practice, we select top 100 documents in the first-round retrieval as D_R and set the number of query-related entities in E_R to 5. Three further parameters need to be adjusted in our MRF-KG model: the number of expansion property terms k, the interpolated coefficient λ in Formulate (3) and the balance parameter α. We sweep over $k \in \{10, 20, ..., 50\}$, $\lambda, \alpha \in \{0, 0.1, 0.2, ..., 1.0\}$ and set $k = 50$, $\lambda = 0.2$ and $\alpha = 0.6$ for two test collections to get the optimal retrieval performance. For the implementation of the baseline model MRF and LCE, we refer to the parameters setting in [6, 7] respectively.

Table 1 shows our evaluation results, from which we make two main observations. Firstly, taking expansion entities or concepts into account, MRF-KG performs significantly better than LCE in ClueWeb12B on all the evaluation metrics and in WT10G on MAP and P@20. Because, in addition to global importance of entities or concepts as LCE does, we also consider local importance in a document, which reflects the degree of entities related to queries more accurately. Secondly, comparing with MRF and LCE, the less perfect appearance of MRF-KG on NDCG@20 and ERR@20 cannot disguise its superiority in overall performance represented by a higher MAP value. In summary, the MRF-KG

Table 1. The retrieval performance comparison of MRF, LCE and MRF-KG. * and + indicate significant improvement ($p < 0.05$ in pair t-test) over MRF and LCE.

	ClueWeb12B				WT10G			
	MAP	P@20	NDCG@20	ERR@20	MAP	P@20	NDCG@20	ERR@20
MRF	0.0365	0.2480	0.1143	0.1415	0.1924	0.2612	0.3208	0.0653
LCE	0.0314	0.2122	0.0893	0.1128	0.1959	0.2755	**0.3293**	**0.0661**
MRF-KG	**0.0400**[*+]	**0.2760**[*+]	**0.1178**[+]	**0.1528**[+]	**0.2251**[*+]	**0.3064**[*+]	0.2977	0.0586

model is more efficient than MRF and LCE with leveraging external knowledge graph, making it richer and more accurate for entity analysis in queries.

4   Conclusions and Future Work

In this paper, we proposed a query expansion approach by using entity distribution based on Markov Random Fields, called MRF-KG. In MRF-KG, two expanded variants, i.e. entities and properties, are integrated with MRF, which can expand the original query to better represent user's information needs. Experimental results on WT10G and ClueWeb12B show that the MRF-KG performs better on web collections than the baseline model MRF and the expansion model LCE. The effectiveness of MRF-KG depends on the accuracy and comprehensiveness of entity annotations in documents. In the future work, we will investigate the relationships between entities, e.g., hypernym/co-hyponym, which is ignored in the current MRF-KG model. Moreover, we will experiment on more large-scale entity annotation resources, such as Wikipedia, to further demonstrate the effectiveness of our MRF-KG.

Acknowledgments. This work is supported in part by Chinese National Program on Key Basic Research Project (973 Program, grant No.2013CB329304, 2014CB744604), and the Natural Science Foundation of China (grant No. 61272265, 61402324).

References

1. Dalton, J., Dietz, L., Allan, J.: Entity query feature expansion using knowledge base links, pp. 365–374 (2014)
2. Diaz, F., Metzler, D.: Improving the estimation of relevance models using large external corpora. In: SIGIR, pp. 154–161 (2006)
3. Evgeniy, G., Michael, R., Amarnag, S.: Facc1: freebase annotation of clueweb corpora (2013)
4. Kotov, A., Zhai, C.: Tapping into knowledge base for concept feedback: leveraging-conceptnet to improve search results for difficult queries. In: WSDM, pp. 403–412. ACM (2012)
5. Lv, Y., Zhai, C.X.: Positional relevance model for pseudo-relevance feedback. In: SIGIR, pp. 579–586. ACM (2010)
6. Metzler, D., Croft, W.B.: A markov random field model for term dependencies. In: SIGIR, pp. 472–479. ACM (2005)

7. Metzler, D., Croft, W.B.: Latent concept expansion using markov random fields. In: SIGIR, pp. 311–318. ACM (2007)
8. Pan, D., et al.: Using dempster-shafer's evidence theory for query expansion based on freebase knowledge. In: Banchs, R.E., Silvestri, F., Liu, T.-Y., Zhang, M., Gao, S., Lang, J. (eds.) AIRS 2013. LNCS, vol. 8281, pp. 121–132. Springer, Heidelberg (2013)

Analysis of Cyber Army's Behaviours on Web Forum for Elect Campaign

Man-Chun Ko and Hsin-Hsi Chen[✉]

Department of Computer Science and Information Engineering,
National Taiwan University, Taipei, Taiwan
jkw552403@gmail.com, hhchen@ntu.edu.tw

Abstract. The goal of cyber army for elect campaign is to promote a certain candidate and denounce his/her rivals for ballots. This paper investigates the cyber army's behaviors with a real case study, 2014 Taipei mayoral race. We analyze the data crawled from the Gossip Forum on the Professional Technology Temple (PTT), Taiwan's largest online bulletin board. The operations of cyber army are shown and discussed.

Keywords: Cyber army · Deception · Opinion spammer

1 Introduction

Referencing public opinions disseminated on the web to make decisions is very common behaviours in daily life. Before purchase like reserving a hotel room and buying a ticket, customers usually consult the experiences about the possible candidates. Due to the influence and usefulness of opinions, how to collect, classify, and summarize opinions from heterogeneous sources is indispensable. On the other hand, opinion spammers, who intend to promote some specific entities and denounce their competitors, may disseminate deceptions, fakes and disinformation. Thus how to identify the misleading information and the information providers becomes another issue.

In elect campaign, camps employ various tools and methods to convince voters to vote their supporting candidates. In Obama's 2008 and 2012 election, his digital team influenced voters on Twitter, Facebook and other social tools. On the other hand, cyber army hired by unscrupulous camps aim at extolling specific candidates and criticizing their opponents. They play the similar roles of opinion spammers, and try to disseminate fake postings to mislead voters and influence the electoral process and outcome. Recently, various approaches have been proposed to detect opinion spams and spammers [1]. However, researches are few on political domain such as elect campaign.

Opinion spam/spammer detection is challenging from two aspects. Firstly, spam postings are often subtle because it will hurt the reputation of the promoted entities when the spam postings and spammers are recognized. Secondly, the opinion spam corpora are difficult to obtain because nobody are willing to recognize themselves as opinion spammers. Fortunately, two opinion spam corpora are disclosed recently. One is 2014 Taipei mayoral race, and the other one is 2013 Samsung Taiwan's event [1]. The former happened in the Gossip Forum on the Professional Technology Temple

© Springer International Publishing Switzerland 2015
G. Zuccon et al. (Eds.): AIRS 2015, LNCS 9460, pp. 394–399, 2015.
DOI: 10.1007/978-3-319-28940-3_32

(PTT), Taiwan's largest online bulletin board, which contains over 20,000 boards covering a multitude of topics. PTT has more than 1.5 million registered users with over 150,000 users online during peak hours and is similar to Reddit in America. A group of cyber army working for one camp in the Taipei mayoral race are disclosed. The latter happened in Mobile01, a web forum in Taiwan, where mobile phones, handheld devices and other consumer electronics are discussed. This paper focuses on the analysis of cyber army's behaviors in the Gossip Forum on PTT.

The structure of this paper is organized as follows. Section 2 surveys the related work in brief. Section 3 describes the dataset used in this study. Section 4 lists some findings from the dataset. Section 5 concludes the remarks.

2   Related Work

Social media brings in significant impact and becomes an essential tool for campaigns. Tumasjan et al. [2] analyze over 100,000 Twitter messages mentioning parties or politicians prior to the German federal election 2009 using LIWC text analysis software. Their results demonstrate that Twitter can be considered as a valid indicator of political opinion, and is indeed used as a platform for political deliberation.

Jindal and Liu [3] are the pioneers in study of opinion spams. Opinion spam detection is formulated as a classification problem. Experimental data are product reviews selected from Amazon. Duplicates or near-duplicates are regarded as fake reviews. Li et al. [4] detect campaign promoters on twitter with Markov Random Fields. Chen and Chen [1, 5] investigate opinion spam and spammer detection on a web forum. The event of Samsung Taiwan over fake web reviews is adopted as a real case study.

3   Dataset

In this paper, we study the cyber army's behaviors with a real case in the 2014 Taipei mayoral race. There are 2 major competitors, abbreviated as SL and KP, in this race. SL's camp hired some cyber army to promote SL or criticize KP on PTT implicitly. It's really difficult to tell whether one is cyber army or not from a post or a comment. In September 16th, a spokesman claimed he owned SL's official account and wanted to communicate with users on PTT. Some keen users launched a cyber-manhunt via IP address of a disclosed account, and discovered other 20 accounts sharing some specific IP addresses with the spokesman. Total 14 of these related accounts posted unusual number of articles from mid-August to mid-September. This research focuses on the operations of these 14 accounts on PTT.

3.1   PTT Corpus

An article has one main content written by an author and several comments from other users. There are usually several passages in a main content, while a comment is just 2 or 3 sentences in at most 30 Chinese characters. PTT provides an article rating mechanism: commendation and criticism. When commenting an article, users can evaluate it by giving a tuei (推) (adding a point, similar to the like button in Facebook), a hsü (噓) (subtracting a point), or an arrow (being neutral). Cyber army posted articles on Gossip Forum from August 18th until they were busted on September 16th. We scraped all the posts from Gossip Forum within this period. Total 49,740 articles and 1,454,994 comments were collected. Tables 1 and 2 show attributes of an article and a comment, respectively.

Table 1. Attributes of an article

Attribute	Description
article_id	Unique id of the article
author_id	Id of the user writing the article
nickname	Nickname of the author
time	Submission time of the article
title	Title of the article
content	Main content of the article
n_positive	Number of the commendation
n_negative	Number of the criticism
n_neutral	Number of the comment keeps neutral

Table 2. Attributes of a comment

Attribute	Description
article_id	Id of the article this comment belong
user_id	Id of the user who commented
position	Position relative to other comments
content	Content of the comment
time	Submission time of the comment
rating	Rating by the user

3.2 Treatment Group

During the period, there are 10,242 users who posted at least one article and 59,200 users who posted at least one comment, but most of them do not active as cyber army. We decide to select active users by measuring activeness of a user, where the activeness is defined by the number of articles a user posted during the period. Each account of cyber army posted at least 18 articles, so we select another 647 users who posted at least 18 articles as our treatment group during the period. The numbers of articles/comments by cyber army and treatment group are 644/293 and 22,569/146,829, respectively. The ratio of articles and comments written by cyber army is 2.198, while the ratio by treatment group is 0.154. It means cyber army tend to actively initiate more discussion. Most normal users in treatment group just passively give comments to posted articles.

3.3   User Posts History and Profile

The number of articles on sub-forum is restricted by PTT system. During the scheduled maintenance, system deletes old articles exceeding the limit. A website, ucptt.com, backs up many hot sub forums on PTT. We scraped post history of the 661 users, including cyber army and treatment group, from ucptt.com. In addition, we scraped profiles of the 661 users.

4   Some Findings

In this section, we analyze the dataset to get a grasp of what kinds of operations cyber army do. Their main goal is to attract young voters. However, it is not easy to gain others' trust for a new user. If a user keeps opposite against others, s/he would hear many acid remarks or even be harassed by others. Therefore, pretending to be one of common users is an important issue. We show how cyber army deal with it implicitly and are not noticed by other users.

4.1   Unusual Number of Articles in Gossip Forum

Recall KP and SL denote the candidates denounced and promoted by cyber army, respectively. The sign of posting unusual number of articles can be discovered by the user history. These 14 accounts of cyber army did not post any articles from May to first half of August. But from August 18th to September 16th, they posted unusual number of articles. In contrast, 96 % (621/647) users in treatment group have posted at least one article before August.

4.2   Working Time of Cyber Army

Figure 1 shows the submission time of articles and comments by cyber army (CA) and treatment group (TG). Their working time is different, because their roles are different on the forum, i.e., cyber army are at work, while treatment group are at leisure.

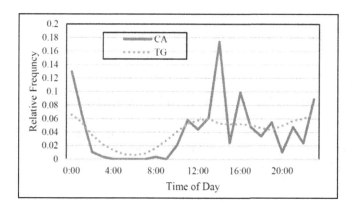

Fig. 1. Post submission time of cyber army and treatment group

4.3   Subtlety of Cyber Army

Posts (i.e., articles and comments) can be divided into the following seven types, where KP and SL denote the candidates denounced and promoted by cyber army, respectively.

(1) NR: posts not related to election.
(2) SL+: posts related to election, and favor SL.
(3) SL−: posts related to election, and disfavor SL.
(4) KP+: posts related to election, and favor KP.
(5) KP−: post related to election, and disfavor KP.
(6) 3rd: post related to election, and favor the 3rd candidate.
(7) NEU: post related to election without any favor.

Most articles by cyber army are not related to the mayoral race. Portion of the race related articles is 10.9 %, which is very close to the portion by treatment group (i.e., 10.3 %). Figure 2 shows the distribution of types of race articles by cyber army. An article may belong to more than one type, e.g., SL+ and KP−. The results demonstrate that 74.29 % favor SL and disfavor KP, 18.57 % are neutral, 5.71 % favor the 3rd candidate, and 1.43 % favor KP.

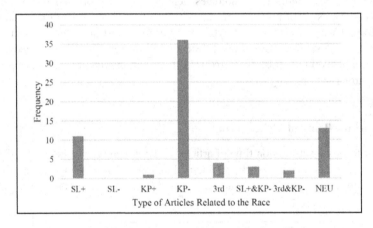

Fig. 2. Distribution of types of race articles by cyber army

4.4   Interaction Among Cyber Army and the Others

Figure 3 shows the distribution of comments by cyber army to all the articles. They interact more with cyber army group themselves than the others. Besides, they like articles of types SL+ and KP−, and dislike articles of types SL− and KP+. In other words, cyber army promote their supporting candidate, and denounce the opponent through the article rating mechanism on PTT.

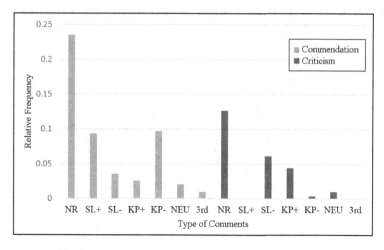

Fig. 3. Distribution of types of comments by cyber army

5 Conclusion

This paper analyzes the operations of a special type of opinion spammers, cyber army in elect campaign. Unusual number of submission within a race period, working time, types of race articles, and types of comments are critical features. The submission time and group behaviors in political opinion spams are similar with those of product opinion spams. However, their sentiment words are different. Political opinions are quite related to the events happening in burst in elect campaign. We will extend the dataset and employ the findings to cyber army detection in the future.

Acknowledgments. Research of this paper was partially supported by Ministry of Science and Technology, Taiwan under the contract MOST104-2221-E-002-061-MY3.

References

1. Chen, Y.R., Chen, H.H.: Opinion spam detection in web forum: a real case study. In: Proceedings of 24th International World Wide Web Conference, pp. 173–183 (2015)
2. Tumasjan, A., Sprenger, T.O., Sandner, P.G., Welpe, I.M.: Predicting elections with Twitter: what 140 characters reveal about political sentiment. In: Proceedings of the Fourth International AAAI Conference on Weblogs and Social Media, pp. 178–185 (2010)
3. Jindal, N., Liu, B.: Opinion spam and analysis. In: Proceedings of the 2008 International Conference on Web Search and Data Mining, pp. 219–230 (2008)
4. Li, H., Mukherjee, A., Liu, B., Kornfieldz, R., Emery, S.: Detecting campaign promoters on Twitter using markov random fields. In: Proceedings of IEEE International Conference on Data Mining, pp. 290–299 (2014)
5. Chen, Y.R., Chen, H.H.: Opinion spammer detection in web forum. In: Proceedings of the 38th International ACM SIGIR Conference on Research and Development in Information Retrieval, pp. 759–762 (2015)

Beyond tf-idf and Cosine Distance in Documents Dissimilarity Measure

Sunil Aryal[1(✉)], Kai Ming Ting[2], Gholamreza Haffari[1], and Takashi Washio[3]

[1] Monash University, Victoria, Australia
{sunil.aryal,gholamreza.haffari}@monash.edu
[2] Federation University, Victoria, Australia
kaiming.ting@federation.edu.au
[3] Osaka University, Osaka, Japan
washio@ar.sanken.osaka-u.ac.jp

Abstract. In vector space model, different types of term weighting schemes are used to adjust bag-of-words document vectors in order to improve the performance of the most widely used cosine distance. Even though the cosine distance with some term weighting schemes result in more reliable (dis)similarity measure in some data sets, it may not perform well in others because of the underlying assumptions of the term weighting schemes. In this paper, we argue that the explicit adjustment of bag-of-words document vectors using term weighting is not required if a data-dependent dissimilarity measure called m_p-dissimilarity is used. Our empirical result in document retrieval task reveals that m_p with the simplest binary bag-of-words representation is either better or competitive to the cosine distance with the best performing state-of-the-art term weighting scheme in four widely used benchmark document collections.

Keywords: Cosine distance · Term weighting · m_p-dissimilarity

1 Introduction

Using the bag-of-words [1] vector representation, a document d_i in a collection of n documents ($i = 1, 2, \cdots, n$) is represented by a r-dimensional vector (where r is the number of terms in the dictionary), i.e., $d_i = \langle d_{i1}, d_{i2}, \cdots, d_{ir} \rangle$, where each entry d_{ij} represents the occurrence frequency of term t_j in d_i. As the most widely used cosine distance [1] estimates the dissimilarity of two vectors using their geometric positions only, it is important to adjust their positions in the space according to the importance of their terms. Two types of term weighting factors are used in the literature to estimate the importance of term t_j in document d_i (w_{ij}) [2]. First, the term frequency (tf) based factor of t_j in d_i (tf_{ij}) can be estimated in different ways: (a) Binary (bin_tf): $tf_{ij} = 1$ if t_j is in d_i and 0 otherwise; (b) Raw term frequency (Raw_tf): $tf_{ij} = d_{ij}$; and (c) Logarithmic (log_tf): $tf_{ij} = \log(1 + d_{ij})$. Second, the inverse document frequency (idf) based weighting factor of a term t_j (idf_j) is estimated using the number of documents in the given collection having term t_j (n_j) as $idf_j = \log\left(\frac{n}{n_j}\right)$.

© Springer International Publishing Switzerland 2015
G. Zuccon et al. (Eds.): AIRS 2015, LNCS 9460, pp. 400–406, 2015.
DOI: 10.1007/978-3-319-28940-3_33

Using tf_{ij} and idf_j, a term-weighted document vector of document d_i is represented as having component $w_{ij} = tf_{ij} \times idf_j$. The dissimilarity between document vectors d_1 and d_2 using the cosine distance is estimated as follows:

$$dist_{cos}(d_1, d_2) = 1 - \frac{\sum_j w_{1j} \times w_{2j}}{\sqrt{\sum_j w_{1j}^2} \times \sqrt{\sum_j w_{2j}^2}} \qquad (1)$$

It has been shown that the above cosine distance is more meaningful than the traditional ℓ_2-norm in text retrieval. The only difference between the cosine distance and ℓ_2-norm is that it uses the length normalized vector which is referred as cosine normalization in the literature.

The ideas of term weighting and cosine normalization are based on the following three monotonic assumptions [3]: (i) Multiple appearances of a term in a document are no less important than single appearance (the tf assumption). (ii) Rare terms are no less important than frequent terms (the idf assumption). (iii) For the same quantity of term matching, long documents are no more important than short documents (the cosine normalization assumption).

Even though these assumptions seem reasonable, the similarity measure biases toward smaller documents, documents with infrequent terms and documents with multiple occurrences of terms which can be disadvantageous in some cases [4]. The cosine distance with term weighted document vectors may perform well in some data sets or domains where the above assumptions hold but it may perform worse in other data sets where the assumptions do not hold (see the experimental results in Sect. 3).

Table 1. Dissimilarity between d_q and other documents in a data set.

Doc	t_1	t_2	t_3	$dist_{cos}$
d_1	5	2	0	0.82
d_2	2	2	0	0.58
d_3	2	0	0	1.00
d_4	1	2	0	0.32
d_5	0	2	1	0.74
d_6	3	2	4	0.93
d_7	1	6	2	0.63
d_q	0	2	0	-

In an example shown in Table 1, the dissimilarity between d_q and each of the documents d_1-d_7 using the cosine distance with $raw\_tf\text{-}idf$ term weighting is provided in the fourth column. Even though d_4 and d_5 have the same occurrences of the common term t_2 with d_q, d_4 is considered to be more similar to d_q than d_5 for no particular reason because of the idf assumption (d_5 is penalized more due to the mismatch in infrequent term t_3). Similarly, d_2 is considered to be more similar to d_q than d_1 because of the cosine normalization assumption (d_2 is shorter than d_1). Even though d_6 has the same occurrences of the common term t_2 with d_q, d_7 is considered to be more similar to d_q than d_6 because of the tf assumption (d_7 has more occurrences of t_2).

Furthermore, in order to use the tf based weights such as raw_tf and log_tf, the frequency of each term in each document is required. However, in some application domains such as legal and medical, it may not be possible to have the exact term frequencies due to privacy issue because it is possible to infer information in the document from its term frequencies [5]. Hence, in some domains, only binary representation of documents is available rather than their raw term frequencies.

In this paper, we investigate a dissimilarity measure that does not require an adjustment of bag-of-words vectors and demonstrate that the recently proposed data-dependent dissimilarity measure called m_p-dissimilarity [6] is one such measure. It uses a similar statistic as used in idf_j but it is used as the measure of dissimilarity directly rather than for vector adjustment in the space. Our empirical evaluation shows that m_p-dissimilarity with the simplest binary representation performs either better than or competitively to the cosine distance with different term weighting schemes in document retrieval tasks. Its performance is more consistent across different data sets than that of the cosine distance with any term weighting scheme.

2 m_p-Dissimilarity in Bag-of-Words Document Vectors

In order to measure dissimilarity between two r-dimensional data points x and y, rather than just relying on the positions of x and y in the space, m_p-dissimilarity [6] (we refer to it as m_p hereafter) considers the probability data mass in the range $R_j(x, y)$ that encloses x and y in each dimension j. It estimates the final dissimilarity as follows [6]:

$$ m_p(x, y) = \left(\frac{1}{r} \sum_{j=1}^{r} \left(\frac{|R_j(x, y)|}{n} \right)^p \right)^{\frac{1}{p}} \tag{2} $$

where $|R_j(x, y)|$ is the number of data points falling in $R_j(x, y)$; n is the total number of data points; and $p > 0$ is a parameter.

Simply by replacing the geometric distance in each dimension by the probability mass in the range, m_p has been shown to provide more reliable nearest neighbours than ℓ_p-norm in high dimensional spaces [6]. But, it is very expensive to compute as it requires a range search to determine how many instances fall in each $R_j(x, y)$. Using a binary search tree, one-dimensional range search can be done in $O(\log n)$ resulting in the run-time complexity of $O(r \log n)$ to measure dissimilarity of a pair of vectors.

In a document collection, only a few terms in the dictionary appear in each document. Many terms do not appear in either of the two documents given for dissimilarity measurement. Since the absence of a term in both the documents does not provide any information about (dis)similarity of documents, those terms should be ignored. Hence, we make a simple modification in the formulation of m_p shown in Eq. 2 by considering only those terms that occur in either of the two documents as follows:

$$m_p(d_1, d_2) = \left(\frac{1}{|T_{1,2}|} \sum_{t_j \in T_{1,2}} \left(\frac{|R_j(d_1, d_2)|}{n} \right)^p \right)^{\frac{1}{p}} \tag{3}$$

where $|T_{1,2}| = |T_1 \cup T_2|$ (T_i is the set of terms that appear in d_i) is the normalization term employed to account for different numbers of terms used for any two documents.

Using the simplest binary representation, where each d_{ij} in a document vector d_i has only two values $\{1,0\}$ indicating whether the term t_j exists in the document d_i, $|R_j(d_1, d_2)|$ can be estimated easily using the total number of documents in the collection (n) and the number of documents where t_j occurs (n_j) as follows:

$$|R_j(d_1, d_2)| = \begin{cases} n & \text{if } d_{1j} \neq d_{2j} \\ n_j & \text{if } d_{1j} = d_{2j} = 1 \end{cases} \tag{4}$$

Note that the case where $d_{1j} = d_{2j} = 0$ is not required because Eq. 3 does not measure dissimilarity of d_1 and d_2 w.r.t a term which does not appear in both d_1 and d_2. n_j can be precomputed for all t_j as a pre-processing; thus, $|R_j(d_1, d_2)|$ can be estimated in $O(1)$ resulting in $O(r)$ complexity to compute m_p-dissimilarity of a pair of documents using Eq. 3 which is equivalent to that of the cosine distance. The pre-processing to compute n_j for all t_j requires $O(rn)$ time and $O(r)$ space complexities. Note that the same complexities are required in computing idf factor.

Note Eq. 3 does not require to adjust the positions of documents in the vector space because it is not using the absolute positions of two vectors in the dissimilarity measure. It estimates dissimilarity w.r.t each term t_j that appears in both the documents based on the distribution of documents (i.e., high dissimilarity if t_j appears in many documents, and low dissimilarity if it appears only in a few documents) and assigns maximal dissimilarity of 1 w.r.t terms that appear in only one of them. Even though a similar statistic as in idf based weighting is used in the case of matching term, it is not used to transform vectors in the space but it is used as a measure of (dis)similarity between two documents w.r.t. t_j directly. In the example shown in Table 1, $m_p(d_q, d_4) = m_p(d_q, d_5)$, $m_p(d_q, d_1) = m_p(d_q, d_2)$ and $m_p(d_q, d_6) = m_p(d_q, d_7)$.

3   Empirical Evaluation

In this section, we present the empirical results of m_p (using the binary representation) and the cosine distance with the six different term weighting schemes as discussed in Sect. 1 (bin_tf, raw_tf and log_tf with and without idf) in document retrieval tasks. Since we want to capture the contrast between two documents with low dissimilarity in a few common terms and maximal dissimilarity w.r.t many terms that appear in either one of them, $p < 1$ is preferred to amplify the effect of low dissimilarities in the average. Hence, we used $p = 0.1$ for m_p (i.e. $m_{0.1}$) in our experiments[1].

Table 2. Data sets

Name	#docs	#terms	#cat
NG20	18,821	5,489	20
R52	9,100	7,379	52
Ohscal	11,162	11,465	10
Wap	1,560	8,460	20

We used 4 different data sets from 4 benchmark document collections that are used in the text mining literature. The data characteristics are provided in Table 2. NG20[2] is the widely used 20 Newsgroup data set and R52 (See footnote 2) is a subset of the another widely used Reuters document collection [7]. Ohscal[3] is a data set from Ohsumed patients' medical information collection and Wap (See footnote 3) is a collection of web pages from Yahoo [8].

Given a query document d_q, documents in a data set were ranked in ascending order of their distance/dissimilarity to d_q; and the first k documents were presented as the relevant documents. For performance evaluation, a document was considered to be relevant to d_q if they have the same category label. A good retrieval system returns relevant documents at the top. Hence, the precision in the top 10 (P@10) retrieved documents was used as the performance measure. The same process was repeated for every document in a data set as a query and the rest of the documents were ranked. The average P@10 over n (the number of documents in a collection) queries of $m_{0.1}$ and cosine with six different term weighting schemes are provided in Table 3. Note that all the differences are statistically significant as they are averaged over n (≥ 1560) queries and the standard error is negligible (up to two decimal places) in each case.

The performance of $m_{0.1}$ was more consistent than the cosine distance with any term weighting scheme across four data sets (see the average result in the last column in Table 3). It was among the top three performers in each data set (and has the best performance in Wap and Ohscal, the second best in R52 and the third best in NG20) whereas none of the term weighting schemes were among the top three performers in all data sets.

---

[1] The parameter p in m_p has the same role as in the case of traditional ℓ_p-norm. The performance of m_p may be changed slightly using different p values in some data sets. Empirically, we observed that $p = 0.1$ is a reasonably good setting.

[2] http://web.ist.utl.pt/acardoso/datasets/.

[3] http://www.cs.waikato.ac.nz/ml/weka/datasets.html.

Table 3. P@10 with average over four data sets in the fourth column (∗: best, †: second best and ‡: third best).

Contenders	NG20	R52	Ohscal	Wap	Avg
raw_tf	0.56	$0.85^{\dagger}$	0.53	0.63	0.64
$raw\_tf\text{-}idf$	0.71	0.81	0.48	0.64	0.66
log_tf	0.70	$\mathbf{0.87^{*}}$	$\mathbf{0.61^{*}}$	0.64	0.71
$log\_tf\text{-}idf$	$\mathbf{0.76^{*}}$	0.81	0.54	$0.67^{\ddagger}$	0.70
bin_tf	0.66	0.84	$0.59^{\dagger}$	0.60	0.67
$bin\_tf\text{-}idf$	$0.75^{\dagger}$	0.79	0.56	$0.68^{\dagger}$	0.70
$m_{0.1}$	$0.74^{\ddagger}$	$0.85^{\dagger}$	$\mathbf{0.61^{*}}$	$\mathbf{0.72^{*}}$	**0.73**

It is interesting to note that the idf based weighting does not always result in good performance as it produced poor results in R52 and Ohscal with any of the three tf representations. Even though log_tf produced the best result in R52 and Ohscal, it did not produce the top three results in the other two data sets. Similarly, $log\_tf\text{-}idf$ produced the best result in NG20 and the third best in Wap but did not produce the top three results in Ohscal and R52. Cosine with $bin\_tf\text{-}idf$ was among the top three performers in two data sets whereas bin_tf and raw_tf in one data set and $raw\_tf\text{-}idf$ did not produce the top three results in any data set.

4   Concluding Remarks

Since the cosine distance measures (dis)similarity solely based on the positions of two vectors, it is important to adjust the positions of document vectors in the space w.r.t the importance of the terms in those documents. In the literature, many term weighting schemes are proposed using the tf and idf factors based on some assumptions. Though these methods perform well in some data sets when the assumptions hold, they could perform poorly when the assumptions do not hold.

Rather than focusing on researching an effective term weighting scheme to improve the performance of the cosine distance, this paper opens a different avenue for research by investigating an alternative dissimilarity measure that does not require the adjustment of document vectors using a term weighting scheme. We show that a 'data-dependent' dissimilarity measure called m_p-dissimilarity is one such effective alternative. It considers (dis)similarity between a pair of documents based on the distribution of documents in each term.

Our empirical result in document retrieval tasks shows that m_p-dissimilarity with the binary bag-of-words representation produces either better or competitive results in comparison to the cosine distance with the state-of-the-art term weighting schemes.

References

1. Salton, G., McGill, M.J.: Introduction to Modern Information Retrieval. McGraw-Hill Inc, New York (1986)
2. Salton, G., Buckley, C.: Term-weighting approaches in automatic text retrieval. Inf. Process. Manage. **24**(5), 513–523 (1988)
3. Zobel, J., Moffat, A.: Exploring the similarity space. SIGIR Forum **32**(1), 18–34 (1998)
4. Polettini, N.: The Vector Space Model in Information Retrieval - Term Weighting Problem, University of Trento, Italy (2004). https://wiki.eecs.yorku.ca/course_archive/2014-15/W/6339/_media/polettini_information_retrieval.pdf
5. Zhu, X., Goldberg, A.B., Rabbat, M., Nowak, R.: Learning bigrams from unigrams. In: Proceedings of ACL 2008: HLT, Association for Computational Linguistics, pp. 656–664 (2008)
6. Aryal, S., Ting, K., Haffari, G., Washio, T.: m_p-dissimilarity: a data dependent dissimilarity measure. In: Proceedings of the IEEE International Conference on Data Mining (ICDM), pp. 707–712. IEEE (2014)
7. Cardoso-Cachopo, A.: Improving Methods for Single-label Text Categorization. Ph.D. Thesis, Instituto Superior Tecnico, Technical University of Lisbon, Lisbon, Portugal (2007)
8. Han, E.-H.(Sam), Karypis, G.: Centroid-based document classification: analysis and experimental results. In: Zighed, D.A., Komorowski, J., Żytkow, J.M. (eds.) PKDD 2000. LNCS (LNAI), vol. 1910, pp. 424–431. Springer, Heidelberg (2000)

Explorations of Cross-Disciplinary Term Similarity

Hanif Sheikhadbolkarim[(✉)] and Laurianne Sitbon

Queensland University of Technology (QUT), Brisbane, Australia
{hanif.sheikhabdolkarim,laurianne.sitbon}@qut.edu.au

Abstract. This paper presents some initial explorations into how to compute term similarity across different domains, or in the present case, scientific disciplines. In particular we explore the concepts of polysemy across disciplines, where the same term can have different meaning across different discipline. This can lead to confusion and/or erroneous query expansion, if the domain is not properly identified. Typical bag-of-words systems are not equipped to highlight such differences as terms would have a single representation. Identifying the synonymy of terms across different domains is also a difficult problem for typical bag-of-words systems, as they use surrounding words that will usually also be different across domains. Yet discovering such similarities across domains can support tasks such as literature discovery. We propose an approach that integrates knowledge based distances into a distributional semantics framework and demonstrate its efficiency on a hand-crafted dataset.

Keywords: Distributional semantics · Cross-disciplinary similarity · Polysemy · WordNet similarity

1 Introduction

Research to support interdisciplinary interactions can be broadly divided into approaches motivated from (1) knowledge engineering, (2) semantic computing and (3) computer-supported cooperative work. Knowledge engineering has proposed systematic approaches to facilitate a common understanding between disciplines, with the development of large ontologies, e.g., the Unified Medical Language System (UMLS[1]). However, ontologies are manually constructed, costly to build, and usually lack complete conceptual coverage of the respective domains of the disciplines. Few publications detail models or tools aimed at exploiting the semantics in ontologies to address the specific challenge of integration.

Modeling connections and disconnections between vocabularies of various disciplines or domains can support cross-disciplinary search with query expansion techniques that include new connections between concepts but also avoid terms that could have a confusing meaning. For instance, the word "cell" has several distinct interpretations including that of "the smallest block of biological unit" in a biology context,

---

[1] http://www.nlm.nih.gov/research/umls/.

© Springer International Publishing Switzerland 2015
G. Zuccon et al. (Eds.): AIRS 2015, LNCS 9460, pp. 407–412, 2015.
DOI: 10.1007/978-3-319-28940-3_34

"web" in architecture and "A basic unit of storage" in computer science. This can also fully automated cross-disciplinary tasks such as literature-based discovery.

One of the basic challenges to model and use such connections between lexical domains is to measure the semantic similarity between terms from these domains. With current approaches to measuring semantic similarity, it would require either very thoroughly hand-crafted knowledge repositories or contextual vectors defined in single or aligned word spaces [2]. The similarity of semantic spaces has been addressed in some studies [4], but is not sufficient to identify polysemy (terms with different meanings across domains).

Thus, the question we address in this paper is how to compare domain concepts whose context vectors are defined in different spaces. We propose an initial approach based on semantic alignment from a generic ontology, and evaluate its potential on a small hand crafted dataset with 2 tasks: identify which terms have different senses and synonyms across 2 disciplines.

2   Related Work

In IR, the notion of domain concept is generally associated with either word senses as defined in some knowledge based repository, or the statistical use of terms in context based on the distributional hypothesis.

Computing relatedness and the degree of similarity between words has often been addressed with corpus based geometric or probabilistic approaches, such as LSA, HAL, LDA and NNLM [8, 10]. These approaches generate a vector (or probabilistic) space in which terms are defined in terms of other terms, effectively dimensions of the semantic space, into what is referred to as context vectors. Similarity between terms can then be established using mathematical tools to compare the vectors, provided that they are defined in the same space. These models cannot directly address cross-domain environments, unless initial connections between domains are provided. WSD and Named Entity Recognition [6] techniques attempt to clarify specific domain terms or senses, where a domain is then considered as an area of knowledge that is widely accepted as unitary, and there is generally no attempt to connect domains.

Knowledge based repositories (e.g. lexical resources) are hand-crafted and are employed for discovering semantic relations between words and as a basis for WSD tasks. WordNet provides some domain specific versions [7], listing domain specific terms. Distributional approaches [5] or graph-based approaches [4] may be used to enhance the estimation of semantic similarity within domain.

Domain adaptation methods seek to learn a common feature representation, by either approximating the posterior distribution of source data on target data [3], using heuristic selections of common pivot features [1], or learning a shared latent sematic space underlying the domains [9]. However, the number of unseen patterns in these methods can impact on final adaptation, these techniques require prior knowledge.

A recent method proposes to use only data labeled in a source domain instead of alignment data, and uses an incremental learning algorithm [3]. In this paper, we make the assumption that common and similar terms between domains seed the notion of alignments.

3   Methodology

Our proposed approach is to first create separate distributional representations for the separate domains, and then connect them. We propose here to generate the initial representations with the HAL approach, which is a distance-sensitive word space.

With a large vocabulary gap between domains, we hypothesise that standard similarity measures will be affected by few identicalterms. We propose to use the semantic similarity provided by external resources to connect the dimensions.

3.1   EHAL: A Syntax-Sensitive Word Space

HAL (Hyperspace Analog to Language) creates a word space (a square term-term matrix) by scanning a corpus of documents, and for each term t encountered, updating its context vector by adding to each dimension corresponding to terms occurring on a window of size W around t a value function of its distance to t. We additionally propose that this function should also account for the syntactic role of these surrounding terms. For a given term t and a context (surrounding) ter.t_y, the weight added to the dimension corresponding to W in the context vector of t would then be:

$$f(t, t_y) = (W - |t - t_y|) \times \frac{\alpha}{|t - t_y|} \times \beta(POS(t_y)) \tag{1}$$

Where α is an additional factor to enhance the effect of distance, and β a weighting factor that is dependent on the syntactic role of the context term

Finally, we normalize the matrix using Positive Pointwise Mutual Information (PPMI), to further emphasise the importance of rare co-occurring terms, and to improve performance of all distributional semantic approaches [2].

3.2   Semantic Similarity Between Context Vectors

For two different domains' concepts for example, "Algorithm" from computer science and "Process" from bio-information, the non-null part of their context vectors may consist of a few dimensions [e.g. step, sequence] and [e.g. stage, chain] respectively and as they don't share any of these non-null dimensions, their similarity would be null. It means the context term "chain" in the context vector of the term "process" could not feature in the context vector of "algorithm". However, "sequence", in the term "algorithm", might be the closest context term according to the WordNet (in this case with a distance of 0.66).

Thus, we propose to leverage the degree of semantic association of the context vectors' non-null dimensions by employing the notion of similarity distance in an external resource, such as WordNet. More precisely, for each context term (non-null dimension) of a source context vector in domain $D1$, we seek the closest context term in the target context vector in domain $D2$. The score we propose is additionally asymmetric, and seeks to measure the similarity by a mapping of the most similar context terms of the second word to the first. The similarity score is the sum of the weighted score for each non-null dimensions of the first vector. This weighted score is the average of the context vector values for that dimension and the most similar non-null dimension in the second vector, multiplied by the distance between the non-null dimensions. This is illustrated by Equation which illustrates how to calculate the semantic association between two words W1 and W2 with the semantic similarity between the context vectors reduced to their non-null values $W1[x_1,\ldots,x_n]$ and $W2[y_1, \ldots,y_m]$ defined in the sub-set of dimensions of domain $A\{a_1,\ldots,a_n\}$ and domain $B\{b_1, \ldots,b_m\}$ respectively. The semantic similarity between the dimensions is defined by an external lexical resource.

$$\text{SemSimilarity}(W1,W2) = \sum_i \frac{x_i + y_j}{2} \times \max_j \big(\text{ExternalSimilarity}(a_i, b_j)\big) \quad (2)$$

4   Evaluation

4.1   Evaluation Dataset

We built a small dataset to evaluate the potential for our approach to similarity measure. To do this, we selected 11 pairs of synonym words, and 11 polysemy words between 2 domains which are biology and computer science as we had experts able to advise on these domains. The synonym pairs are (Process, Algorithm), (Analyse, Compute), (Study, Survey), (Reproduction, Replication), (Duplicate, Clone), (Sequence, Chain), (Reserve, Store), (Produce, Develop), (Flaw, Imperfect) and the polysemous words are (Cell, Gene, Host, Synthesize, Clone, Synthetic, Mutant, Object, Genetic, Taxonomy, Recovery, System, Benchmark).

4.2   Implementation

Two different corpora belonging two different domains were implemented by crawling two publicly available corpora to build the models for our evaluation. There are PubMed[2] for the Biology domain and ACM[3] for the computer science domain.

We first pre-processed the corpora (stemming and removing stop words and low frequency terms). HAL and the EHAL were implemented with window size 20 (10 words each side of the target term). In EHAL, we set $\alpha = 5$ and $\beta = 2$ if $POS(t_y) = $ *verb and noun, else* 0.8 on the basis of intuition. α set to 5 to emphasise context words in a short distance to the target, while the β function is based on the intuition that emphasising noun-noun and noun-verb contexts is more likely to provide the significant semantic context across domains.

We used a path-distance measure of WordNet as the external similarity measure [5], which is the minimum number of edges that separate two words.

---

[2] http://www.ncbi.nlm.nih.gov/pmc/tools/ftp/#Data_Mining ≈Documents.

[3] http://dl.acm.org/, https://www.comp.nus.edu.sg/~ sugiyama/SchPaperRecData.html.

4.3   Rank-Based Evaluation

We first investigated the appropriateness of the similarity scores provided by each method to differentiate between synonyms and polysemous terms expressed in different domains. More precisely, we measured the proportion of synonyms that had a similarity score higher than the highest score between variants of polysemous terms, and higher than the second highest score, respectively.

The results presented in Table 1 show that our proposed combined approach provides the most accurate distinctions. It also shows that it is reasonably stable when we look further than the highest score and is not only relying on chance. This suggest that a threshold could be estimated to provide a reliable way of distinguishing between words. It is important to note however that the detection of synonyms is still reasonably low and these results recognize that it is a difficult task.

Table 1. Accuracy of the methods when using a 0.5 threshold for each of the tasks.

Task	HAL+cosine	EHAL+cosine	EHAL+SemSimilarity
Higher than 1st	0 %	9 %	45 %
Higher than 2nd	9 %	18 %	45 %

4.4   Threshold-Based Evaluation

In this second evaluation, we considered a threshold for separating synonyms from non-synonyms. We first proposed to apply a standard threshold of 0.5 on both WordNet and cosine similarity measures and measured the accuracy for each task. We also investigated what an oracle threshold would lead to, and reported the range in which it would have to be for each method to maximize the overall accuracy across both tasks. The results are reported in Tables 2 and 3. All methods use PPMI.

Table 2. Accuracy of the methods when using a 0.5 threshold for each of the tasks.

Task	HAL+cosine	EHAL+cosine	EHAL+SemSimilarity
Detect Polysemy	91 %	91 %	100 %
Detect Synonyms	0 %	0 %	46 %

Table 3. Accuracy of the methods with oracle thresholds optimized for overall accuracy

	HAL+cosine	EHAL+cosine	EHAL+SemSimilarity
Threshold range	[0.08-0.09]	[0.1-0.11]	[0.35-0.46]
Detect polysemy	55 %	59 %	77 %
Detect synonyms	64 %	82 %	100 %
Overall accuracy	45 %	36 %	55 %

We observe that not only did our approach always outperform the comparisons, but also that the range of an oracle threshold is much wider than for the other methods, suggesting a much better separation between synonyms and unrelated words.

The results also show that all methods could identify differences between polysemous words, which suggests that indeed the vectors built from the two different corpora are quite different for these words. However, standard approaches are not able to draw enough similarities between the vectors of synonyms across the different domains, and the results suggest that using external resources is helpful in that regard.

5 Conclusion and Future Work

Our proposed approach can successfully distinguish similarity relations, especially between nouns, even though, the proposed implementation cannot fully account for highly specific terms in context vectors, as they would be missing from the WordNet ontology. However the results on our initial dataset suggest that our method is promising to reduce the semantic gap in how the terms are contextually defined differently across different domains. The approach to measure the semantic similarity could be experimented on as a symmetric measure, by simply adding or averaging both asymmetric measures. Additionally, it would be interesting to trial the semantic measure on a regular word association task or one that does require asymmetric associations to be taken into account, within a single domain. We will also grow the evaluation dataset by recruiting more experts, and evaluate our approach with other representations of the word spaces (e.g. NNLM) and a range of distance measures.

References

1. Blitzer, J., Mcdonald, R., Pereira, F.: Domain adaptation with structural correspondence learning. In: Proceedings of EMNLP, pp. 120–128 (2006)
2. Bullinaria, J.A., Levy, J.P.: Extracting semantic representations from word co-occurrence statistics: stop-lists, stemming, and SVD. Behav. Res. Methods **44**, 890–907 (2012)
3. Gopalan, R., Li, R., Chellappa, R.: Domain adaptation for object recognition: an unsupervised approach. In: Proceedings of ICCV, pp. 999–1006 (2011)
4. Hu, X., Cai, Z., Graesser, A.C., Ventura, M.: Similarity between semantic spaces. In: Proceedings of CogSci 2005, pp. 995–1000 (2005)
5. Kamps, J., Marx, M., Mokken, R.J., De Rijke, M.: Using wordnet to measure semantic orientations of adjectives. In: Proceedings of LREC, pp. 1115–1118 (2004)
6. Kim, S.N., Cavedon, L.: Classifying domain-specific terms using a dictionary. In: Proceedings of ALTA, p. 57 (2011)
7. Bentivogli, L., Forner, P., Magnini, B., Pianta, E.: Revising the wordnet domains hierarchy: semantics, coverage and balancing. In: Proceedings of the Workshop on Multilingual Linguistic Ressources, pp. 101–108. ACL (2004)
8. Mikolov, T., Zweig, G.: Context dependent recurrent neural network language model. In: Proceedings of SLT, pp. 234–239 (2012)
9. Pan, S.J., Kwok, J.T., Yang, Q.: Transfer learning via dimensionality reduction. In: Proceedings of AAAI, pp. 677–682 (2008)
10. Turney, P.D., Pantel, P.: From frequency to meaning: vector space models of semantics. J. Artif. Intell. Res. **37**(1), 141–188 (2010)

A Signature Approach to Patent Classification

Dilesha Seneviratne[1]([✉]), Shlomo Geva[1], Guido Zuccon[1], Gabriela Ferraro[2],
Timothy Chappell[1], and Magali Meireles[3]

[1] Queensland University of Technology (QUT), Brisbane, Australia
d.dwmhakmanawalawwe@hdr.qut.edu.au,
{s.geva,g.zuccon,timothy.chappell}@qut.edu.au
[2] NICTA and Australian National University, Canberra, Australia
gabriela.ferraro@nicta.com.au
[3] Pontifical Catholic University of Minas Gerais (PUC Minas), Belo Horizonte, Brazil
magali@pucminas.br

Abstract. We propose a document signature approach to patent classification. Automatic patent classification is a challenging task because of the fast growing number of patent applications filed every year and the complexity, size and nested hierarchical structure of patent taxonomies. In our proposal, the classification of a target patent is achieved through a k-nearest neighbour search using Hamming distance on signatures generated from patents; the classification labels of the retrieved patents are weighted and combined to produce a patent classification code for the target patent. The use of this method is motivated by the fact that intuitively document signatures are more efficient than previous approaches for this task that considered the training of classifiers on the whole vocabulary feature set. Our empirical experiments also demonstrate that the combination of document signatures and k-nearest neighbours search improves classification effectiveness, provided that enough data is used to generate signatures.

1 Introduction

Patents are legal documents issued by governments for giving rights of exclusivity and protection to inventors. Patents play a significant role in helping inventors and organisations to protect their intellectual property. The number of patent applications filed every year is increasing rapidly. For example, in 2012 the World Intellectual Property Organization (WIPO) reported an increase of 9.2 % from previous years. Patents are organised in a classification system, called International Patent Classification (IPC), which provides for a hierarchical taxonomy where patents are classified according to the areas of technology to which they pertain. IPC contains about 120 classes and about 630 subclasses. This taxonomy is complex, large and nested (hierarchical), adding to the complexity of the patent classification task.

Given the increasing rate at which patents are filed, the current practice of manually classifying patents is unsustainable due to the time and resource burden it presents [10]. Automated classification systems have therefore emerged;

© Springer International Publishing Switzerland 2015
G. Zuccon et al. (Eds.): AIRS 2015, LNCS 9460, pp. 413–419, 2015.
DOI: 10.1007/978-3-319-28940-3_35

see for example Chakrabarti et al. (multi level Bayesian classifiers) [3,4], Tikk et al. (hierarchical classifiers) [12], Larkey (k-nearest neighbour) [11], Cai and Hofmann (hierarchical classifiers based on SVM) [2], and Chen and Chang (three phase classification) [5]. Of interest to this paper is the work of Fall et al. [7] who have evaluated a number of machine learning classifiers, including support vector machines (SVM), naive Bayes (NB), and k-nearest neighbour (kNN) classifiers, using bag of words as feature set. Their results suggest that SVM and NB have similar effectiveness when considering the highest level of the IPC hierarchy (class level), while kNN had lower effectiveness. When considering the lower level classification (subclass level), instead, SVM was found to outperform the other classification methods. We shall use the methods explored by this work as a benchmark for our document signature approach.

However, in previous work, little attention has been paid to the *efficiency* of automated methods for patent classification, with improvements in classification effectiveness taking the lion's share of the research efforts. The use of classification techniques such as support vector machines (SVM), however, does not scale to the increasing amount of patents being filed every year. In this paper we address this concern by examining an approach to patent search that is well-known for its efficiency: signature search [8].

Signatures are lengthy bit strings of words that are often created using an hash function. Signatures are used to quickly identify potentially relevant documents. We exploit signatures for patent classification by performing a signature search for patent signatures that are similar to a target patent that is provided as a query for classification. The patents in the k-nearest neighbourhood of the (query) target patent are considered to determine the target's classification code; this is obtained by weighting the classification codes of the patents in the neighbourhood. This approach guarantees extreme efficiency, specifically because of its capacity to scale to very large collections given that indexing is linear with the size of the collection (like inverted file search engines) and searching time increases linearly at a lower rate than the increase in collection size [1].

2 Patent Classification with Signatures

Early approaches used to generate document signatures are based on the bitwise OR composition of binary signatures associated with terms in documents [6,8]. Further refinements of the signature generation process have been proposed. A recent approach, called TopSig [6,9], uses random indexing for compressing the standard term-document matrix, followed by aggressive numerical precision reduction to maintain only the sign bits of the projected term-document matrix. This approach has been shown to be superior to standard signature approaches: we thus rely on the TopSig method to generate patent signatures.

Patent signatures are generated from snippets of text extracted from the patents. Specifically, in the experiments of Sect. 3, we consider the effectiveness of signatures generated either from patent titles, from patent abstracts, from claim texts, or from the first 300 words[1] of patent text.

---

[1] Previous work has also used the first 300 words extracted from each patent: this setting has in fact shown strong promise [7].

To perform automated patent classification with signatures, we first construct a signature for the patent requiring classification (target patent). The document signatures are formed through the successive summing of pseudo-randomly generated term vectors created from the patents' text. The resulting document vector is then flattened into a binary signature which functions as a locality-sensitive hash (LSH). We then use this signature to query a collection of patent signatures derived from a training set (in the same manner as the query signature), where patents are labelled with their correct IPC code. This process results in a ranking of patents, ordered in decreasing similarity to the target patent. Similarity in the signature space is measured according to the Hamming distance, i.e., the number of bits in which the two signatures differ.

To determine the first level of classification (section), the k-nearest neighbour classification algorithm is employed. The k-nearest neighbourhood to the target patent signature is formed by selecting the top-k patents from the ranking. A classification label is then produced for the target patent by a majority vote of its neighbours, with the label being selected from the class most common amongst its k nearest neighbours measured by the distance function $w(p) = \frac{1}{\sqrt{rank(p)}}$, where $rank(p)$ is the rank at which patent p has been retrieved in answer to the query formed by the target patent's signature. This procedure is akin to a simple voting process, where each training patent in the k-neighbourhood votes for its label, votes for the same label cast by different patents are accumulated and modulated by a weight $w(p)$ inversely proportional to the rank of the voting patent in the neighbourhood ranking.

To determine further levels of classification (class, subclass, group), the voting process is iterated but considering only subsets of patents in the k-neighbourhood that share the same higher level label as that assigned to the target patent. For example, when determining the class label for a target patent, the training patents that are considered are only those in the k-neighbourhood that share the section label assigned by our method to the target patent are considered.

3   Experiment Settings

3.1   Dataset

To evaluate the proposed approach to patent classification based on signatures we use the WIPO-alpha collection, a standard collection for patent classification used also by previous work. The WIPO-alpha collection (WIPO in short) consists of over 75,000 patent applications that have been submitted to WIPO under the Patent Cooperation Treaty (PCT). The collection is split into train and test sets which consist of 46,324 and 28,926 patents respectively.

3.2   Evaluation Measures

In line with previous work [7,12], to evaluate the effectiveness of the automated patent classification approaches, we analyse the number of correct guesses made by the classifiers when compared with the ground truth (precision). The (micro-average) precision is computed according to three settings: (1) the *top prediction* made by each classifier, where the classifier prediction with highest score is compared with the classification label recorded in the ground truth; (2) the top *three guesses* made by each classifier, where the classifier predictions with the three highest scores are compared with the classification label recorded in the ground truth, and success is recorded if one of these prediction does match with the ground truth; and (3) the *All Categories* method, where the top prediction of the classifier is compared with all the categories recorded in the ground truth, in case one match is found, the classification is deemed successful. Note that generally patents are assigned to multiple classification codes.

3.3 Approaches and Settings

For the signature based approach, we first preprocessed the patents by removing characters that are not alphabetic; we also removed stop words and applied the Porter stemmer. To create signatures, we set the signature width to 4096 bits. For the kNN classifier, we set k to 30 following the benchmark approach [7].

The effectiveness of the signature based approach is compared to that achieved by the classifiers investigated by Fall et al. [7] because they also used the WIPO-alpha collection and considered all classification codes (rather than limiting to particular sections of the IPC hierarchy). Moreover, that work explored the effectiveness of the kNN algorithm and thus we can directly compare the benefits of using signatures over bag of words.

4 Results

4.1 Effectiveness

Table 1 reports the results obtained by benchmarks and proposed approach at the IPC class level classification, while the results obtained at the IPC subclass level evaluation are reported in Table 2. The results were obtained by considering different text snippets to create representations of patents, either based on bag of words (for the benchmark methods) or document signatures (for the proposed method); these are: title of patent, abstract of patent(including titles, inventors, applicants) first 300 words of the patent text (titles, inventors, applicants, abstracts, and descriptions).

The results highlight that all classification methods are less effective when classifying at lower granularity levels (subclass) than at higher granularity (class). For the kNN method with document signatures, this is because classifications for low granularity levels are affected by those obtained at higher granularities: thus, if an error is made at class level, the error is propagated to subclass and group level. For the benchmark methods, loss in effectiveness is instead generally due to less training data being available for subclass level classification than at class level classification.

Table 1. Classification results at IPC class level.

Indexing field	Evaluation measures	NB Fall et al. [7]	k-NN Fall et al. [7]	SVM Fall et al. [7]	k-NN proposed method
Title	Top-prediction	45 %	33 %	Not reported	**40 %**
First 300 words	Top-prediction	55 %	51 %	55 %	**56 %**
Title	Three-guesses	66 %	52 %	Not reported	**63 %**
First 300 words	Three-guesses	79 %	77 %	73 %	**81 %**
Title	All-categories	52 %	38 %	Not reported	**46 %**
First 300 words	All-categories	63 %	58 %	62 %	**63 %**

Table 2. Classification results at IPC subclass level.

Indexing field	Evaluation measures	NB Fall et al. [7]	k-NN Fall et al. [7]	SVM Fall et al. [7]	k-NN proposed method
Abstract	Top-prediction	28 %	26 %	34 %	**32 %**
First 300 words	Top-prediction	33 %	39 %	41 %	**42 %**
Abstract	Three-guesses	47 %	45 %	52 %	**53 %**
First 300 words	Three-guesses	53 %	62 %	59 %	**67 %**
Abstract	All-categories	35 %	32 %	41 %	**38 %**
First 300 words	All-categories	41 %	46 %	48 %	**50 %**

When signature and benchmark methods are compared, we observe that the signature method is always more effective than its direct counterpart in the bag-of-words space, i.e., the kNN classifier of Fall et al. [7]. When other benchmark methods are considered instead, SVM and NB are found to be more effective that the signature based kNN when title snippets are used to generate patent representations. However, if longer snippets are used, as is the case when using the first 300 words of a patent, then the classification precision increases (for both proposed and benchmark methods); more importantly, the effectiveness of the proposed approach reaches (and can even outperform) that of benchmark methods. This suggests that signature approaches are comparable (or sometimes superior) to bag-of-words approaches in terms of classification effectiveness if enough evidence is used to produce the patent representations. This is more evident when considering top three predictions (Tables 1 and 2). Moreover, the results suggest that if more than one classification could be assigned to a patent, then effectiveness increases on average of approximately 20 %. Finally, when we analysed the effectiveness of the classifiers with respect to each IPC section (not reported for brevity), we found that the section where all classifiers delivered the lowest precision was section G (Physics); previous work had reported similar findings [7].

4.2   Efficiency and Scalability

The use of document signatures as an alternative to bag-of-word features for patent classification was motivated by the fact that document signatures provide significant advantages in terms of search times and scalability. Table 3 reports the

D. Seneviratne et al.

Table 3. Time required to index and search a patent collection.

Collection	Field	Indexing time	Searching time (Avg per query)
WIPO Train-46,324 Test- 28,926	Abstracts Title First 300 words	4.43 s 1.40 s 9.14 s	2.8×10^{-2} s 1.6×10^{-4} s 6.9×10^{-2} s
USPTO (2006-2013) Train-1,358,908 Test-51,324	Abstracts	68.96 s	4.5×10^{-1} s

time required to index the WIPO-alpha collection and that required to search in order to perform classification[2]. These results highlight that signatures allow searching patent collections within milliseconds and that the increase in the amount of text that is represented by a signature does not result in a large increment in time required to search for similar signatures. To study whether the document signature approach is scalable to larger patent collections, we replicated the classification experiments (for abstract only) for the USPTO collection, a dataset of more than 1.4 million patents (thus three orders of magnitude larger than WIPO). The runtime results (Table 3) highlight the scalability of the signature approach, since querying time increased of only one order of magnitude while the collection increased of three orders of magnitude.

5 Conclusions

In this paper we have investigated the use of document signatures for patent classification. Our initial empirical experiments have provided a number of interesting insights on the use of document signatures for this classification task and have opened avenues for future work. The results continued that the signature approach provides an efficient and scalable solution for this problem, and this is highly comparable in terms of effectiveness with other approaches to patent classification like SVM (which in turn do not scale to large patent collections). Moreover, our initial experiments have shown that the selection of which part of the patent is used to generate signatures is fundamental for the effectiveness of the classifiers. More research is however needed to understand what is the best content/part of the patent that should be used to generate signatures: titles, abstracts and the first 300 words gave correspondingly different results.

---

[2] No publicly available implementation of Fall et al.'s methods was available and our re-implementation did not lead to effectiveness comparable to the reported one. We were therefore unable to obtain efficiency figures for the benchmark methods. Similarly, we were unable to test for significant differences.

References

1. Chappell, T., Geva, S., Zuccon, G.: Approximate nearest-neighbour search with inverted signature slice lists. In: Hanbury, A., Kazai, G., Rauber, A., Fuhr, N. (eds.) ECIR 2015. LNCS, vol. 9022, pp. 147–158. Springer, Heidelberg (2015)
2. Cai, L., Hofmann, T.: Hierarchical document categorization with support vector machines. In: Proceedings of CIKM 2004, pp. 78–87 (2004)
3. Chakrabarti, S., Dom, B., Agrawal, R., Raghavan, P.: Using taxonomy, discriminants, and signatures for navigating in text databases. VLDB 97, 446–455 (1997)
4. Chakrabarti, S., Dom, B., Agrawal, R., Raghavan, P.: Scalable feature selection, classification and signature generation for organizing large text databases into hierarchical topic taxonomies. VLDB J. 7(3), 163–178 (1998)
5. Chen, Y.-L., Chang, Y.-C.: A three-phase method for patent classification. Inf. Process. Manage. 48(6), 1017–1030 (2012)
6. De Vries, C.M., Geva, S.: Pairwise similarity of topsig document signatures. In: Proceedings of ADCS 2012, pp. 128–134 (2012)
7. Fall, C.J., Törcsvári, A., Benzineb, K., Karetka, G.: Automated categorization in the international patent classification. SIGIR Forum 37(1), 10–25 (2003)
8. Faloutsos, C.: Signature-based text retrieval methods: a survey. Data Eng. 13(1), 25–32 (1990)
9. Geva, S., De Vries, C.M.: TopSig: topology preserving document signatures. In: Proceedings of CIKM 2011, pp. 333–338 (2011)
10. Kim, J.-H., Choi, K.-S.: Patent document categorization based on semantic structural information. Inf. Process. Manage. 43(5), 1200–1215 (2007). Patent Processing
11. Larkey, L.S.: A patent search and classification system. In: Proceedings of DL 1999, pp. 179–187 (1999)
12. Tikk, D.: A hierarchical online classifier for patent categorization, pp. 244–267 (2007)

The Impact of Using Combinatorial Optimisation for Static Caching of Posting Lists

Casper Petersen[(✉)], Jakob Grue Simonsen, and Christina Lioma

University of Copenhagen, Copenhagen, Denmark
{cazz,simonsen,c.lioma}@di.ku.dk

Abstract. Caching posting lists can reduce the amount of disk I/O required to evaluate a query. Current methods use optimisation procedures for maximising the cache hit ratio. A recent method selects posting lists for *static* caching in a greedy manner and obtains higher hit rates than standard cache eviction policies such as LRU and LFU. However, a greedy method does not formally guarantee an optimal solution. We investigate whether the use of methods guaranteed, in theory, to find an approximately optimal solution would yield higher hit rates. Thus, we cast the selection of posting lists for caching as an integer linear programming problem and perform a series of experiments using heuristics from combinatorial optimisation (CCO) to find optimal solutions. Using simulated query logs we find that CCO yields comparable results to a greedy baseline using cache sizes between 200 and 1000 MB, with modest improvements for queries of length two to three.

Keywords: Posting list · Caching · Combinatorial optimisation

1 Introduction

A *posting list* consists of a term t and $n \geq 1$ postings, each containing the ID of a document where t occurs, and other information required by the search engine's scoring function, e.g. the frequency of t in each document [6]. Posting list caching can reduce the amount of disk I/O involved [13,14] in query processing, affords higher cache utilisation and hit rates than result caching [12], and can combine terms to answer incoming queries.

Our Contribution: We show that static caching of posting lists can be modelled in a principled manner using constrained combinatorial optimisation (CCO), a standard method that has yielded great improvements in many fields [9, Chap. 35], and we provide a principled investigation of whether CCO would yield better solutions (preferably using modest extra computational resources) than greedy methods. Using simulated query logs for a range of cache sizes, we perform a sequence of experiments that show that results using combinatorial optimisation is comparable to the greedy baseline of Baeza-Yates et al. using 200-1000 MB cache sizes, with some modest improvements for queries of length two to three.

© Springer International Publishing Switzerland 2015
G. Zuccon et al. (Eds.): AIRS 2015, LNCS 9460, pp. 420–425, 2015.
DOI: 10.1007/978-3-319-28940-3_36

2   Related Work

Much prior work has been devoted to caching posting lists [3–6,10,11,13–15]. Zhang et al. [15] benchmark five posting list caching policies and find LFU (least frequently used – cache members are evicted based on their infrequency of access) to be superior, and that cache hit rates for static posting list are similar to the LFU, but with less computational overhead. An integrated cache that merges posting lists of frequently co-occurring terms to build new posting lists in the inverted index is used by Tolosa et al. [14]. Using a cost function that combines disk lookup and CPU time, the integrated cache improves performance over standard posting list caching by up to 40 %. Combinatorial optimisation for caching has not been investigated to the same degree: Baeza-Yates et al. [5] cache query terms based on their frequencies in a query log, and obtain ≈20 % reduction in memory usage without increasing query answer time. Baeza-Yates et al. [3] extend this approach by caching query terms using (i) their frequency in a query log weighted by (ii) their frequency in a collection. Posting lists with the highest weight are then cached. This method obtains higher hit rates than their approach in [5], dynamic LRU (least recently used) and dynamic LFU for all cache sizes. We propose an extension of [3] which uses a principled method to select posting lists for static caching. Next, we describe the original method by Baeza-Yates et al., and our extension.

3   Posting Lists Caching

Greedy Posting Lists Caching. Consider a list of queries, each of which consists of one or more terms and a cache of finite capacity. Let $F_q(t)$ denote the number of queries that contain term t in some query log Q_L and $F_d(t)$ the number of documents that contain t in some collection C. A *greedy* strategy to posting list selection chooses the query terms (representing posting lists) with the highest $F_q(t)$ until cache space is exhausted as in [5]. However, Baeza-Yates et al. [3] observe a trade-off between terms with high $F_q(t)$ and high $F_d(t)$ as these have long posting lists that consume substantial cache space. They address this trade-off by using the ratio $F_q(t)/F_d(t)$, called QTFDF, to select terms for static caching by (i) calculating QTFDF of each $t \in Q_L \cap C$, (ii) sorting terms in decreasing value of QTFDF and (iii) caching the terms with the highest QTFDF until cache space is exhausted. The method of [3] is thus a clever variation of the *profit-to-weight ratio* approach first used by Dantzig [8].

Selecting which posting lists to load into the cache is a *0-1 knapsack problem* [3,4]: given a knapsack with capacity c and n items $c_1, \ldots, c_n$ having values $v_1, \ldots, v_n$ and weights $w_1, \ldots, w_n$, take the items that maximise the total value without exceeding c. An item can only be selected once and fractions of items cannot be taken. As the knapsack optimisation problem is NP-hard and cannot in general be solved optimally using a greedy strategy [7, Chap. 16], we next describe how to formulate posting list selection as a combinatorial optimisation problem which, in theory, would find an approximately optimal solution.

Combinatorial Optimisation for Posting Lists Caching. We formalise the observation of [3] that a trade-off exists between $F_q(t)$ and $F_d(t)$ as follows: terms should be cached that yield the highest possible $F_q(t)$ subject to the constraint that the total size of the posting lists of cached terms should not exceed cache size. This is a classic CCO problem (a fact already noted by [3], but without formalisation or reported experiments). We cast posting list selection as an *integer linear program* of the form:

$$\text{max} \quad \sum_{i=1}^{n} v_i x_i \tag{1}$$
$$\text{subject to} \quad \sum_{i=1}^{n} w_i x_i \leq c \tag{2}$$
$$x_i \in \{0, 1\}, 1 \leq i \leq n \tag{3}$$

where $\sum_{i=1}^{n} v_i x_i$ is the *objective function*, $\sum_{i=1}^{n} w_i x_i \leq c$ and $x_i \in \{0, 1\}, 1 \leq i \leq n$ are *constraints* where x_i represents a term t_i (a posting list). A *solution* is a setting of the variables x_i; a *feasible* solution is a solution that satisfies all constraints; and an *optimal* solution is a feasible solution with maximal value of the objective function. We consider only optimal solutions here. Equation (2) states that the total weight of the selected terms cannot exceed c, and Eq. (3) that each term is either selected or discarded. We set $v_i = F_q(t_i)$ and $w_i = F_d(t_i)$, and refer to the method described here as the *CCO* method.

We emphasise two points. First, the CCO method maximises the chance of a query term cache hit, but does not consider disk I/O a factor. We can do this using a *multi*-objective CCO problem where one objective function seeks to minimize disk I/O (using the length of the posting list of x_i as values v_i) and a second objective function that seeks to maximise the number of cache hits. Second, if a term is selected its *entire* posting list is loaded. Another approach is to allow fractions of posting lists to be loaded and access the main index as needed. This may be useful if e.g. each posting list is sorted so access to the main index is reduced. We leave both topics as future work.

4   Simulating Queries

Query logs from large search engines are typically not publicly available in large numbers. Instead, we construct simulated query logs using (i) the method of Azzopardi et al. [2] and (ii) random sampling from a large synthetic query log.

Known-item Queries. We construct synthetic query logs containing known-item queries using the method of [1,2] as follows: We first select a document d_k from the collection (with uniform probability), then select a query length l and then select l terms $t_{1,...,l}$ from the document language model (LM) of d_k with probability $p(t_i|\Theta_d)$ and add t_i to q. $p(t_i|\Theta_d)$ is a mixture of (i) the maximum likelihood estimate of a term occurring in a document and (ii) a background model $p(t)$ (maximum likelihood estimate of t in the collection). Estimating (i) is done using one of two LMs [1]. The *popular* LM is given by

$$p(t_i|d_k) = n(t_i, d_k) / \sum_{t_j \in d_k} n(t_j, d_k) \tag{4}$$

where t_i, t_j are terms in d_k and $n(t_i, d_k)$ is the term-frequency of t_i in d_k. The *discriminative* LM is given by

$$p(t_i|d_k) = b(t_j, d_k)/p(t_i) \cdot \sum_{t_j \in d_k} b(t_j, d_k)/p(t_j) \qquad (5)$$

where $b(t_j, d_k) = 1$ if term t_j occurs in d_k.

Sampling from a Large Query Log. We use the anchor text query log from ClueWeb09[1] as starting point, which contains 500M triplets of the form <URL, anchor text, fq> where fq is the frequency of the tuple <URL, anchor text>. From this query log, we sample with replacement to generate new query logs.

5   Experiments

We describe how we simulate repeated queries and how we measure performance. We evaluate the CCO method against the greedy baseline of Baeza-Yates et al. [3], using the number of cache hits as our cache performance measure.

Simulating Repeated Queries. The method of Sect. 4 generates queries occurring exactly once. To generate *repeated* queries in the synthetic query sets we do as follows: after simulating a query, we generate a random number r in the interval $(0; 1)$ and compare it to a threshold τ. If $r > \tau$ we duplicate the query. We fix $\tau = 0.44$ meaning that $\sim 56\%$ of the queries have multiple occurrences [4]. We simulate queries of length $l = 1, 2, 3$ and generate $m = 5$ queries from each document. For query logs simulated using the method in Sect. 4, we cannot control repeated queries.

Experimental Settings. We experiment with cache sizes of 200, 600 and 1000 MB (cache sizes can vary between 100 MB to 16 GB [14]) and fix the size of a posting to 8 bytes. We use ClueWeb09 cat. B. – a domain-free crawl of ca. 50 million web pages in English – indexed using INDRI 5.8 with no stemming and with stop words removed as collection. We simulate query logs of 1M, 5M and 10M queries using each of the two LMs from Sect. 4 and the method from Sect. 4 with the anchor text as queries. As in [3], we estimate $F_q(t)$ from each query log and $F_d(t)$ from the collection. Each CCO problem is solved using SYMPHONY[2] (extensive experiments and tuning using LP_SOLVE[3] gave no consistent improvements). We count a cache hit for a query iff *at least one* of its terms is found in the cache (see [15] for alternative definitions). A single hit is sufficient for efficient retrieval as we need only traverse that term's posting list, and scan the forward index of each document to determine if remaining query terms are found. Counting query hits using this linear scan approach is less efficient than posting list intersection, but in this preliminary work, it allows us to test the merit of our method.

---

[1] http://lemurproject.org/clueweb09/anchortext-querylog/.
[2] https://projects.coin-or.org/SYMPHONY.
[3] http://lpsolve.sourceforge.net/5.5/.

Table 1. Results for the Discriminative and Popular 5M and 10M query log for query lengths = 1,2,3 and cache sizes: 200,600 and 1000 MB. x/y means *CCO / baseline*. OC is the *overlap coefficient* $= \frac{|X \cap Y|}{\min(|X|,|Y|)}$. CT is the number of *cache terms*. CH is the number of *cache hits*. *Diff* is the difference in CH. Entries where *Diff*>0 (boldfaced).

		Simulated 5M					
		Discriminative			Popular		
		qlen=1	qlen=2	qlen=3	qlen=1	qlen=2	qlen=3
OC	200M	0.851	0.884	0.923	0.990	0.973	0.977
	600M	0.962	0.934	0.899	0.997	0.987	0.999
	1000M	0.952	0.952	0.942	0.995	0.994	0.998
CT	200M	24938/24951	24922/24920	24973/24974	9311/9310	13799/13782	16841/16845
	600M	74648/74483	74725/74740	74780/74752	13804/13803	21483/21473	27568/27557
	1000M	114269/115850	122655/123358	124525/124546	16309/16307	25704/25696	33194/33209
CH	200M	217626/217626	228183/228266	228692/228877	19185/19185	28655/28671	35242/35238
	600M	546566/546566	596812/596790	600376/600733	28537/28537	44579/44573	57352/57351
	1000M	765793/765793	859142/857718	880740/880132	33768/33767	53531/53529	69370/69366
DIFF	200M	0	-83	-185	0	-16	**4**
	600M	0	**22**	-357	0	**6**	**1**
	1000M	0	**1424**	**608**	**1**	**2**	**4**

		Simulated 10M					
		Discriminative			Popular		
		qlen=1	qlen=2	qlen=3	qlen=1	qlen=2	qlen=3
OC	200M	0.949	0.957	0.865	0.961	0.929	0.977
	600M	0.890	0.909	0.885	0.989	0.982	0.985
	1000M	0.910	0.891	0.957	0.999	0.989	0.999
CT	200M	24988/24958	24955/24926	24954/24935	13892/13886	19404/19335	22775/22799
	600M	74589/74537	74468/74591	74642/74562	21623/21614	32962/32919	41095/41170
	1000M	124251/124127	123925/124091	124632/124350	25872/25876	40122/40096	51247/51247
CH	200M	240056/240056	246439/246440	246117/246293	28634/28634	40152/40145	48251/48258
	600M	629405/629405	662230/662148	664890/664515	44710/44710	68199/68203	86191/86154
	1000M	957964/957963	1033190/1033109	1044362/1044852	53631/53629	83379/83374	107333/107332
DIFF	200M	0	-1	-176	0	**7**	-7
	600M	0	**82**	**375**	0	-4	**37**
	1000M	**1**	**81**	-490	**2**	**5**	**1**

6   Findings

We show results for the 5M and 10M query logs generated using the method from Sect. 4 in Table 1. Results for all other query logs are qualitatively similar. We do not report CPU or memory consumption as this cost is likely minimal compared to indexing and retrieval costs. Across all query logs, query lengths (qlen) and cache sizes, the overlap coefficient is > 85 % and both CCO and the baseline cache contain approximately the same number of terms. For qlen=1, CCO and the baseline perform nearly identically for all query logs. For qlen=2, the discriminative query log gives rise to the largest differences between CCO and the baseline though these differences are negligible relatively to the total number of cache hits. For the popular query log, the differences are substantially smaller. The observations for qlen=3 are identical to those for qlen=2.

7   Conclusions and Future Work

We have investigated static posting list caching as a constrained combinatorial optimisation (CCO) problem and have evaluated this theoretically principled method against the greedy method of Baeza-Yates et al. [3]. We found both methods performed similarly for all cache sizes, with some modest gains for the CCO method. The high values (>85 %) of the overlap coefficient in all experiments suggest that both methods mostly identify the *same* high-frequency query terms and that differences in cache hits can be attributed to a small set of infrequent terms. However, while combinatorial optimisation gives, in theory, optimal solutions, in practice the quality of the solution also depends on the problem, the solver and the settings of the solver's parameters. In future work, we will investigate (i) how this impacts posting list selection, (ii) if CCO can obtain consistent performance improvements for domain-specific query logs, and (iii) the use of multi-objective CCO to balance disk I/O with cache hits.

References

1. Azzopardi, L., de Rijke, M.: Automatic construction of known-item finding test beds. In: SIGIR, pp. 603–604 (2006)
2. Azzopardi, L., de Rijke, M., Balog, K.: Building simulated queries for known-item topics: an analysis using six european languages. In: SIGIR, pp. 455–462 (2007)
3. Baeza-Yates, R., Gionis, A., Junqueira, F., Murdock, V., Plachouras, V., Silvestri, F.: The impact of caching on search engines. In: SIGIR, pp. 183–190. ACM (2007)
4. Baeza-Yates, R., Gionis, A., Junqueira, F.P., Murdock, V., Plachouras, V., Silvestri, F.: Design trade-offs for search engine caching. TWEB 2(4), 20 (2008)
5. Baeza-Yates, R., Saint-Jean, F.: A three level search engine index based in query log distribution. In: Nascimento, M.A., de Moura, E.S., Oliveira, A.L. (eds.) SPIRE 2003. LNCS, vol. 2857, pp. 56–65. Springer, Heidelberg (2003)
6. Broder, A.Z., Carmel, D., Herscovici, M., Soffer, A., Zien, J.: Efficient query evaluation using a two-level retrieval process. In: IKM, pp. 426–434. ACM (2003)
7. Thomas, H.C., Charles, E.L., Ronald, L.R., Clifford, S.: Introduction to Algorithms. MIT Press, Cambridge (2001)
8. Dantzig, G.B.: Discrete-variable extremum problems. Oper. Res. 5(2), 266–288 (1957)
9. Grotschel, M., Lovász, L.: Combinatorial optimization. Handb. Comb. 2, 1541–1597 (1995)
10. Liu, Z., Nain, P., Niclausse, N., Towsley, D.: Static caching of web servers. In: PWEI, pp. 179–190. ISOP (1997)
11. Long, X., Suel, T.: Three-level caching for efficient query processing in large web search engines. WWW 9(4), 369–395 (2006)
12. Papadakis, M., Tzitzikas, Y.: Answering keyword queries through cached subqueries in best match retrieval models. In: JIIS, pp. 1–40 (2014)
13. Saraiva, P.C., Silva de Moura, E., Ziviani, N., Meira, W., Fonseca, R., Riberio-Neto, B.: Rank-preserving two-level caching for scalable search engines. In: SIGIR, pp. 51–58. ACM (2001)
14. Tolosa, G., Becchetti, L., Feuerstein, E., Marchetti-Spaccamela, A.: Performance improvements for search systems using an integrated cache of lists+intersections. In: Moura, E., Crochemore, M. (eds.) SPIRE 2014. LNCS, vol. 8799, pp. 227–235. Springer, Heidelberg (2014)
15. Zhang, J., Long, X., Suel,T.: Performance of compressed inverted list caching in search engines. In: WWW, pp. 387–396. ACM (2008)

Improving Difficult Queries by Leveraging Clusters in Term Graph

Rajul Anand and Alexander Kotov[✉]

Department of Computer Science, Wayne State University,
Detroit, MI 48226, USA
{rajulanand,kotov}@wayne.edu

Abstract. Term graphs, in which the nodes correspond to distinct lexical units (words or phrases) and the weighted edges represent semantic relatedness between those units, have been previously shown to be beneficial for ad-hoc IR. In this paper, we experimentally demonstrate that indiscriminate utilization of term graphs for query expansion limits their retrieval effectiveness. To address this deficiency, we propose to apply graph clustering to identify coherent structures in term graphs and utilize these structures to derive more precise query expansion language models. Experimental evaluation of the proposed methods using term association graphs derived from document collections and popular knowledge bases (ConceptNet and Wikipedia) on TREC datasets indicates that leveraging semantic structure in term graphs allows to improve the results of difficult queries through query expansion.

Keywords: Difficult queries · Term graphs · Graph clustering · Knowledge bases

1 Introduction

Vocabulary mismatch between documents and queries, when the searchers and authors of relevant documents use different terms to refer to the same concepts, is one of the major causes of poor initial results for some queries (or difficult queries). Due to the lack of positive relevance signals in the initial retrieval results, improvement of retrieval accuracy for such queries cannot be achieved by employing standard techniques, such as pseudo-relevance feedback, and requires utilization of additional resources, such as term graphs. Term graphs are weighted directed graphs, in which the nodes correspond to the basic lexical units (terms or phrases) and the weighted edges represent the strength of semantic relatedness between a pair of such units. Term graphs are rich sources of terms for query and document expansion and can be constructed either manually or automatically. Automatically constructed term graphs (or statistical term association graphs) are derived from document collections by calculating a term co-occurrence based information-theoretic measure, such as mutual information (MI) [8], for each pair of distinct terms in the vocabulary of a collection.

© Springer International Publishing Switzerland 2015
G. Zuccon et al. (Eds.): AIRS 2015, LNCS 9460, pp. 426–432, 2015.
DOI: 10.1007/978-3-319-28940-3_37

Besides term association graphs, expansion LMs can also be derived from manually curated knowledge repositories, such as ConceptNet [6]) (large semantic network) or DBpedia (Wikipedia infoboxes represented as an RDF graph).

Automatically constructed term association graphs have been recently applied to address the problem of vocabulary mismatch in ad-hoc information retrieval through document and query expansion. In particular, Karimzadehgan and Zhai [2] leveraged the MI-based term association graph to estimate translation model for document expansion, Kotov and Zhai [3] used term association graphs for interactive query disambiguation and Bai et al. [1] experimented with using different number of top-k related terms from statistical term association graphs for query expansion. All these methods expand a given query or document term with either the *top-k or all related terms from the term association graph*. We, however, hypothesize that *unstructured and indiscriminate utilization of term association graphs results in suboptimal retrieval performance*, since statistical term association graphs are usually fairly noisy. To overcome this problem, we propose to capture semantic structure in term association graphs through graph clustering and leverage the identified clusters to derive more precise and robust query expansion language models (LMs) to improve the retrieval results of difficult queries.

We illustrate our approach with an example in Fig. 1. This example shows a fragment of a term graph, which includes the query term "greek" from the TREC topic 433 "Greek philosophy, stoicism" and the 8 terms that are most strongly associated with it. Previously proposed methods [1,2] include all these related terms into the resulting expansion LM. Instead, we propose to apply graph clustering methods to term graphs to first determine a set of clusters (connected components) with an intuition that such components correspond to sets of semantically coherent expansion terms. Given a query, our method would then include only those related terms from a term graph that are in the same clusters with the query terms. We hypothesize that such filtering allows to effectively discard spurious term associations and improve the retrieval effectiveness of resulting expansion LMs. Applying the proposed method to our example, the query term "greek" will contribute the terms "greece", "cyprus", "cypriot" and "athens" to the query expansion LM (shown in gray shade).

The key difference between our proposed approach and the previously proposed clustering-based retrieval methods [5,7] is that our method leverages

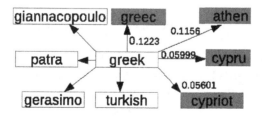

Fig. 1. Constructing more robust query (document) expansion LM by filtering out the terms that are not in the same term graph cluster as the query (document) term.

clusters in a term graph, rather than a document collection. The main contributions of this work are two-fold:

- we propose a method to derive more robust query expansion LMs based on leveraging clusters in term graphs and experimentally demonstrate that the query expansion LMs constructed using the proposed method are more effective in improving the accuracy of difficult queries than the query expansion LMs obtained by including all related terms;
- we compare the retrieval effectiveness of term association graphs with term graphs derived from popular knowledge repositories (DBpedia and ConceptNet). Although ConceptNet [4] and Wikipedia [10] have both been individually utilized for different IR tasks, our approach is the first to leverage the clusters in term graphs derived from these knowledge repositories.

2   Method

2.1   Term Graph Construction and Clustering

The proposed method uses mutual information (MI) [8], a co-occurrence based information-theoretic measure, to captures semantic relatedness between the nodes in the term graph. Infomap [9] is a *state-of-the-art, non-parametric* algorithm for finding communities in large networks, which utilizes information-theoretic measures and models stochastic graph flow to obtain the optimal clusters. The algorithm uses hierarchical map equation to measure the per-step average code length necessary to describe a random walker's movements on a graph, given its hierarchical partition, and finds the partition that minimizes the code length. For a term graph with n nodes divided into m modules, the lower bound on the code length is defined by the map equation:

$$L(M) = \sum_{i=1}^{m} Q_i \log \sum_{i=1}^{m} Q_i - 2 \sum_{i=1}^{m} Q_i \log Q_i - \sum_{j=1}^{n} p_j \log p_j$$
$$+ \sum_{i=1}^{m} (Q_i + \sum_{j \in i} p_j) \log(Q_i + \sum_{j \in i} p_j)$$

where Q_i is the probability of the random walk to exit the partition i and p_j is the frequency of node j.

Table 1. Statistics of experimental datasets.

Dataset	# docs	size (MB)	# tops	# hard
AQUAINT	10,033,461	3,042	50	17
ROBUST	528,155	1,910	250	75
GOV	1,247,753	18,554	225	147

2.2 Datasets

For all experiments in this work we used AQUAINT, ROBUST and GOV TREC collections, various statistics of which are summarized in Table 1. For each experimental dataset, we constructed a term association graph using MI as a similarity measure. DBpedia term graph was constructed by treating DBpedia 3.9[1] extended abstracts, which contain all the words in the first section of Wikipedia articles, as a document collection and using MI as a similarity measure. Concept-Net term graph was constructed by removing all non-English terms and negative associations from the core ConceptNet 5 term graph. We considered two versions of the ConceptNet term graph. The first version uses original weights of edges provided with ConceptNet 5 (**CNET**)[2], while in the second version, the weights between the concepts are calculated for each collection using MI (**CNET-MI**). We further customized Wikipedia and ConceptNet term graphs for each experimental collection by removing all the nodes that do not occur in the index of that collection.

Our proposed methods can be divided into three categories. Methods using collection term association graphs include **COL-ALL**, which uses all related terms to construct query expansion LM and **COL-INFO**, which selects expansion terms based on Infomap clustering. Similarly, **WIKI-ALL, CNET-ALL, CNET-MI-ALL** use all related terms from the corresponding term graph, while **WIKI-INFO, CNET-INFO, CNET-MI-INFO** filter expansion terms based on Infomap clustering from Wikipedia and ConceptNet term graphs, respectively.

2.3 Retrieval Model and Query Expansion

We used the KL-divergence retrieval model with Dirichlet prior smoothing [11] (**KL-DIR**), according to which the retrieval task involves estimating Θ_Q, a query language model (LM) for a given keyword query $Q = \{q_1, q_2, \ldots, q_k\}$, and document language models Θ_{D_i} for each document D_i in the document collection $\mathcal{C} = \{D_1, \ldots, D_m\}$. We define a query as difficult (or hard), if the average precision of results retrieved with the **KL-DIR** retrieval model is less than 0.1.

In language modeling approach to IR, query expansion is typically performed via linear interpolation of the original query model $P(w|Q)$ and query expansion LM $P(w|\hat{Q})$ with parameter λ:

$$P(w|\tilde{Q}) = \lambda P(w|Q) + (1 - \lambda)P(w|\hat{Q}) \tag{1}$$

Estimating query expansion LM $P(w|\hat{Q})$ using clusters in a term graph involves finding a set of semantically related terms $\mathcal{E}_{q_i}$ for each query term q_i (i.e. all direct neighbors of the query term q_i in the term graph that are in the

---

[1] http://wiki.dbpedia.org/Downloads39.
[2] http://conceptnet5.media.mit.edu/downloads/20130917/associations.txt.gz.

same term graph cluster C_{q_i} as q_i) and normalizing the probabilities using the following formula:

$$p(w|\hat{Q}) = \frac{\sum_{i=1}^{k} p(w|q_i)}{\sum_{i=1}^{k} \sum_{w \in C_{q_i}} p(w|q_i)} \tag{2}$$

3   Experiments

We pre-processed each dataset by removing stopwords and stemming (using Porter stemmer). To construct collection term association graphs, we removed all the terms that either occur in less than five documents or in more than 10 % of all documents in a given collection. We used the following settings of Infomap parameters: self link teleportation probability was set to 0.1, node teleportation probability was to 0.01 and random seed to 111222333. The optimal value for self link teleportation probability was determined empirically to reduce the number of very small clusters (which include less than 5 terms). We used the **KL-DIR** retrieval model and document expansion using translation model based on MI term graph (**TM**) [2] as the baselines. The reported results are based on the optimal settings of the Dirichlet prior μ, interpolation parameter λ that were empirically determined for all methods and the baselines. Summary of retrieval performance of the proposed methods and the baselines on each experimental dataset is provided in Tables 2, 3 and 4. The entries corresponding to the highest and second highest values were highlighted in boldface and italics. We performed statistical significance testing of MAP values using Wilcoxon signed rank test (▲ and • represent statistically significance difference ($p < 0.05$) relative to **KL-DIR** and **TM** baselines, respectively).

Several conclusions can be drawn from experimental results. First, cluster-based filtering of query expansion LMs derived from both statistical term association graphs and knowledge base term graphs improves retrieval performance

Table 2. Summary of retrieval performance on AQUAINT collection for difficult topics.

Method	MAP	P@5	GMAP
KL-DIR	0.0474	0.1250	0.0386
TM	0.0478	0.1250	0.0386
COL-ALL	0.0476	0.1375	0.0393
COL-INFO	0.0482	0.1375	0.0397
WIKI-ALL	0.0528▲•	0.1850	0.0452
WIKI-INFO	0.0501▲	0.1750	0.0405
CNET-ALL	0.0504▲	0.1875	0.0440
CNET-INFO	**0.0531▲•**	**0.1950**	**0.0471**
CNET-MI-ALL	0.0496▲•	0.1875	0.0422
CNET-MI-INFO	0.0527▲•	**0.1950**	0.0416

Table 3. Summary of retrieval performance on ROBUST collection for difficult topics.

Method	MAP	P@5	GMAP
KL-DIR	0.0410	0.1544	0.0261
TM	0.0458	0.1646	0.0267
COL-ALL	0.0429▲	0.1594	0.0273
COL-INFO	0.0463▲	0.1949	0.0279
WIKI-ALL	0.0503▲•	0.1848	0.0301
WIKI-INFO	0.0535▲•	0.1870	0.0271
CNET-ALL	0.0559▲•	0.1899	*0.0334*
CNET-INFO	*0.0580*▲•	*0.1924*	**0.0344**
CNET-MI-ALL	0.0560▲•	**0.1949**	0.0326
CNET-MI-INFO	**0.0582**▲•	0.1899	0.0301

Table 4. Summary of retrieval performance on GOV collection for difficult topics.

Method	MAP	P@5	GMAP
KL-DIR	0.0114	0.0233	0.0103
TM	0.0128	0.0248	0.0107
COL-ALL	0.0120	0.0243	0.0105
COL-INFO	0.0125	0.0245	0.0112
WIKI-ALL	0.0123	0.0242	0.0112
WIKI-INFO	0.0121	0.0236	0.0104
CNET-ALL	0.0128▲	*0.0258*	0.0121
CNET-INFO	**0.0196**▲•	**0.0290**	0.0124
CNET-MI-ALL	0.0176▲•	0.0242	*0.0129*
CNET-MI-INFO	*0.0195*▲•	0.0255	**0.0131**

in majority of cases on all 3 experimental collections. This indicates that graph clustering is effective at capturing semantically strong associations in the context of a given collection, while discarding the spurious ones. Secondly, query expansion LMs based on filtered term graphs derived from Wikipedia and ConceptNet (**WIKI-INFO, CNET-INFO, CNET-MI-INFO**) generally outperformed query expansion LMs derived from association term graphs (**COL-ALL** and **COL-INFO**) as well as document expansion based on translation model (**TM**) according to all metrics on all datasets. Finally, term graphs derived from ConceptNet generally outperformed the ones derived from Wikipedia on all 3 collections, which highlights the importance of commonsense knowledge in IR besides entity information in DBpedia.

4   Conclusion

In this paper, we proposed a method to derive more accurate query expansion LMs by clustering term graphs and experimentally demonstrated that applying this method to statistical term association graphs and term graphs derived from knowledge bases translates into more accurate retrieval results for difficult queries.

References

1. Bai, J., Song, D., Bruza, P., Nie, J.-Y., Cao, G.: Query expansion using term relationships in language models for information retrieval. In: Proceedings of CIKM 2005, pp. 688–695 (2005)
2. Karimzadehgan, M., Zhai, C.: Estimation of statistical translation models based on mutual information for ad hoc information retrieval. In: Proceedings of SIGIR 2010, pp. 323–330 (2010)
3. Kotov, A., Zhai, C.: Interactive sense feedback for difficult queries. In: Proceedings of CIKM 2011, pp. 59–66 (2009)
4. Kotov, A., Zhai, C.: Tapping into knowledge base for concept feedback: leveraging conceptnet to improve search results for difficult queries. In: Proceedings of WSDM 2012, pp. 403–412 (2012)
5. Kurland, O.: The opposite of smoothing: a language model approach to ranking query-specific document clusters. In: Proceedings of SIGIR 2008, pp. 171–178 (2008)
6. Liu, H., Singh, P.: Conceptnet - a practical commonsense reasoning tool-kit. BT Technol. J. **22**(4), 211–226 (2004)
7. Liu, X., Croft, B.: Cluster-based retrieval using language models. In: Proceedings of SIGIR 2004, pp. 186–193 (2004)
8. Manning, C., Schütze, H.: Foundations of statistical natural language processing. The MIT Press, Cambridge (1999)
9. Rosvall, M., Bergstrom, C.: Maps of information flow reveal community structure in complex network. PNAS **105**, 1118–1123 (2008)
10. Xu, Y., Jones, G., Wang, B.: Query dependent pseudo-relevance feedback based on Wikipedia. In: Proceedings of SIGIR 2009, pp. 59–66 (2009)
11. Zhai, C., Lafferty, J.: Document language models, query models, and risk minimization for information retrieval. In: Proceedings of SIGIR 2001, pp. 111–119 (2001)

Demonstrations

EEST: Entity-Driven Exploratory Search for Twitter

Chao Lv, Runwei Qiang, Lili Yao, and Jianwu Yang[✉]

Institute of Computer Science and Technology,
Peking University, Beijing 100871, China
{lvchao,qiangrw,yaolili,yangjw}@pku.edu.cn

Abstract. Social media has become a comprehensive platform for users to obtain information. When searching over the social media, users' search intents are usually related to one or more entities. Entity, which usually conveys rich information for modeling relevance, is a common choice for query expansion. Previous works usually focus on entities from single source, which are not adequate to cover users' various search intents. Thus, we propose EEST, a novel multi-source entity-driven exploratory search engine to help users quickly target their real information need. EEST extracts related entities and corresponding relationship information from multi-source (i.e., Google, Twitter and Freebase) in the first phase. These entities are able to help users better understand hot aspects of the given query. Expanded queries will be generated automatically while users choose one entity for further exploration. In the second phase, related users and representative tweets are offered to users directly for quickly browsing. A demo of EEST is available at http://demo.webkdd.org.

Keywords: Entity driven · Twitter search · Real-time exploratory search

1 Introduction

Social media such as Twitter has become a comprehensive platform for users to obtain information. When searching over the social media, users' initial interest is usually vague. However, their search intents are usually linked to an entity. As related entities can reflect different aspects of a topic, users often choose them to expand their queries. Previous studies mainly focus on how to use related entities as a feedback process. However, they simply adopt entities from single source such as DBpedia, which is not adequate to cover various search intents from our point of view.

Hence, we introduce EEST, a multi-source entity-driven exploratory search engine. For a certain query posted from users, related entities are offered in the first phase, to help users better understand hot aspects of the initial topic. According to the nature of selected source, two kinds of related entities, i.e., real-time entities and historical entities, are adopted. Real-time entities extracted

© Springer International Publishing Switzerland 2015
G. Zuccon et al. (Eds.): AIRS 2015, LNCS 9460, pp. 435–439, 2015.
DOI: 10.1007/978-3-319-28940-3_38

from real-time source provide latest aspects while historical entities extracted from historical source offer global aspects of the query. Those related entities as well as their relationship will be presented in a graph view, and users can choose a certain entity for further exploration. For each selected entity, EEST will retrieve related users and tweets from Twitter. Moreover, considering the redundancy problem of retrieved tweets, we also apply TTG[1] (tweet timeline generation) to generate a summary that captures relevant information.

2   System Architecture

EEST can be divided into two phases, as illustrated in Fig. 1.

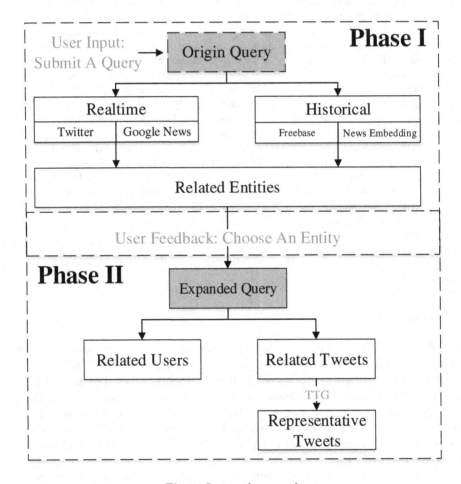

Fig. 1. System framework

---

[1] https://github.com/lintool/twitter-tools/wiki/TREC-2014-Track-Guidelines.

2.1   Phase I: Entity Extraction

In the first phase, EEST accepts a certain query submitted from a user, and we denote it as *OriginQuery*. To help users better understand hot aspects about the given query, related entities are extracted from multi-source. Extracted entities could be divided into categories of real-time entities and historical entities. Google News and Twitter are chosen as real-time entities source while Freebase and News Embedding are selected as historical entities source.

– **Google News** is our first choice as news is more formal and brief compared to normal text. *OriginQuery* is retrieved in Google News, and the latest related news is returned. We extract related entities from these news text by using named entity recognition (NER) [3]. In particular, three kinds of entities are involved, i.e., Person, Location and Organization.
– **Twitter** is a popular application for users to share and discuss information. Just like Google News, related tweets are retrieved from real-time Twitter steam, then related entities are separated out.
– **Freebase** is a practical, scalable tuple database used to organize general human knowledge [1]. We take advantage of the summary description information in Freebase to get a descriptive text about *OriginQuery*. Similarly, we extract entities from the description content via NER.
– **News Embedding** Currently, the distributed word representations (i.e. word embedding) have attracted more attention in text understanding. The word embedding allows to explicitly encode various semantic relationships as well as linguistic regularities and patterns into the new embedding space [2]. For this purpose, we downloaded pre-trained vectors trained on part of Google News dataset[2]. Then we can compute the cosine similarity distance of *OriginQuery* and other terms in the vocabulary. The top k scored terms are regarded as related entities to the given query.

A new *ExpandedQuery* is generated by combining *OriginQuery* with the chosen entity. This *ExpandedQuery* is going to be transmitted to second phase as input.

2.2   Phase II: Result Presentation

ExpandedQuery is submitted to the Twitter Search[3] in the second phase, and related users and related tweets are obtained. For related users, profile images and lots of statistical information are provided, including followers number, following number, tweets number, etc. For related tweets, TTG is conducted on them with the goal of noise elimination and representative tweets selection. We adopt a star clustering algorithm proposed in [4] as our TTG core algorithm. After that, a clear and representative tweets list will be unfolded in front of users.

---

[2] https://code.google.com/p/word2vec/.
[3] https://dev.twitter.com/rest/public/search.

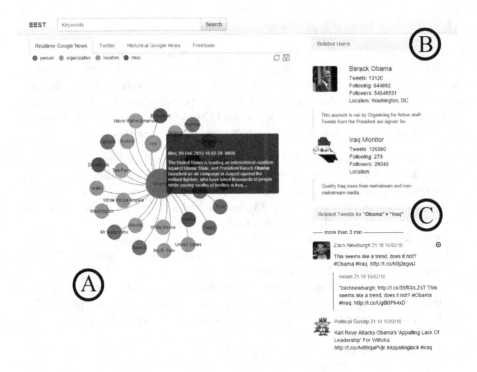

Fig. 2. An example for query "Obama" in EEST.

3 Demonstration

We take a query "Obama" as an example to discuss the main modules of EEST briefly, indicated with circled letters in Fig. 2.

3.1 A - Related Entities Module

Related entities extracted from multi-source are expressed as a graph in this module. As we can see, "Iraq", "White House" and "European" are highly related to "Obama" in Google News recently. Related news will appear above corresponding entities for users to see what's happening between them. Let us assume that user is interested in entity "Iraq" and choose it for further exploration.

3.2 B - Related Users Module

After the user choose entity "Iraq" for further exploration, two Twitter account, "Barack Obama" and "Iraq Monitor", are extracted and displayed in this module. Statistical information and Twitter account links are offered for users to easily navigate to their homepages, including followers number, following number, tweets number, etc.

3.3 C - Related Tweets Module

Related tweets talking about "Obama" and "Iraq" are displayed in this module. Redundant tweets are clustered to their representative tweets, which makes the related tweets list readable and clear.

4 Conclusions

In this paper, we have described a demonstration of a multi-source entity-driven search engine for Twitter, called EEST. We presented our initial motivation and the proposed methods, as well as the main functionality of the system. With the help of related entities, users are able to better understand their information need. At the same time, summarized information can save users' time for browsing the retrieval results.

Acknowledgment. The work reported in this paper was supported by the National Natural Science Foundation of China Grant 61370116.

References

1. Bollacker, K., Evans, C., Paritosh, P., Sturge, T., Taylor, J.: Freebase: a collaboratively created graph database for structuring human knowledge. In: Proceedings of the 2008 ACM SIGMOD International Conference on Management of Data, pp. 1247–1250. ACM (2008)
2. Mikolov, T., Sutskever, I., Chen, K., Corrado, G.S., Dean, J.: Distributed representations of words and phrases and their compositionality. In: Advances in Neural Information Processing Systems, pp. 3111–3119 (2013)
3. Nadeau, D., Sekine, S.: A survey of named entity recognition and classification. Lingvisticae Investigationes **30**(1), 3–26 (2007)
4. Zeng, H.J., He, Q.C., Chen, Z., Ma, W.Y., Ma, J.: Learning to cluster web search results. In: Proceedings of the 27th Annual International ACM SIGIR Conference on Research and Development in Information Retrieval, pp. 210–217. ACM (2004)

Oyster: A Tool for Fine-Grained Ontological Annotations in Free-Text

Hamed Tayebikhorami[1,2(✉)], Alejandro Metke-Jimenez[1], Anthony Nguyen[1], and Guido Zuccon[2]

[1] Australian E-Health Research Centre (CSIRO), Herston, Australia
{hamed.tayebikhorami,alejandro.metke,anthony.nguyen}@csiro.au
[2] Queensland University of Technology, Brisbane, Australia
g.zuccon@qut.edu.au, hamed.tayebikhorami@connect.qut.edu.au

Abstract. Oyster is a web-based annotation tool that allows users to annotate free-text with respect to concepts defined in formal knowledge resources such as large domain ontologies. The tool has been explicitly designed to provide (manual and automatic) search functionalities to identify the best concept entities to be used for annotation. In addition, Oyster supports features such as annotations that span across non-adjacent tokens, multiple annotations per token, the identification of entity relationships and a user-friendly visualisation of the annotation including the use of filtering based on annotation types. Oyster is highly configurable and can be expanded to support a variety of knowledge resources. The tool can support a wide range of tasks involving human annotation, including named-entity extraction, relationship extraction, annotation correction and refinement.

1 Introduction

The development of techniques and tools for information extraction and named-entity recognition from free-text cannot prescind from the collection of annotated data, either for training such techniques (e.g. when using supervised methods) or for evaluating them. A number of tools have been developed for the acquisition of such annotated data. For example, BRAT is a popular web-based annotation tool that provides a user-friendly interface and the possibility to collect rich structured annotations [2]. WordFreak is an annotation tool that supports both human-provided and computer-generated annotation; active learning plugins are available that can learn from labelled data and suggest annotations for unseen data that can in turn be manually corrected by the annotators [3]. Other annotation tools for free-text data are publicly available, e.g., Knowtator[1], MMAX2[2] and GATE[3]; they often share similar functionalities but are all somehow limited in the annotation classes and tasks that are supported.

---

[1] http://knowtator.sourceforge.net/.

[2] http://mmax.eml-research.de/.

[3] http://gate.ac.uk/.

© Springer International Publishing Switzerland 2015
G. Zuccon et al. (Eds.): AIRS 2015, LNCS 9460, pp. 440–446, 2015.
DOI: 10.1007/978-3-319-28940-3_39

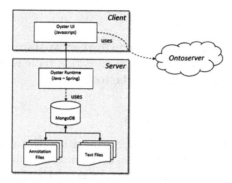

Fig. 1. The architecture of Oyster.

Despite the successful adoption of the tools mentioned above, tools like BRAT are difficult to use when numerous target annotation classes are considered. This is even more so when annotations need to be expressed against a large ontology, as it is the case when annotating clinical free-text data with the formal concepts defined in medical terminologies such as SNOMED CT. This paper introduces Oyster, an annotation tool that allows the fine-grained annotation of text with respect to large knowledge resources such as ontologies; it does so by supporting (both automated and manual) searching through the concepts defined within the knowledge resource. The version of Oyster described here is configured to provide annotations based on the SNOMED CT terminology via the Ontoserver platform [1], a terminology web server for clinical ontologies, and it is therefore tailored to the annotation of medical free-text data, such as discharge summaries, nurse handovers, radiology reports, etc. Nevertheless, Oyster can be easily extended to support the use of annotations from any knowledge resource, e.g. from Freebase (https://www.freebase.com/).

2   Oyster's Architecture

Figure 1 shows the architecture of Oyster, which is divided into the client side and the server side. The client side is responsible to query the knowledge resource and identify candidate annotations within the resource itself. The server side is for serving the data and collecting annotations from the client and save it as JSON files.

The architecture is based on Spring MVC framework using JPA to connect with MongoDB database that is responsible for storing the free-text data. This data requires annotation, the annotations that are in turn collected, and the details associated to the users that provide the annotations. The server-side component is written in Java while the client-side component uses JavaScript, JSP and HTML, including libraries such as JQuery, Bootstrap and AnnotationJS, with this last library responsible for supporting the core annotation functionalities.

3 Features in Oyster

Next we describe the features developed in Oyster; we start from the key features that differentiate Oyster from existing annotation tools and we then turn our attention to other features present in the tool.

3.1 Ontology-Based Annotations

This is the most significant feature that distinguishes Oyster from other annotation tools. The tool allows the provision of annotations with respect to named-entities (concepts) as defined in a reference knowledge resource, such as an ontology. The current version of Oyster integrates the SNOMED CT terminology as a target reference knowledge resource for annotations, but Oyster can be easily configured to point to other knowledge resources, making it a general purpose tool.

Automated Annotation Suggestions Based on the Ontology. Another feature developed within Oyster is the automatic suggestion of candidate concept annotations derived from the reference knowledge resource. This feature aims to simplify the annotation efforts of human coders and therefore speed up the annotation process. When a text span is highlighted, Oyster suggests the best matching candidates from the reference knowledge resource for the user to select the appropriate entity (or entities if multiple ones are appropriate), see Fig. 2. To produce candidate suggestions, Oyster currently uses the Ontoserver web service [1], issuing as query the text span for which an annotation is to be assigned to.

Manual Annotation Search on the Ontology. If the highlighted textual content does not retrieve the desired annotation, the user is allowed to search manually on the ontology. A list of candidate annotations will be updated with the new search results. If the desired match still cannot be found, the user can map the annotation to a high-level parent concept and flag it as a supertype.

3.2 Annotating Multiple Spans

Oyster allows users to select and annotate multiple textual spans by holding the ctrl key while selecting the desired text spans. As can be seen in Fig. 2, the two disjoint text spans "Fecal occult blood" and "negative" were selected for annotation[4]. After selecting the desired entity (or entities) from the suggested list or manually searched list, the user may save the annotation using the "Save" button (Fig. 2).

3.3 Relationship

Oyster can link annotations to each other based on entity relationships in the ontology. For instance, consider this sentence from a medical report: "the patient

---

[4] Note that this is only feasible in the Firefox browser as other browsers do not support multiple selections.

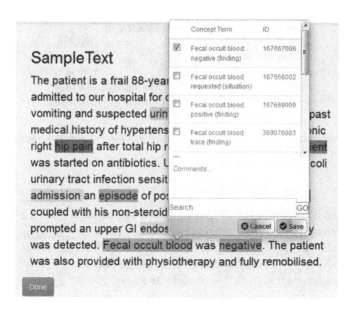

Fig. 2. The Oyster annotation process returning the best matching candidates from the reference knowledge resource for the user highlighted text spans.

is **suspected** to have **urinary track infection**". The two bolded spans refer to two SNOMED CT concepts. The user can relate the two concepts by creating an annotation for each of them and then linking the "suspected" annotation to the "urinary track infection" annotation through a relationship called "clinical context" which is a SNOMED CT concept itself.

3.4 Category Colour Coding

Oyster allows administrators to choose color codes for ontology categories. Once the user creates an annotation, its span will be highlighted with the color code of the corresponding category. Figure 2 shows the highlighted spans that have colours corresponding to each top level hierarchy in SNOMED CT.

3.5 View and Edit

After creating annotations for specific spans, the user can view the details of annotations by hovering the mouse on the highlighted span. Icons in the tooltip allow for the revision or deletion of the annotations (Fig. 3).

3.6 Filtering

Users may filter their annotations based on the category or categories that they wish to view. This is handled by the filtering feature that allows users to filter annotations by categories (Fig. 4).

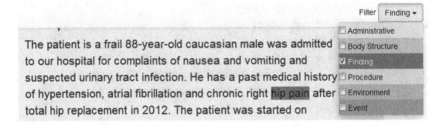

urinary tract infection sensitive to trimethoprim. During admission an episc No Comment ✎ ✕ vomiting coupled with his no Fecal occult blood negative 167667006 use prompted an upper (finding) bnormality was detected. Fecal occult blood was negative. The patient was also provided with physiotherapy and fully remobilised.

Fig. 3. Viewing annotations. Annotations can be updated or deleted suing the appropriate icons in the tooltip.

Filter | Finding ▾

The patient is a frail 88-year-old caucasian male was admitted to our hospital for complaints of nausea and vomiting and suspected urinary tract infection. He has a past medical history of hypertension, atrial fibrillation and chronic right hip pain after total hip replacement in 2012. The patient was started on

☐ Administrative
☐ Body Structure
☑ Finding
☐ Procedure
☐ Environment
☐ Event

Fig. 4. Filtering annotations.

3.7   Timing

The time to annotate each file is measured in the application's background. Timing information can be used, for example, to study the complexity of certain annotation tasks and for estimating the duration of subsequent annotation tasks. Oyster measures the annotation time for each file without involving the user in time controls such as start, pause and resume (see Sect. 3.9 for more details).

3.8   File Manager and User Access

Oyster enables an administrator to upload documents (in plain text format; otherwise they are ignored) using drag and drop multiple file selection. Then the administrator can assign the uploaded text files to existing users. The files can be shared or private. If the file is shared, users can contribute together and view or edit each other's annotations, while if private, then each user has their own independent annotations.

3.9   Text File Life Cycle

After logging into the system, users can see a list of documents that have been assigned to them by the administrator. These documents or text files are shown along with three icons: not commenced (O), in progress (Ⓒ) and completed (✔). By selecting any document from the in-progress list, the time for that particular file will start (or resume if it has already started before). This time will be paused if another text file is selected or if the user has no activities (mouse or keyboard events) in the Oyster page for a pre-specified idle time. The timing will resume by selecting a document again or resuming activities in the Oyster page. The user can complete the annotation process by pushing the "Done" button which stops the timing and moves the document from the in-progress list to the completed list. The completed documents can only be viewed, which means that the user cannot edit or delete any annotations while the text file is completed. The user can continue editing a completed file by pushing the "Edit" button which moves the document back to the in-progress list (Fig. 5).

Fig. 5. Document list.

4   Conclusion

In this paper we have presented Oyster, a tool for supporting the collection of annotation from free-text data. Unlike other tools for data annotation, Oyster allows annotations to be defined with respect to named entities contained in large formal knowledge resources; we showed examples that used Oyster to annotate text using SNOMED CT concepts. One of the key features of Oyster are the automated suggestion of candidate annotations based on named-entities' descriptions contained in the reference knowledge resource and the possibility of manually searching for alternative named-entities to be used for annotation. Future work will consider implementing plugins for Oyster to allow annotation using a number of different popular knowledge resources as reference target, e.g. Freebase. In addition, we plan to extend the current search functionalities to allow (1) searching through the assigned annotations (e.g., for consistency checking or further analysis), and (2) loading pre-annotated information along with the input documents, e.g. for annotation checking.

References

1. McBride, S., Lawley, M., Leroux, H., Gibson, S.: Using australian medicines terminology (AMT) and snomed CT-AU to better support clinical research. Stud. Health Technol. Inform. **178**, 144–149 (2012)
2. Stenetorp, P., Pyysalo, S., Topić, G., Ohta, T., Ananiadou, S., Tsujii, J.: BRAT: a web-based tool for NLP-assisted text annotation. In: Proceedings of the Demonstrations at the 13th Conference of the European Chapter of the Association for Computational Linguistics, pp. 102–107. Association for Computational Linguistics (2012)
3. Morton, T., LaCivita, J.: Wordfreak: an open tool for linguistic annotation. In: Proceedings of the Conference of the North American Chapter of the Association for Computational Linguistics on Human Language Technology, vol. 4, pp. 17–18. University of Pennsylvania (2003)

TopSig: A Scalable System for Hashing and Retrieving Document Signatures

Timothy Chappell$^{(\boxtimes)}$ and Shlomo Geva

Queensland University of Technology, Brisbane, Australia
qut@timothychappell.com, s.geva@qut.edu.au

Abstract. There are a large number of overlapping problems within information retrieval that involve retrieving objects with certain features or objects based on their similarity to other objects. If the features that define these objects can be extracted, these objects can be reduced to a common representation that maintains pairwise similarity but discards all other data in order to facilitate compact storage and scalable retrieval. In this paper we introduce TopSig, an open-source tool for hashing and retrieving topology-sensitive document signatures.

Keywords: Document signatures · Hamming distance · Clustering · Locality-sensitive hashing

1 Introduction

The searching of objects other than the standard text files that have long been the main focus of information retrieval is a multi-stage process: the features that identify those objects must be identified and extracted, those features must then be stored in a representation allowing for efficient retrieval, and finally, a search engine must be developed or reconfigured to work with the new representation. For this reason we have seen that there are advantages to addressing the common aspects of this problem: representation and searching.

The intermediary format chosen to act as a representation of these searchable objects is the topology-sensitive document signature [5]. These document signatures function as semantically-aware digests of the objects they represent, with pairwise similarity relationships between signatures maintained. These signature representations have a number of properties that make them more desirable to store and search from an information retrieval perspective than the original objects they represent. They are usually much smaller in size, meaning they are easier to store in memory and replicate across cluster nodes. They are also of fixed length, which makes them easy to process efficiently.

2 Document Signatures

The document signatures used as an intermediary retrieval format by this software are functionally *locality-sensitive hashes*; that is, rather than exhibiting the

© Springer International Publishing Switzerland 2015
G. Zuccon et al. (Eds.): AIRS 2015, LNCS 9460, pp. 447–452, 2016.
DOI: 10.1007/978-3-319-28940-3_40

traditional properties of hashes (where two different documents will return two completely different hashes and none of the properties of the original document will be reflected in the hash), these more similar the original documents are to each other, the more similar their hashes will be in terms of Hamming [7] distance (the number of bits that differ between the two hashes). As a result, these document signatures can be used as a proxy for calculating the similarity between documents even when the original documents are unavailable.

The TopSig software uses the TOPSIG [5] approach to generate signatures, which is a modified version of the Sahlgren [10] random indexing method, where the terms or features that make up the original document are hashed and the hashes summed together. Figure 1 shows this process used to build an 8-bit document signature from a sentence; the terms are hashed into vectors of positive and negative values which are weighted with the term's inverse document frequency [11], summed together, then reduced to a binary representation (Table 1).

The resulting signatures can be used in a number of information retrieval tasks, and because the representation is fixed, they can be used in this manner regardless of the format of the original documents. One of the most natural retrieval models this lends itself to is "more of the same"-style queries, where one or multiple known documents is available and the user wants to find more documents in the collection that are similar. These representations also lend themselves to classification and clustering tasks; we find that documents that share categories tend to occupy a Hamming space that is closer together than documents that do not (Fig. 1). This property holds for non-text data as well; Fig. 2 shows the results of utilising TopSig in conjunction with spatial smoothing to classify the hyperspectral 145×145 Indian Pines remote sensing scene with a 6.1 % error rate.

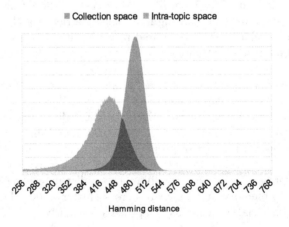

Fig. 1. Histogram of pairwise Hamming distances in both the whole collection and the intra-topic space of the Wikipedia collection [4], using relevance judgements from the 2009 INEX ad hoc track [6]

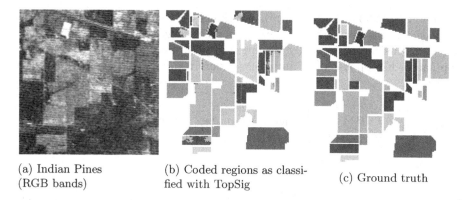

(a) Indian Pines
(RGB bands)

(b) Coded regions as classified with TopSig

(c) Ground truth

Fig. 2. Using TopSig with spatial smoothing to code the Indian Pines scene (6.1 %
error rate)

3 Retrieval

A historic criticism of signature-based retrieval approaches is that queries, which
require iterating over every signature in the collection in order to perform a
search, are not as scalable as their inverted file counterparts [13]. When applied
more generally to the problem of calculating many Hamming distances and finding those hashes that fall closest to the query, this is referred to as the Hamming
distance problem [8]. TopSig supports both exhaustive Hamming distance queries
(which exhibit the Hamming distance problem), as well as the generation and
searching of inverted signature slice lists [3] (ISSLs).

ISSLs are an inverted representation designed specifically to work with
topology-aware document signatures and other locality-sensitive hashes, and are
designed for the situation where a known signature is available for use as a query,
the user wishes to retrieve signatures that are close to the query signature (and
is less interested in signatures that are more distant) and where 100 % accuracy is not required. The signatures are divided up into slices and each of these
slices is inverted for storing in a table, where slices and their neighbours can
be looked up in constant time. This slicing is functionally equivalent to the use
of multiple independent hashing functions in LSH approaches like Minhash [1].
The approach can be configured to search a specified Hamming radius of slices
adjacent to the query slice, and this can be expanded to bring in more results.
Exhaustive Hamming distance comparisons can then be performed on only those
results with the greatest potential.

As Table 2 shows, with an appropriate slice width selected, ISSLs provide a great improvement to search speed, allowing the 500 million document
ClueWeb09 English collection [2] to be searched in under 57 ms on one machine,
something that would not be possible with exhaustive Hamming distance queries
(due to factors including memory bandwidth limitations). The ISSL table needs
to be generated beforehand and stored in memory during searching, compared
to raw Hamming distance queries which can be performed on the signature

collection as-is; however, the fact that it can easily scale to large collections easily makes up for this. While the ISSL approach is geared towards approximate retrieval, this makes it highly flexible and performance-efficient compared to approaches like Spotsigs [12] and Multi-Index Hashing [9] that guarantee exact retrieval.

TERM	the	quick	brown	fox	jumped	over	dogs
WEIGHT	0	0.13	0.15	1	0.3	0.1	0.01

Table 1. Example signature calculation utilising inverse document frequency in addition to term frequency to weight the terms

the	0	0	0	0	0	0	0	0
quick	0	0	**+0.13**	0	0	−0.13	0	**+0.13**
brown	−0.15	0	0	**+0.15**	0	−0.15	**+0.15**	0
fox	0	−1	0	0	0	0	**+1**	0
jumped	0	0	0	0	−0.3	0	**+0.3**	−0.3
over	**+0.01**	0	−0.01	0	0	0	0	0
the	0	0	0	0	0	0	0	0
dogs	0	**+0.1**	0	**+0.1**	0	−0.1	0	0
SUM	−0.14	−0.9	**+0.12**	**+0.25**	−0.3	−0.38	**+1.45**	−0.17
SIGNATURE	0	0	1	1	0	0	1	0

4   Performance

TopSig is written in C and the core indexing and search tools have been heavily optimised and developed to take full advantage of multi-threaded systems. While support for multi-node clusters is not explicitly built into the software, the highly configurable nature of TopSig and the fact that signatures produced from multiple sources can be aggregated means that the time-consuming task of generating signatures from a large collection can be easily distributed over many machines with a standard job control system.

5   Demonstration

In our demonstration we will show how TopSig can be used in a variety of information retrieval contexts that can benefit from the highly scalable and efficient architecture it provides. TopSig is designed to fit into any part of the information retrieval pipeline where it can be of service; as a result, the demonstration will be geared towards showing multiple scenarios where the software can be used.

Table 2. Searching ClueWeb09 English (500 million documents) with inverted signature slice lists and exhaustive Hamming search. CDR (Cumulative distance ratio) [3] gives an approximate idea of the different levels of accuracy that can be expected from various slice width settings.

Slice width	Search time	Memory (MB)	CDR@10
23-bit	199.738 ms	180029.43	0.925
24-bit	112.783 ms	177417.01	0.915
25-bit	66.753 ms	176260.59	0.902
26-bit	56.955 ms	177346.38	0.894
Exhaustive	2843.619 ms	92266.03	1

6 Conclusion

This paper provides a short introduction of the TopSig software, the algorithms it implements and the situations in which it can be used. While the software is still under active development, the main signature indexing and retrieval engines have been solid for the last few years and have already found their use in a variety of different researcher and student projects.

7 Availability

TopSig is currently distributed under the GNU General Public License and the source code is freely available to access, modify and redistribute. It is available for download from http://www.topsig.org.

References

1. Broder, A.Z.: On the resemblance and containment of documents. In: Proceedings of Compression and Complexity of Sequences, 1997, pp. 21–29. IEEE (1997)
2. Callan, J., Hoy, M., Yoo, C., Zhao, L.: Clueweb09 data set, January 2009. http://boston.lti.cs.cmu.edu/Data/clueweb09/
3. Chappell, T., Geva, S., Zuccon, G.: Approximate nearest-neighbour search with inverted signature slice lists. In: Hanbury, A., Kazai, G., Rauber, A., Fuhr, N. (eds.) ECIR 2015. LNCS, vol. 9022, pp. 147–158. Springer, Heidelberg (2015)
4. Denoyer, L., Gallinari, P.: The Wikipedia XML Corpus. SIGIR Forum (2006)
5. Geva, S., DeVries, C.M.: Topsig: topology preserving document signatures. In: Proceedings of the 20th ACM International Conference on Information and Knowledge Management, pp. 333–338. ACM (2011)
6. Geva, S., Kamps, J., Lehtonen, M., Schenkel, R., Thom, J.A., Trotman, A.: Overview of the INEX 2009 ad hoc track. In: Geva, S., Kamps, J., Trotman, A. (eds.) INEX 2009. LNCS, vol. 6203, pp. 4–25. Springer, Heidelberg (2010)
7. Hamming, R.W.: Error detecting and error correcting codes. Bell Syst. Tech. J. **29**(2), 147–160 (1950)

8. Manku, G.S.,. Jain, A., Sarma, A.D.: Detecting near-duplicates for web crawling. In: Proceedings of the 16th International Conference on World Wide Web, pp. 141–150. ACM (2007)

9. Norouzi, M., Punjani, A., Fleet, D.J.: Fast exact search in hamming space with multi-index hashing. IEEE Trans. Pattern Anal. Mach. Intell. **36**(6), 1107–1119 (2014)

10. Adibi, S.: Introduction. In: Adibi, S. (ed.) Mobile Health. SSBN, vol. 5, pp. 1–8. Springer, Heidelberg (2015)

11. Salton, G., Wong, A., Yang, C.-S.: A vector space model for automatic indexing. Commun. ACM **18**(11), 613–620 (1975)

12. Theobald, M., Siddharth, J., Paepcke, A.: Spotsigs: robust and efficient near duplicate detection in large web collections. In: Proceedings of the 31st Annual International ACM SIGIR Conference on Research and Development in Information Retrieval, pp. 563–570. ACM (2008)

13. Zobel, J., Moffat, A., Ramamohanarao, K.: Inverted files versus signature files for text indexing. ACM Trans. Database Syst. (TODS) **23**(4), 453–490 (1998)

Author Index

Printed in the United States
By Bookmasters